Werner Schmidt

**Optical Spectroscopy
in Chemistry and Life Sciences**

Further Titles of Interest

Stefan Berger/Siegmar Braun

200 and More NMR Experiments

A Practical Course

Mai 2004
ISBN 3-527-31067-3

Yong-Cheng Ning

Structural Identification of Organic Compounds with Spectroscopic Techniques

2005
ISBN 3-527-31240-4

Horst Friebolin

Basic One- and Two-Dimensional NMR Spectroscopy

2004
ISBN 3-527-31233-1

Helmut Günzler/Hans-Ulrich Gremlich

IR Spectroscopy

2002
ISBN 3-527-28896-1

Heinz W. Siesler/Yukihiro Ozaki/
Satoshi Kawata/H. Michael Heise (Eds.)

Near-Infrared Spectroscopy

Principles, Instruments, Applications

2001
ISBN 3-527-30149-6

Isao Noda/Yokihiro Ozaki

Two-Dimensional Correlation Spectroscopy

2004
ISBN 0-471-62391-1

Werner Schmidt

Optical Spectroscopy in Chemistry and Life Sciences

WILEY-VCH

WILEY-VCH Verlag GmbH & Co. KGaA

Author

Prof. Dr. Werner Schmidt
Raiffeisenstraße 2
78479 Konstanz
Germany

Universität Konstanz
Fakultät für Biologie
Postfach M645
78457 Konstanz
Germany

Universität Marburg
Fachbereich Biologie/Botanik
Karl-von-Frisch-Straße 8
35032 Marburg
Germany

Cover Picture
Reproduction of the "optical bench" in part
with courtesy of Ocean Optics, Inc.

Library of Congress Card No.: Applied for
British Library Cataloguing-in-Publication Data:
A catalogue record for this book is available
from the British Library.

**Bibliographic information published by
Die Deutsche Bibliothek**
Die Deutsche Bibliothek lists this publication
in the Deutsche Nationalbibliografie;
detailed bibliographic data is available in the
Internet at <http://dnb.ddb.de>.

© 2005 WILEY-VCH Verlag GmbH & Co.
KGaA, Weinheim

Composition   hagedorn kommunikation,
Viernheim
Printing   betz-druck GmbH, Darmstadt
Bookbinding   J. Schäffer GmbH i.G.,
Grünstadt
Coverdesign   SCHULZ Grafik-Design,
Fußgönheim

Printed in the Federal Republic of Germany.
Printed on acid-free paper.

ISBN-13:   978-3-527-29911-9
ISBN-10:   3-527-29911-4

Contents

Optical Spectroscopy in Chemistry and Life Sciences. W. Schmidt
Copyright © 2005 WILEY-VCH Verlag GmbH & Co. KGaA, Weinheim
ISBN 3-527-29911-4

Preface

The first edition of this book, written in German *Optische Spektroskopie, Eine Einführung für Naturwissenschaftler und Techniker* was published in 1994. It was written in response to a need that I first encountered when I was faced with various problems in Optical Spectroscopy during my practical biophysical work as a Ph. D. student at the University of Freiburg/Breisgau, Germany. When working as a postdoc with the late Warren Lee Butler (see the photograph in Chapter 1) at the University of California in San Diego, CA, USA, I acquired the broad basics required for practical work in a biologically/chemically oriented laboratory. Of course, there are many excellent textbooks on specific themes such as atomic spectroscopy, NIR spectroscopy, photoacoustics, fluorescence spectroscopy or polarization spectroscopy, and quantum mechanics, optical design, radiation detectors or data acquisition, but volumes covering all these aspects simultaneously in a limited space are still rare.

I am aware of the problems of trying to cover the areas of many specialists and of provoking their surely often justified disapproval and contradiction. Nevertheless, I do not intend to presume that the principles of Optical Spectroscopy can only be conveyed as an endless succession of mathematical equations, whose comprehension and manipulation demand a greater knowledge of mathematics than the ordinary student of science already possesses. Thus, in this introductory text the aim is to help students develop an interest in Optical Spectroscopy: (i) by ensuring that no mathematical techniques are employed that are beyond their immediate comprehension, hence various necessary equations have not been explicitly derived but are simply offered as a "proven entity"; (ii) by displaying physical and/or mathematical phenomena "in words" rather than in the "correct" physical/mathematical terms; and (iii) by providing pragmatic examples (sometimes from my own published work), which can be easily adapted by the reader to their own particular problems. As with the second German edition of the book *Optische Spektroskopie, Eine Einführung* (2000), this first English edition is the result of substantial rewriting of the previous versions.

To this end, the book's contents falls into "natural" sequential sections that are virtually self-contained. Where a particular issue bridges more than one chapter, cross-references are included in the text so that the sections can be selected at will. In order to understand the problems and terminology of a specific topic it is useful

Optical Spectroscopy in Chemistry and Life Sciences. W. Schmidt
Copyright © 2005 WILEY-VCH Verlag GmbH & Co. KGaA, Weinheim
ISBN 3-527-29911-4

to know something of the history behind it. However, during my research I discovered that there is no comprehensive book on the "History of Optical Spectroscopy", so I had to scan numerous different texts (Chapter 1). The quantum-mechanical, theoretical principles of Optical Spectroscopy, including the nature of light, the structure of matter and the interaction of light and matter are condensed into Chapter 2, possibly the most demanding chapter for a beginner. Chapter 3 focuses on practical issues and describes radiometric and photometric units, light sources, spectroscopic optics, monochromators and light detectors. Chapter 4 discusses the spectroscopy of atoms as the simplest systems that can be investigated opto-spectroscopically. Classical atomic absorption spectroscopy (AAS) using flame and electrothermal methods with different correction procedures of background absorption is presented. The principles of atomic emission spectroscopy [inductively coupled plasma (ICP) spectroscopy] are outlined.

The first specialized chapter, Chapter 5, deals with the absorption spectroscopy of molecules and the derivation of the most important Bouguer–Lambert–Beer law, and describes various monochromators, the general absorption properties of molecules and – most significantly in this era of personal computers – the mathematical treatment of absorption spectra. The second specialized chapter, Chapter 6, focuses on fluorescence spectroscopy. The mechanisms of fluorescence are described, followed by discussions of fluorimeters as well as fluorescence measurement and different fluorescence parameters that lead to various molecular properties, such as polarization, including inter-molecular and molecular interactions with the microenvironment. Discussions of phosphorescence and chemo- and photo-bioluminescence conclude Chapter 6. Photoacoustic spectroscopy (PAS) is treated separately in Chapter 7, even though this method still only occupies an analytical niche rather than being a generally applicable technique. Again, the highly demanding theory is introduced, followed by experimental procedures and several selected practical examples demonstrating the drawbacks and advantages of PAS. Chapter 8, on Scattering, Diffraction and Reflection, covers a very important section of optical spectroscopy that enables valuable information from various large molecules and their dynamics to be extracted in a reasonably easy and cheap manner. Attenuated Total Reflection Spectroscopy (ATRS) allows – similar to Photoacoustic Spectroscopy (Chapter 7) – real absorption spectra of extremely small layers ($<10~\mu m$) and highly opaque samples to be scanned. Circular Dichroism (CD) and Optical Rotatory Dispersion (ORD) are the theme of Chapter 9. These methods reveal structural and dynamic information, particularly of larger enantiomeric molecules. Finally, Chapter 10 is dedicated to Near-infrared Spectroscopy (NIRS). Similar to the ultraviolet spectral range, near-infrared adjoins the "true visible wavelength range" and is usually treated using the same techniques. Even though there are no electronically excited states for the molecules involved but only vibrational overtones, it is still regarded as being an area of Optical Spectroscopy. However, its analytical benefits do not really depend on the classical spectroscopical techniques but on chemometrics, statistical methods which have only become applicable with the advent of personal computers.

Although primarily intended as an introduction to Optical Spectroscopy for students of biology/biochemistry and chemistry, I hope that this book will prove helpful to students of general science including medicine and pharmacy, and that it will also be of interest to their more experienced colleagues.

I am first and foremost greatly indebted to the late Prof. Warren Lee Butler who introduced me to the practice and theory of the engaging field of Optical Spectroscopy. I was lucky enough to work in his lab (1974) with one of the first computerized spectrophotometers in the world. In 1976 I moved to the Plant Research Laboratory of Prof. Ken Poff at Michigan State University, a former graduate researcher of Prof. Butler, where I was involved in early work on another similar computer-based spectrophotometer. In particular Prof. Poff is thanked for teaching me the fast and at the time absolutely essential programming in assembler language.

I wish to express my thanks to Dr. H. F. Ebel at VCH-Verlagsgesellschaft (as it was known at the time) who originally (1991) encouraged me to write a book such as this, and particularly to write it in English. Following his suggestion, I am pleased that finally after 14 years I have succeeded in accomplishing this. My particular thanks go to Judith Egan-Shuttler and her husband Dr. Ian Shuttler. Dr. Shuttler (PerkinElmer Inc.) provided me with the latest information on atomic absorption spectroscopy, including Table 4.2 and Figure 4.9 (the two-dimensional Echellogram), and Judith Egan-Shuttler copy-edited the whole manuscript, written in the modest English of a non-native speaker, to produce an English textbook suitable for publication. Last, but not least, I am indebted to the three colleagues at Wiley-VCH for their excellent cooperation and patience during the preparation of this book: Dr. Waltraud Wüst, Dr. Uschi Schling-Brodersen and Dr. Gudrun Walter.

Konstanz
May 2005 *Werner Schmidt*

1
Introduction to Optical Spectroscopy

Bene docet, qui bene distinguit
(Horace)

Good choices means good teaching

1.1
Overview

The term *Optical Spectroscopy* (OS) in this book covers all types of qualitative and quantitative analytical methods that are based on the interaction of light with living and non-living matter.

Here the term "light" includes electromagnetic radiation covering the spectral range from the far-ultraviolet (UV), followed by visible light (VIS) up to the near-infrared region (NIR). For more than 200 years optical spectroscopy has been utilized in various fields of science, industry and medicine, particularly in (bio-) chemistry, biology, physics and astronomy. OS is highly specific, each substance is discernible from all others by its spectral properties. Samples can be qualitatively and quantitatively analyzed. In contrast to other spectroscopic methods such as NMR (nuclear magnetic resonance), ESR (electron spin resonance), Mößbauer or mass spectroscopy, the requirements of the samples to be analyzed are not particularly restrictive. Measurements of various optical parameters as a function of wavelength/energy ("spectrum") or time ("kinetics") provide valuable insights that are not, or not readily, attainable by other analytical methods.

Optical spectral analysis is a well developed method. However, there is still a rapidly growing market for spectrophotometers and in addition to routine analyses many novel applications are being continuously reported. Depending on their required specifications, spectrophotometers differ markedly in size, form and usability and, last but not least, in price. They can be specifically adapted to the envisaged purpose. Therefore, the current trend is to use highly specific, moderately priced spectrophotometers rather than installing and using the most expensive "all-purpose machines" that have the best specifications available for more diverse applications.

Optical Spectroscopy in Chemistry and Life Sciences. W. Schmidt
Copyright © 2005 WILEY-VCH Verlag GmbH & Co. KGaA, Weinheim
ISBN 3-527-29911-4

The specifications of modern optical spectrophotometers are now approaching the limits set by physics; the advantages compared with other analytical methods are many, as follows.

- OS is not destructive or invasive.
- Remote measurements are feasible, i.e., measurements ranging from distances of a few millimeters up to long distances, such as from airplanes or satellites, without any physical contact with the sample. Hazardous or unattainable objects are easily analyzed.
- In principle liquid samples as well as solids and gases are acceptable, independent of their "optical quality". Transparent samples as well as highly light scattering, turbid and opaque samples are measurable.
- The photometric as well as kinetic time scales covered are extra-ordinarily wide and are unmatched by other analytical methods.
- OS (for example, femtosecond laser spectroscopy) allows measurements of extremely fast reactions down to a femtosecond time scale (10^{-15} s).
- Minute samples in the micrometer range (microspectroscopy) or very rare events can be investigated using single photon counting (measurement of a few photons per second).
- Extremely low concentrations down to 10^{-18} moles are detectable by luminescence methods.
- Radioactive markers for the investigation of biochemical and molecular-biological processes are increasingly being replaced by cheaper and easy to use luminescence markers and low cost luminescence spectrometers.

In this book we will discuss the essential features of optical spectroscopy with minimum descriptions of complex theories, various spectrophotometric procedures and apparatus. This is not an easy task in light of the significant progress that has been made in optoelectronics, laser and microcomputer techniques in recent years. However, the book is not intended to be a "how to use" manual for a specific spectrophotometer. For this purpose the reader should consult the manufacturer's manual. In addition, spectrophotometer manufacturers offer practical and theoretical introductory seminars; however, they are often rather expensive, time consuming and only focus on their own products.

After a short journey through the history of optical spectroscopy in the current chapter, Chapter 2 offers a short orientation and introduction to quantum theoretical concepts that are a prerequisite to the understanding of the spectroscopic properties of matter. Starting with the hydrogen atom as the simplest quantum mechanical system we then focus on the spectroscopic properties of small, diatomic molecules and finally turn our attention to complex (bio-)molecules.

Optical spectroscopy is intimately connected to optical physics. Thus, in Chapter 3 we will establish the necessary fundamental knowledge and become acquainted with the diverse optical elements: starting with the radiometric and photometric definition of light units, the basics of geometric, wave and particle optics, and finally the generation of light and its measurement will be described. We will discuss the properties of optical components such as filters, mirrors, lenses, optical fibers, integrating spheres as well as methods for dispersing light by prisms, gratings and interferometers.

In the fourth chapter we will utilize the theoretical and practical knowledge obtained from the first three chapters to describe atomic absorption (AAS) and atomic emission spectrometry (AES) as two variants of atomic spectroscopy for theoretical and practical discussions.

In the fifth chapter we will discuss in more detail the theoretical and practical aspects of molecular absorption spectroscopy, the most widely utilized analytical tool in various fields. We will discuss different versions of absorption spectrophotometers and also focus on diverse methods of spectral evaluation in order to extract "hidden" information from spectra that may not be immediately apparent at first glance.

The sixth chapter deals with luminescence spectroscopy. While absorption or extinction is the only measured parameter to be determined in absorption spectroscopy, luminescence spectroscopy provides a number of measurable parameters for the molecular system of interest.

Photoacoustic spectroscopy (PAS), where heat emission rather than light is detected, will be discussed in the seventh chapter. Supplementing absorption and luminescence spectroscopy, PAS allows conclusions to be drawn about thermodynamic relationships, particularly of complex molecular systems in biology, chemistry and medicine. Moreover, it allows spectral scans to be made of non-visible, subsurface layers of sample materials such as skin or fruit shells.

Measuring methods and phenomena that depend on various scattering effects such as Rayleigh, Mie, Fraunhofer or Raman scattering, including the related reflection and ATR spectroscopy (*attenuated total reflection*), will be discussed in the eighth chapter.

Because of the importance of the chiral and stereochemical properties of biomolecules, we introduce ORD (*optical rotational dispersion*) and CD (*circular dichroism*) spectroscopy in the ninth chapter, even though both methods are derivatives of absorption spectroscopy and are related to Chapters 5 and 8. Finally, ellipsometry utilizes the phase shift from the surface reflection of polarized light, an effect that is applicable for the highly sensitive analysis of biologically relevant molecules.

With the availability of low cost, fast, on-line microcomputers with large memories, near-infrared spectroscopy (NIR) has experienced an unprecedented boom in recent years, particularly in routine industrial analysis. NIR spectroscopy has become indispensable in the pharmaceutical industry, medicine and the food industry, and in petrochemical, geological and environmental analysis. Further technical developments in these areas are anticipated, also extending into other application areas of industrial research. Therefore, this topic is treated in Chapter 10.

The most important physical constants as used in this book can be found in the Appendix. The Periodic Table of the elements, including detailed electron distributions, may serve as a quick overview. Another table allows the easy conversion

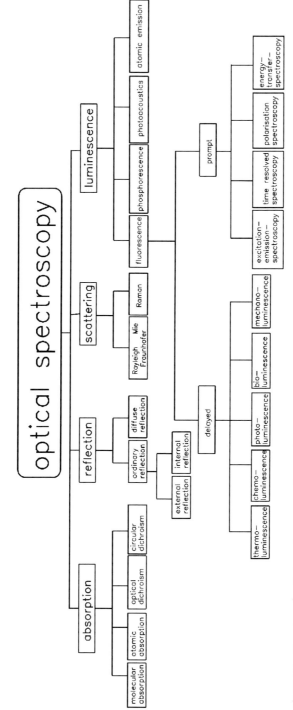

Figure 1.1 Overview of optical spectroscopy as discussed in the present book. We distinguish absorption, reflection, scattering and luminescence spectroscopy, with secondary disciplines as shown. For simplicity, photoacoustic spectroscopy, a type of emission spectroscopy (emission of heat), is attributed here to luminescence spectroscopy.

of various units of energy. As an "interface" between the reader and various manufacturers and distributors of optical spectrophotometers and components of all types, important addresses including email and URL addresses are listed. Supplementary literature is given at the end of each chapter. Figure 1.1 depicts the whole field of optical spectroscopy as discussed in the present book.

1.2
History of Optical Spectroscopy

Table 1.1 presents the highlights in the history of optical spectroscopy.

1666	Sir Isaac Newton is the first to use a glass prism for spectral dispersion
1675	Newton's famous book on Optics ("Optiks")
1758	Markgraf: Analysis by flame coloration (without spectral dispersion)
1800	Herschel discovers IR radiation (England)
1801	Ritter discovers UV radiation (Jena, Germany)
1802	Wollaston uses a slit and lens in flame spectroscopy for the first time
1807	Young: Experiment with double slits to prove wave-nature of light, interprets as the first "colors of thin layers" and action of optical gratings properly
1808	Malus recognizes polarization of light (by reflection)
1814	Fraunhofer: Invention of the grating (1500 lines in the sun's spectrum)
1817	Young solves the problem of "polarization by reflection": united wave- and corpuscle picture; transversal polarization
1818	Fresnel and Huygens unite wave character and Huygens' principle
1834	Talbot distinguishes spectrally "strontium-red" and "lithium-red", birth of analytical optical spectroscopy
1848	Foucault: sodium in the electrical arc (absorption/emission)
1859	Kirchhoff: absorption = emission wavelength, discovers sodium in the sun's spectrum
1860	Kirchhoff and Bunsen discover cesium and rubidium
1868	Ångström lists lines of the sun's spectrum
1868	Kirchhoff and Bunsen discover helium in the sun
1885	Balmer discovers Balmer series lines in hydrogen spectrum
1887	Michelson and Morley experiment, end of ether-theory
1887	Henry A. Rowland revolutionizes experimental spectroscopy with his concave grating; improvement of Ångström's experiments of 1868
1888	Rowland publishes the first sun atlas ("Rowland circle")
1891	Rowland compares spectra of all known elements with the sun's spectrum and finds several new lines
1893	Rowland publishes the standard spectral atlas (reference: sodium)
1894	Helium discovered on earth (1868 in the sun's spectrum)
1896	Balmer develops his first series formula

1897	Thompson discovers the electron
1897	Rubens succeeds with the first isolation of an infrared band utilizing multiple reflections at metal layers
1911	Rutherford discovers the atomic nucleus
1920	H. v. Halban: first determination of concentration of molecules utilizing "light electric apparati"
1925	G. Scheibe, W. Gerlach and E. Schweitzer: first real quantitative spectral analysis
1926	G. Hansen builds the first recording double beam spectrophotometer following the plans of P. P. Koch (1912)
1941	The famous prism photometer, Model DU, from Zeiss comes to the market; for the first time this spectrophotometer covers the UV range, resolution 1 nm, measured values are read from a scale with a needle pointer
1972	W. L. Butler uses a monochromator (Cary 14) on-line with a minicomputer (PDP11, Digital Equipment Corp. AF-01) thereby pioneering optical spectroscopy in biological/biochemical research (see photograph of Butler)

Table 1.1 History of optical spectroscopy

Warren Lee Butler (28.1.1925–21.6.1984), co-discoverer of the plant pigment phytochrome and one of the pioneers of optical spectroscopy in biological/biochemical research (photograph taken by the author)

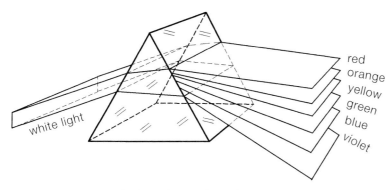

red
orange
yellow
green
blue
violet

white light

Figure 1.2 Newton was the first to discover that a glass prism will disperse parallel sunlight into its components, i. e., its spectral colors.

The word "spectrum" is derived from the Latin and implies "apparition" or "scheme". It was first introduced, in the sense of today's science, by Isaac Newton in 1666 in his classic volume "Optiks". He was the first to use a prism to disperse sunlight into its spectral colors (Figure 1.2). In 1758 Markgraf was the first to use the coloration of flames to identify substances by visualization with the bare eye. In 1802 the English physicist William H. Wollaston resumed Newton's prism experiment, improved the set-up and thus provided the first observation of numerous dark lines in the sun's spectrum. In the same way J. Herschel and W. H.T. Talbot analyzed the light from a flame. In 1834 Talbot, using the same method, succeeded in distinguishing unequivocally the red flames of lithium and strontium on a spectral basis, which could be regarded as the birth of spectral chemical analysis.

With this new technique, which has enriched optical spectroscopy up to the present day, an apprentice who originally learned to grind mirrors, who was self-taught and later became Professor of Physics in Munich, Josef Fraunhofer, succeeded in the dispersion of sunlight (1814).

Fraunhofer's family were very poor and as a young boy he became an "apprentice". For three years he was taught how to grind lenses and mirrors, and thus he became very interested in optics; but as a boy from the lowest social class he had no chance or opportunity to study, and his application to study was turned down. Thus, he taught himself secretly, and invented the optical grating, which later "enriched" science tremendously. At the beginning of these activities he was not accepted by the German scientific/academic community. Nevertheless, after many years they had to accept his work and finally he even became Professor of Physics in Munich and a highly respected person.

On the basis of the optical grating he invented – and which is still used in modern spectral apparatus today – Fraunhofer developed a high dispersive optical spectroscope, which can now be found in the Deutsches Museum in Munich, Germany (Figure 1.3). Using such an apparatus, in 1859 Kirchhoff detected sodium in the sun's spectrum (Figure 1.4), and on this basis derived the famous *Kirchhoff law*:

Joseph von Fraunhofer (1787–1826)

"At a given wavelength and temperature the magnitude of the spectral emission of any object equals the magnitude of its absorption".

Prior to the 20th century there were no theories that could explain satisfactorily the complex behavior exhibited by all substances. The most important founda-

Figure 1.3 First grating spectroscope, Fraunhofer, 1821 (Deutsches Museum, Munich, Germany).

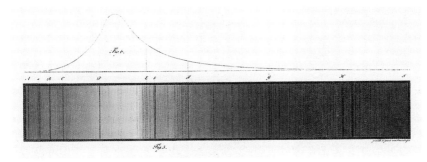

Figure 1.4 Fraunhofer lines in the sun's spectrum. Copper engraving, 1814, taken from: *Denkschriften der kgl. Akademie der Wissenschaften zu München*, 1814 and 1815, Vol. 5, Table II, Munich 1817. The solid line (cf. Figure 3.4) shows the sensitivity curve of the human eye (Deutsches Museum, Munich, Germany).

Gustav Robert Kirchhoff (1824–1887)

Max Planck (1858–1947)

tions which led to today's understanding of spectra were laid down by the following scientists.

In 1885 the Swiss scientist J. J. Balmer discovered the "Balmer" series of lines in the spectrum of hydrogen. In 1897, the English physicist J. J. Thompson discovered the electron. In 1911 his fellow countryman Ernest Rutherford discovered the atomic nucleus. In 1900 *Max Planck* formulated the first principles of quantum theory. Werner Heisenberg (1932) and Erwin Schrödinger (1933) won the Nobel Prize for their pioneering work on quantum mechanics. These early quantum theory concepts were further advanced by Paul A. M. Dirac and Wolfgang Pauli (1945), for which both researchers were honored with the Nobel Prize (Dirac in 1933, together with Schrödinger). Today quantum theory is well established and its development is in a sense complete.

As the history of science is entwined with the history of measurements and analysis, the history of optical spectroscopy is largely reflected in the history of

Werner Karl Heisenberg
(1901–1976)

Erwin Schrödinger
(1887–1961)

Wolfgang Pauli
(1900–1958)

astronomy and thus in the history of atomic spectroscopy. Only at the end of the 19th century did *molecular spectroscopy* become important as a powerful analytical method. For example, forensic investigations could now solve murders through just tiny spots of blood, using spectrophotometers that were capable of measuring the characteristic "bands" of hemoglobin and thus distinguishing between blood and red dyes.

For many decades, as a result of the simple tungsten glow lamps, prisms, gratings and light detectors, optical spectroscopy was confined to the narrow visible wavelength range between 500 and 700 nm.

In the 40 years at the beginning of the 20th century only a few commercial spectrophotometers were available as they were manufactured in limited number, and, in addition, were difficult to operate (General Electric Hardy Spectrophotometer, Cenco "Spectrophotelometer", Coleman Model DM). At that time the "measurement" of extinction for the determination of concentration was performed visually by comparison of two visual fields next to each other (similar to the Nagel anomaloscope used today for testing color vision) by the naked eye. The famous Pulfrich-Photometer (some thousand units were produced by Zeiss in Jena, Germany) also worked in this way, relying heavily on a series of so-called S-filters in the visible range (interference filters with half-widths of only 15–20 nm). By 1941 there were already more than 800 publications on the determination of the concentration of the clinically important components of blood and other body fluids using this very spectrophotometer.

However, even by the 1930s the great significance of the ultraviolet spectral range (UV) had been recognized, particularly for the quantitative determination of vitamin A (which has an absorption at 320–330 nm). Indeed, this turned out to be the key factor in the subsequent development of UV–VIS (visible) spectrophotometers. Under the leadership of H. H. Cary and A. O. Beckman, the famous Model D spectrophotometer, with a selling price in 1941 of US$ 723, brought about a breakthrough in optical spectroscopy, as it was the first commercial instrument to include the UV region. A hydrogen lamp as the only UV light source was developed for this purpose. The quartz glass required for the manufacture of UV prisms was difficult to obtain, as it was used primarily by the military in quartz-stabilized oscillators for radio transmitters. A photomultiplier for the UV range down to 220 nm had to be developed (the available cesium oxide types were only good for wavelengths >600 nm). The DU/DU1 series of instruments from Beckman, probably the most successful series worldwide ever, was developed continuously up till 1976, and all together more than 30000 units were sold.

In parallel with the development of UV–VIS spectrophotometers, infrared spectrometers were developed. However, only after World War II did the market for spectroscopic analytical instruments widen rapidly, and various manufacturers developed the techniques further. Owing to the better resolution and low stray light compared with a prism, dispersing gratings were chosen, and automatic scanning double monochromators yielded baseline-corrected spectra, thereby leading to use for routine analytical work. Stray light was lowered dramatically thereby increasing the detection capability of spectrophotometers by 4–5 orders of magnitude. Specialized photometers, e. g., for radiometry, colorimetry or dual wavelength analysis, were developed for the market. A significant push came at end of the 1970s with the advent of powerful, low cost minicomputers, which accelerated still further in the early 1980s with the appearance of the first microcomputers leading to today's PC (personal computer) based technology. The PC not only lightens the routine workload, but in addition, enables on-line measurements and complex analyses to be made, which had previously only been possible with large computers. Old theories such as that of Kubelka–Munk on scattering analysis, Fourier analysis or diverse correlation and fitting procedures or chemometrics as utilized in near-infrared spectroscopy, which depend heavily on calculation power, are now available to everyone and are performed readily.

1.3
Selected Further Reading

Beckmann, A. O., Gallaway, W. S., Kaye, W., Ulrich, W. F. History of Spectrophotometry at Beckman Instruments, Inc. *Anal. Chem.* **1977**, *49*, and references cited therein.

Brand, J. C.D. *Lines of Light: The Sources of Dispersive Spectroscopy, 1800–1930*, Gordon & Breach Science, New York, **1995**.

Hearnshaw, J. B. *The Analysis of Starlight: Two Centuries of Astronomical Photometry*, Cambridge University Press, Cambridge, **1996**.

Laitinen, H. A., Ewing, G. W., eds., *A History of Analytical Chemistry*, American Chemical Society, Washington, D. C., **1977**.

Lindon, J. C., Tranter, G. E., Holmes, J. L., eds., *Encyclopedia of Spectroscopy and Spectrometry*, Academic Press, London, Sydney, Tokyo, **1999**.

Ninnemann, H. A Scientific Homage to W. L. Butler, *Photochem. Photobiol.* **1985**, *42*, 619–624.

Sawyer, R. A. *Experimental Spectroscopy*, Dover Publishers, New York, **1963** (the first chapter largely covers the history of optical spectroscopy).

Wilde, E. *Geschichte der Optik*, Sändig Reprint Verlag, H. R. Wohlwend, Vaduz/Lichtenstein **1985**. (History of optics, from ancient times to 1838, an excellent source of the history of optical sciences; in German).

2
Fundamental Principles

Contra principia negantem non est diputandum
(Horace)

It is impossible to discuss something with somebody who doesn't
understand the principles

2.1
The Nature of Light

In general, optical phenomena can be considered from three different points of view. Geometric optics is the straightforward view that is the easiest to visualize, interpreting light in the form of rays, and describes the rules that determine how light propagates and how optical images are formed. However, it does not describe the interaction of light and matter. In wave optics light is interpreted as periodic oscillations of electric and magnetic fields in time and space (Figure 2.1). Building on this concept as first described by Faraday (1831), in 1864 the English physicist James Clark Maxwell formulated the basic equations of electrodynamics, which "reveal the structure of the electromagnetic (light) fields" (Albert Einstein). In many respects light waves are comparable to mechanical waves in liquids, although this analogy has its limit. However, only in this consideration are the phenomena such as refraction, reflection, interference and polarization "understandable". In the third point of view, via quantum optics, light is described as a stream of mass-less particles, known as "photons". It is only the quantum optical interpretation that is capable of describing light absorption and emission as the foundations of optical spectroscopy.

Even though these three points of view appear to be contradictory, they do not exclude each other, and in fact the contrary is the case: they simply characterize the "real nature" of light, which cannot be conceived by our intuition. This complementary view is known by the term "dualism". Depending on the individual experiment and requirements we will adopt the most appropriate of these three descriptions: that is light as a ray, a wave or a particle stream.

Optical Spectroscopy in Chemistry and Life Sciences. W. Schmidt
Copyright © 2005 WILEY-VCH Verlag GmbH & Co. KGaA, Weinheim
ISBN 3-527-29911-4

James Clerk Maxwell (1831–1879)

Albert Einstein (1879–1955)

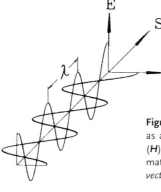

Figure 2.1 According to Maxwell's theory, light is described as an electromagnetic transversal wave where the magnetic (**H**) and electric components (**E**) oscillate in phase. Their mathematical product **S** = **E** × **H** is the so-called *Poynting vector* and follows the direction of energy flow.

2.2
Electromagnetic Radiation

Spectroscopy is usually defined as an analytical procedure dependent on the separation of electromagnetic radiation corresponding to the wavelength, i. e., energy. According to Einstein the energy E of electromagnetic radiation is proportional to the frequency ν, i. e., $E \sim \nu$. With the proportionality constant $h = 6.626 \times 10^{-34}\,\text{J s}$, which is known as Planck's constant, we obtain the equation

$$E = h \times \nu \tag{2.1}$$

As the propagation velocity of electromagnetic radiation in a vacuum is a constant $c = 2997925 \pm 3\ \text{m s}^{-1}$ (cf. Appendix A2), wavelength and frequency are correlated according to

$$\nu = \frac{c}{\lambda} \quad \text{and}$$

$$\tag{2.2}$$

$$E = h \times \frac{c}{\lambda}$$

Therefore, the energy of electromagnetic radiation can be expressed as a reciprocal length, which is referred to as wavenumber, k:

$$k = \frac{1}{\lambda} \qquad (2.3)$$

The energy of one individual light quantum is

$$h \times \nu = k \times 1.24 \times 10^{-4} \text{ eV cm} \qquad (2.4)$$

Because scanning by a grating monochromator is essentially proportional to wavelength (cf. Chapter 5), spectra in optical molecular spectroscopy are usually plotted on a wavelength scale, even if this presentation is less meaningful than the linear energy scale, i.e., wavenumber scale. If required, both spectral presentations can easily be converted into each other using Eq. (2.3).

Electromagnetic radiation includes the whole spectral range between an alternating current of 50 Hz up to high energy γ-radiation (Figure 2.2). For technical reasons the various spectral ranges require different types of spectrophotometers; even the "visible" spectrum, which is defined as the range approximately between 200 and 2000 nm, including the UV range between 200 and 400 nm and the near-infrared range (NIR) above 800 nm, cannot be measured continuously but needs different spectral components. However, with modern spectrophotometers the user is often freed from this task and the spectrum is scanned automatically across the whole range.

Quantization of light does not have the same meaning as quantization of matter, because the frequency, i.e., the energy, can be chosen arbitrarily. With respect to wavelength, light is generally inhomogeneous: light "consists" of quanta of various energies where the energy distribution typically exhibits a maximum with a specific half-width. Both energy distribution and maximum depend on the specific process of light generation, e.g., by an incandescent lamp, a gas discharge lamp, a laser or a solid state emitter such as an LED (*light emitting diode*), or a laser diode. A stream of light quanta with a small energy distribution, or in the ideal case with only one wavelength, is called monochromatic.

In the classical picture, emission of light is described by vibrating, elastically bound electrons, the energy of which decrease exponentially according to the laws of classical physics. Assume N atoms are in an excited state j. Then the intrinsic lifetime is defined by the equation:

$$-\,dN_j\,/dt = \; N_j \sum_{i<j} A_{ji}$$

where A_{ji} are the so-called Einstein coefficients for the spontaneous transition from the excited state j to the ground state i. Integration of this equation yields

$$N_j(t) \;=\; N_j(0)\,e^{\frac{-t}{\tau_j}}$$

with the lifetime of emission being τ_j.

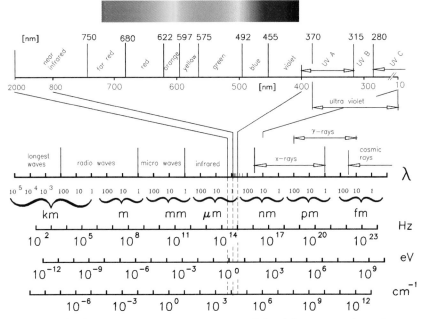

Figure 2.2 The electromagnetic spectrum covers more than 24 decades. The spectral range that is important in optical spectroscopy, including the infrared and ultraviolet radiation, is expanded and the corresponding color experienced by the human eye indicated. In the lower part some typical energy units are added.

$$\tau_j = \frac{1}{\displaystyle\sum_{i<j} A_{ji}}$$

The lifetime of free excited atoms or molecules that do not interact with each other is defined as the time after which the emission intensity declines to $1/e$ of the starting value. Typical values are in the order of 10^{-7} to 10^{-9} s. Values that deviate significantly will be discussed in Chapter 6 on luminescence spectroscopy. Taking the velocity of light c into account, a finite length of the wave packet (wave group) of 3 m is calculated, which is termed the coherence length. The smaller the lifetime, the greater the linewidths, and vice versa. Following Heisenberg's uncertainty principle with the abbreviation $\hbar = h/2\pi = 1.0546 \times 10^{-34}$ J s we obtain

$$\Delta E \times \Delta t \geqslant \frac{\hbar}{2} \tag{2.5}$$

For example, a lifetime of 10^{-8} s yields a natural linewidth of 1.18×10^{-5} nm.

Owing to this extraordinary line sharpness, in 1960 the standard for a meter in the form of a rod made from a platinum–iridium alloy was abandoned and the electronic $2p^{10} \rightarrow 5d^5$ transition of krypton atoms was chosen instead as the international definition of length (cf. Section 2.3.2.2 for definitions of atomic terms). The wavelength of this orange spectral line is exactly $\lambda = 605.780211$ nm in vacuum and 605.612525 nm in air.

Coherence length, emission lifetime and linewidth concomitantly depend on each other. Coherence is a macroscopic definition and is simply defined by the fact that interference patterns, such as Newton's rings, can be observed. Microscopically this means that the individual wave packets have a fixed phase relationship. The laser, which is an important tool in spectral analysis, fulfills the requirements of coherence in an almost ideal manner (Section 3.4.5).

Strong atomic absorption transitions exhibit A_{ji} values of 10^8 to 10^9 s^{-1}, i. e., their lifetimes are 1 to 10 ns. The lifetime can be significantly shortened by colli-sion or by induced emission. The natural (intrinsic) linewidth (i. e., with external influences being excluded) of an excited state is given according to Heisenberg's uncertainty principle:

$$\Delta E_j = \frac{h}{2\pi\tau_j}$$

or

$$\Delta E_j = \frac{h \sum\limits_{i<j} A_{ji}}{2\pi}$$

With $E = h\nu$ it follows that

$$\Delta\nu = \sum\limits_{i<j} A_{ji}$$

where $\Delta\nu$ is the linewidth, in frequency units, of the transition from the excited to the ground state. As the ground state has an infinite lifetime and therefore repre-sents a so-called δ-function (infinitely narrow peak), the width immediately mir-rors the width of the excited state.

However, the measured linewidth is always greater than the actual linewidth, which is derived from the natural lifetime. We have to take into account a further broadening of the linewidth, i. e., a reduction in the lifetime, by collision, stimu-lated emission (as in lasers, cf. section 3.4.5) or by Doppler shift as caused by the relative motion of the molecules and the observer. Broadening of the linewidth by collision and induced emission, as homogeneous processes, lead to a Lorentzian shape, while Doppler broadening, as an inhomogeneous process, yields a Gaus-sian line shape. The combination of both line shapes is described by the so-called "Voigt" profile.

2.3

Interaction of Light and Matter

By the 19th century spectroscopists had already collected sufficient data pertaining to the interaction of light and matter. The essential observation was that atoms and molecules were able to absorb or emit light only at specific wavelengths or wavelength bands. As mentioned previously, Kirchhoff had discovered that for a given atom the frequencies of absorption and emission were identical (which, however, is not true for molecules; cf. Chapter 6). The breakthrough in quantum theory, and thus in spectral analysis, is attributed to Albert Einstein when he was working as an examiner in the Patent Office in Bern, Switzerland, in 1905. He discovered the emission of electrons from the surfaces of solids on incidence of light of a suitable wavelength. In his spare time, he developed a theory for the photoelectric effect. Planck had assumed that the oscillators comprising a black-body radiator were quantized, but Einstein extended this concept to suggest that the radiation itself was quantized. Thus, the light incident on a probe was composed of quanta $h\nu$. When they struck the surface of a metal, some of the energy was used to overcome the attractive potential, w, which held the electrons in the metal and the remaining energy was converted into the kinetic energy of the ejected electrons. Thus, according to Einstein

$$E_{max} = h\nu - w \tag{2.6}$$

The Einstein equation for the photoelectric effect has been confirmed many times experimentally and provides a convenient method for evaluating Planck's constant.

2.3.1

Fundamental Principles of Optical Spectroscopy: The Hydrogen Atom

The basic structure of atoms was established only after many failures of the famous scattering experiments with β-rays by Lenard (1903; textbook of *German Physics*!) and with α-rays by Rutherford (1911). According to these workers, the positively charged atomic nucleus with a diameter of approximately 10^{-4} Å (1/10000th of the atom diameter) carries almost the total mass of the atom in the form of positively charged protons and electrically neutral neutrons. The remaining mass (approximately 1/2000th of the total mass) is confined to the electron shell of the atom, which defines the form, size and, above all, its chemical properties. To avoid a collapse of the negatively charged electrons into the positively charged nucleus the then 28 year old Niels Bohr boldly assumed that electrons move in closed orbits around the nucleus in such a manner that the forces of attraction and centrifugal force are balanced in equilibrium:

$$(Z\,e)\,\frac{e}{r^2} = \frac{m\,v^2}{r} \tag{2.7}$$

Niels Hendrik David Bohr (1885–1962)

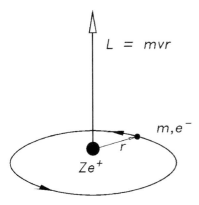

Figure 2.3 Visualization of the vector of angular momentum $L = mvr$ of an electron of mass m and charge e^- which revolves around a nucleus of charge Ze^+ on a circular track with radius r.

where Z is the number of protons, e the electrical charge and r the radius of the orbit around the nucleus (Figure 2.3).

According to the laws of electrodynamics, any rotating charge (i. e., in the vertical projection, an oscillating) should emit radiation energy; consequently the electron would be slowed down and finally collapse into the nucleus. Then Bohr came up with his famous *ad hoc Bohr's theory*, the so-called quantum condition. This stated that an electron is non-radiative and its orbit is stable, when the angular momentum (mvr) just equals an integer multiple n of $\hbar$, where n is the principal quantum number, with the abbreviation $\hbar = h/2\pi$:

$$mvr = n\hbar \tag{2.8}$$

With Eqs. (2.7) and (2.8) we obtain the formula for the radii of the allowed electron orbits in the central field:

$$r = \frac{n^2\,\hbar^2}{m\,Z\,e^2} \tag{2.9}$$

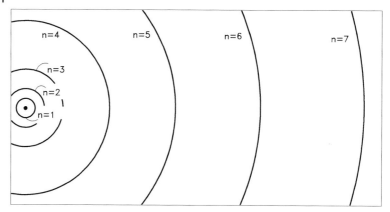

Figure 2.4 Geometrical representation of Bohr's circular
tracks of an electron in a central field with quantum number *n*
= 1, 2, ... The radius of a circular track increases with the square
of the quantum number *n*.

For the special case of $Z = 1$ and $n = 1$, which describes the hydrogen atom in its
ground state we obtain the so-called Bohr radius:

$$r = a_0 = \frac{\hbar^2}{m\, e^2} = 5.29177 \pm 7 \times 10^{-11}\ \mathrm{m} \tag{2.10}$$

For higher quantum numbers of n we obtain a square dependency of the radius of
the hydrogen atom in the form of $r_n = n^2 \times a_0$ (Figure 2.4).

A discovery by the French physicist L.-V. de Broglie in 1924 meant that Bohr's
theory, Eq. (2.7), gained acceptance. He described wave properties of electrons as
elementary particles with velocity v. Similarly, a monochromatic light beam is associated with a specific wavelength [cf. Eq. (2.2)]. The wavelength of an electron
beam equals

$$\lambda = \frac{h}{p} \tag{2.11}$$

where $p = mv$ is the momentum of the particle.

Louis-Victor, Prince de Broglie (1892–1987)

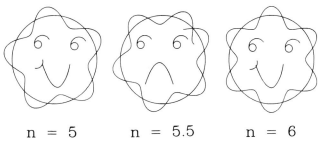

$$n = 5 \qquad n = 5.5 \qquad n = 6$$

Figure 2.5 Visualization of the quantum condition discovered by Bohr. Only integer quantum numbers lead to standing (matter) waves that allow radiationless, stable paths.

Bohr's theory states that the only possible electron orbitals are those that consist of an integer number of de Broglie waves; i. e., the de Broglie wave is a standing wave (Figure 2.5). Thus the possible electron orbitals around the nuclei are discrete and have quantized energy.

The total energy (E_{tot}) of an electron is calculated as the sum of the kinetic (E_k) and potential energy (E_p):

$$E_{tot} = E_k + E_p = \frac{m \, v^2}{2} - \frac{Z \, e^2}{r} \tag{2.12}$$

With the abbreviation

$$R_y = \frac{2 \, \pi^2 \, m \, e^4}{h^2} = 1.31225 \times 10^6 \text{ J mol}^{-1} = 3.288 \times 10^{15} \text{ s}^{-1} = 13.59 \text{ eV}$$

we obtain the expression

$$E_{tot} = R_y \times \frac{Z^2}{n^2} \tag{2.13}$$

However, this term cannot be verified experimentally and is only of theoretical interest. The main task in optical spectroscopy, however, is the determination of *differences* in energy ΔE, either in the form of absorption, i. e., elevation of a lower energy state with quantum number m to a higher orbital with quantum number n, or emission, which occurs with concomitant lowering of the principal quantum number.

The energy difference ΔE is calculated according to Eq. (2.13):

$$\Delta E = R_y \times Z^2 \left(\frac{1}{m^2} - \frac{1}{n^2} \right) \tag{2.14}$$

where m and n are the principal quantum numbers of the lower and upper energy levels of the electron transition involved. This equation was established experimentally in 1890 by J. R. Rydberg, on the basis of the hydrogen spectrum. It is valid experimentally up to a seven-fold ionized oxygen ion (8 protons, 1 electron

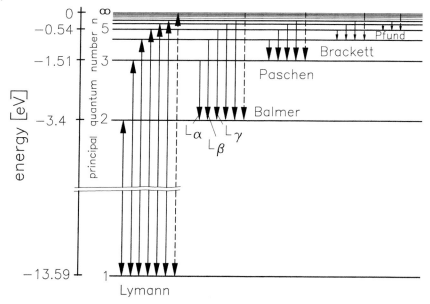

Figure 2.6 Term scheme for a hydrogen atom with the corresponding series of transitions according to the Rydberg formula. The Lymann series is observed typically under absorption and emission conditions, all others are usually only as emission lines. Only the Balmer series is found in the visible wavelength range and thus was discovered first.

in the shell). From every level of a given quantum number m, a series of spectral lines are attributed to transitions to or from higher energy states with quantum number n (Figure 2.6). The so-called Balmer series arise from radiative transitions from the upper energy levels to the $n = 2$ level, and is the only series in the visible wavelength range (and was hence discovered first).

The Rydberg equation is true in the strictest sense only for the so-called *central field* in which a single electron moves around a spatially fixed nucleus with the charge $Z \times e$. However, for the analysis of all other atomic or molecular spectra, Bohr's atomic model is insufficient, because of the effects of the numerous electrons in atoms with a higher atomic number than hydrogen. Moreover, the Bohr model requires only a single quantum number n, essentially taking only one spatial dimension into account, while in reality we expect at least three quantum numbers, corresponding to the Cartesian coordinate systems. The assumption of plane orbitals for moving electrons does not reflect the three-dimensional properties of space. Bohr himself did not believe in plane orbitals for long.

In 1926 the Austrian Erwin Schrödinger formulated quantum mechanics for atomic systems with his famous equation as he took (following the concept of de Broglie) the wave properties of matter into account. Here we will describe only the essential ideas that allow us to understand the spectroscopic concepts. We will encounter some statements that seriously contradict our macroscopic view of nature, a problem which in principle cannot be avoided. Firstly, what is it that vibrates in de Broglie's concept of the de Broglie wave of an electron? For electrons in an atom we describe the probability, ψ^2, of finding an electron in a given volume element dV. ψ is the eigenfunction, or wavefunction and is commonly referred to as an orbital. From the mathematical viewpoint ψ is a complex function with a real and an imaginary part, which describes any (microscopic) system completely in space and time, an entity which carries a sign (however, only in a few exceptional cases is ψ known exactly). The density of probability is, according to the laws of physics, defined by the square ψ^2 (i.e., a real number), the probability of finding an electron in the volume element dV is obtained by $\psi^2 dV$. The probability of finding the electron somewhere in space is of course 100%, i.e., $\int \psi^2 dV = 1$ (*normalization* of the eigenfunction). The Schrödinger equation is analogous to the equation of motion in classical mechanics which, for example, describes the vibration of a one-dimensional string with its fundamental and over-tones. The equation is a partial differential equation that connects ψ, V, kinetic (E_k) and potential energy (E_p) of a system.

The short form of the Schrödinger equation does not tell the beginner much: $H\psi = E\psi$, where H is the Hamiltonian operator, E the energy and ψ the wavefunction or eigenfunction. The Hamiltonian operator, as a mathematical entity, "extracts" the energy value out of the corresponding eigenfunctions ψ.

Equivalent to the mechanical vibrations of a plucked string, there are an infinite but discrete number of solutions to this equation for ψ, where each solution, the *eigenfunction*, corresponds to a specific value of the total energy $E = E_k + E_p$, the *eigenvalue*. While in classical mechanics the vibration of a one-dimensional string is characterized by essentially one value, we expect three discrete parameters for the three-dimensional system: the principal quantum number n, which we are already familiar with from the discussion of Bohr's model, the orbital momentum quantum number l and the magnetic quantum number m_l. A specific eigenvalue belongs to each eigenfunction ψ_{n,m,m_l}.

In a sense the electron is spread out over the whole space. It forms a type of "electron cloud". This uncertainty is a basic property of matter, which, however, is only manifested in a microcosm. It was quantified by Werner Heisenberg in the well known *uncertainty principle*, mentioned previously. According to the so-called *conjugated* entities such as locus and velocity v of a particle of mass m (specifically momentum $p = m \times v$) or energy and time [Eq. (2.5)] are not defined or measurable *simultaneously*. Again Heisenberg's uncertainty principle is valid, but with an uncertainty of momentum and locus.

$$\Delta p \times \Delta d \geqslant \frac{h}{2} \tag{2.15}$$

The more precisely the position is determined, the less precise is the determination of the momentum, and vice versa. As precise calculations show, the maximum probability of the presence of the 1s electrons in the hydrogen atom is at a_0, the classical Bohr radius which confirms the simple concept proposed by

Bohr. However, the electron is only within this sphere for 32 % of the time; but the probability of its presence inside a sphere of radius $5 \times a_0$ is as high as 99.7 %. Nevertheless, the simple model of Bohr where the electron moves like a point of mass in an orbit obviously contradicts the Heisenberg uncertainty principle. Schrödinger's equation provides, in an unrestricted manner, the atomic orbitals and their eigenvalues, which are both specified by the three quantum numbers already mentioned. These are integers and are mutually limited as the appropriate quantum mechanical calculations show: The principal quantum number runs from $n = 1, 2, 3...$, the orbital angular momentum quantum number l at a given value of n will run from $0, 1,...$ to $n - 1$. Finally, the magnetic quantum number m_s with given values for n and l will take on the values $-l, -l + 1, ...,0 ,... l - 1, l$. Using the orbital angular momentum quantum number l, the angular momentum L itself is calculated to be

$$L = \hbar \times \sqrt{l\,(l + 1)} \tag{2.16}$$

The product in the form of $l\,(l + 1)$ is known as the "quantum mechanical square" as this entity, following the classical derivation, is l^2. This finding alone reflects the difference between classical and quantum mechanics. For larger quantum numbers both points of view converge.

For historical reasons electrons with orbital angular momentum quantum numbers $l = 1, 2, 3, 4$ are called s-, p-, d- and f-electrons. In addition to the quantum numbers discussed so far, a fourth quantum number, namely the spin quantum number m_s (projection onto the z-direction), is indispensable, and is explained theoretically in the relativistic quantum theory of Paul Dirac. It takes account of the manifold of space and time and describes the intrinsic quantized angular momentum of the electron spin and possesses only 2 possible values: $m_s = \pm 1/2$. Even if there is no physical explanation that an electron rotates, this picture is capable of visualizing the spin, including the dimension of an angular momentum. The angular momentum of spin is calculated in a manner analogous to Eq. (2.16): $\hbar\,[1/2(1/2 + 1)]^{1/2}$. In Table 2.1 all possible combinations of the four quantum numbers are listed, and Figure 2.7 shows the energy levels ("terms") of the hydrogen atom together with the corresponding nomenclature. Now it becomes clear why Bohr's model verified the experimental data for the hydrogen atom so well. In a central field, such as in a hydrogen atom, we only have to take two particles, such as the proton and the electron, into account, and all energy levels are degenerate with respect to the three quantum numbers l, m_l and m_s. This means that different orbitals with the same principal quantum number possess identical eigen-, that is, energy-values E. The most important atomic orbitals (more precisely ψ^2, i.e., the density distribution) including the correct terminology can be visualized in Figure 2.8.

Table 2.1 The possible combinations of the four quantum numbers and their allocation to the atomic shells.

Atomic shell	n	l	m_l
K	1	0	0
L	2	0	0
		1	−1, 0, 1
M	3	0	0
		1	−1, 0, 1
		2	−2, −1, 0, 1, 2
N	4	0	0
		1	−1, 0, 1
		2	−2, −1, 0, 1, 2
		3	−3, −2, ,1, 0, 1, 2, 3

General relationship of quantum numbers:

$$n = 1, 2, \dots n$$
$$m_l = 0, \pm 1, \pm 2, \pm 3 \dots \pm l$$
$$l = 0, 1, 2, \dots (n-1)$$
$$s = \pm 1/2, 2e^- \text{ per orbital}$$

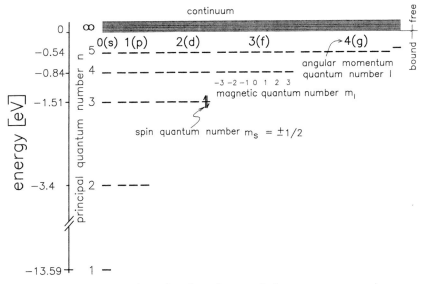

Figure 2.7 Extensive term scheme for solving the Schrödinger equation of a hydrogen atom. For a better understanding, in contrast to Figure 2.6, the distribution of the four quantum numbers is included (cf. Table 2.1), even if we have to take into account a strong degeneration. On excitation with light energies exceeding 13.59 eV the atom is ionized (transition to the continuum).

a)

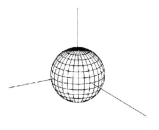

1 s-orbital (1 piece)

2 p-orbitals (3 pieces)

b)

c)

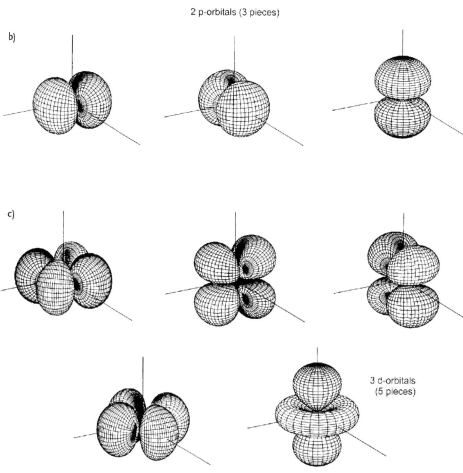

3 d-orbitals
(5 pieces)

Figure 2.8 Spatial representation of the s-, p- and d-atomic orbitals with the lowest possible quantum number. Because it is not feasible to depict a function that depends on the three space coordinates *x*, *y* and *z*, in these diagrams the function which corresponds to a fixed or- bital value of ±0.01 is drawn. As only real fig- ures can be depicted, we have shown linear combinations ψ_{nlm} of $\psi_{nl(+m)}$ and $\psi_{nl(-m)}$ (Brickmann et al. 1978, *Chemie in Unserer Zeit* *1*, 23–26, with permission of *Verlag Chemie*).

2.3.2
Atoms with Several Electrons

In 1928 Dirac made the provocative statement that "chemistry is now reduced to mathematics". However, the physics of larger molecules is much too complex to be solved by rigorous mathematics. The calculations of energetic states of even small, simple biomolecules, such as a sugar or an amino acid, only yield relative values when explaining experimental results. Thus on such a basis, hypotheses are only verified or contradicted. Even though modern supercomputers allow the theoretician to calculate information on relatively complex biomolecules with some rigor, for example, by *ab initio* quantum mechanical methods, analytical spectral analysis will remain, and represents an indispensable experimental tool for the study of biomolecules.

2.3.2.1   Shells and Periods

In hydrogen-like atoms all states with the same principal quantum number n are degenerate, i. e., they have the same energy.

However, this holds true only in the first approximation; with high resolution spectrophotometers the individual lines show a distinct fine structure, which can also be explained on the basis of a more refined theory.

If we now turn our attention to atoms with several electrons, degeneracy with respect to the orbital angular momentum quantum number l arises. The degeneracy with respect to the magnetic quantum number m_l and the spin quantum number m_s remain. Elements with the same principal quantum number n belong to the same shell (n = 1, 2, 3, 4... correspond to the K-, L-, M-, N-...shells), where each shell accommodates a maximum of $2n^2$ elements, as is easily calculated on the basis of the possible combinations of the quantum numbers. Elements that are grouped energetically belong to one period. Periods encompass successively 2, 8, 8, 18, 18, 32, 32 elements as is immediately deduced from the scheme in Figure 2.9, and which is reflected by the Periodic Table (Appendix A4). This is an *energetic* order, which, essentially, is deduced empirically. A period is *not* defined by the principal quantum number as is sometimes assumed: elements with the same quantum number n are localized on "diagonals" of the term scheme and may belong to different periods.

At the first sight it appears to be difficult to describe an atom with numerous electrons and four quantum numbers, each with a common spectroscopic scheme. How does the atomic system behave as a whole? The solution is surprisingly simple if we are dealing with atoms within the range of atomic numbers of less than 40, which is usually the case in chemistry and the bio-sciences. The so-called *Russel–Saunders coupling* applies (in contrast to the so-called *j–j coupling* for atoms with atomic numbers larger than 40). Each atom has a specific total angular momentum $J = S + L$, which is composed of the vector sum S of all angular

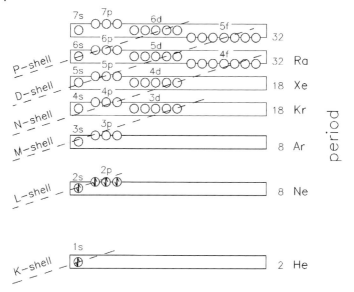

Figure 2.9 Schematic representation of the electron distribution in the periods of the Periodic Table of the elements. Shells correspond to the principal quantum number and typically cover several periods. This clearly shows how the number of electrons, i.e., elements, increases with increasing periods. The closed shells correspond to the noble gases indicated.

momenta of all individual spins *s* and the sum **L** of all individual angular momenta *l* of all shells that are not completely filled.

"Lower" lying, completely filled shells do not contribute to optical spectra and therefore can be ignored (the relevant spectroscopy of inner electrons is not the subject of this textbook). The sum of all wavefunctions of a completely filled shell is precisely spherical in shape.

Of course, the resulting total angular momentum **J** is again an integer or half-integer multiple of the (elementary) angular momentum ℏ. Only the partly filled "outer" shells determine the chemical and therefore the opto-spectroscopic behavior of atoms; the same holds true for molecules as we will see later.

2.3.2.2 Vector Model, Rules of Pauli and Hund

As discussed previously for each orbital angular momentum quantum number *l*, $2l + 1$ orbitals always exist, which differ only in their magnetic quantum number m_l, i.e., their spatial orientations. These differently oriented orbitals are degenerate. The degeneracy in m_l is expected to be removed when we apply an external linear magnetic field (conventionally drawn in the *z*-direction), as the *x*-, *y*- and *z*-directions are no longer equivalent: As a result, we observe a splitting of the spectral lines. A vector picture is very useful to obtain a clear overview of the quantization in space: the three p- or the five d-orbitals are depicted by arrows of length "1" and "2", respectively, where the projections onto a fixed direction in space (usually the vertical) correspond to integer lengths of the unit $\hbar$. Figure 2.10 shows carbon (C) and iron (Fe) as examples. Each arrow corresponds to an orbital ψ, which holds a maximum of 2 electrons of different spin quantum numbers $m_s = \pm 1/2$, as the other three quantum numbers are identical. This rule holds for all elementary particles with spin 1/2 (fermions) and is known as "Pauli's exclusion principle":

- All electrons in an atom or molecule are distinguishable by at least one quantum number.

The scheme of *how* individual electrons fill the available orbitals is defined by the rules given by Hund:

- The first Hund rule says that the ground state of an atom (or molecule) is the state with the largest total angular momentum **L**.
- The second rule postulates concomitantly the largest possible total spin momentum **S**.
- Finally, the third rule relates the vectorial connection between the angular momentum **L** and spin angular momentum **S** in the ground state of a system, thus allowing a classification of atomic spectra. If an electron shell is less than half filled we obtain **J** = **L** − **S**, if it is more than half filled **J** = **L** + **S**, and if it is just half filled than **J** = **L** + **S** = **S** because **L** = 0.

Correspondingly we attribute a so-called spectral term to an atom, which defines the spectral properties:

$$^{2S+1}L_J \tag{2.17}$$

The superscript index $2S + 1$ denotes the *multiplicity*, L the angular momentum and the subscript the *total angular momentum* (Figure 2.10). As a preview to the molecular spectroscopy to be discussed later, we will emphasize here two important cases of multiplicity. If all electrons are paired, i.e., all orbitals are doubly occupied and the total spin is $S = 0$, than we talk about a *singlet state*, because an external magnetic field does not cause a splitting of the spectral lines. For $S = 1$, however, the state is a *triplet state* because it is split into three closely spaced levels when an external magnetic filed is applied.

For carbon atom the vector sum of the angular momentum **L** = $1\hbar$ and the total spin momentum **S** = $1\hbar$ yields three possible quantum numbers for the total angular momentum, where the ground state is defined by **J** = 0. For iron with L = 2 and S = 2 there are five possible quantum numbers for the total angular momentum **J** with J = 4 in the ground state. Table 2.2 lists terms for more chemically and biologically relevant atoms in the ground state.

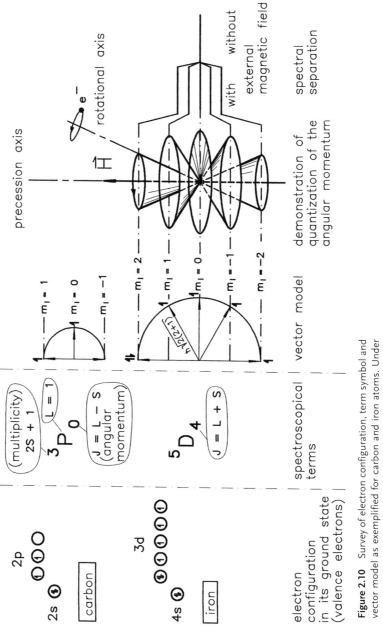

Figure 2.10 Survey of electron configuration, term symbol and vector model as exemplified for carbon and iron atoms. Under isotropic conditions there is no directional quantization, the corresponding states are degenerate. If a linear magnetic field *H* is applied we observe spectral splitting.

Table 2.2 Term symbols of the elements in the first three periods of the Periodic Table. Vertically arranged elements have the same number of s- and p-electrons and thus the same term symbols, i.e., similar spectra (cf. Figure 2.13).

^{2}S	^{1}S	$^2P_{1/2}$	$^3P_0$	^{4}S	$^3P_2$	$^2P_{3/2}$	^{1}S
H							He
Li	Be	B	C	N	O	F	Ne
Na	Mg	Al	Si	P	S	Cl	Ar
K	Ca	Ga	Ge	As	Se	Br	Kr

Sc $^2D_{3/2}$	Ti $^3F_2$	V $^4F_{3/2}$	Cr ^{7}S	Mn ^{6}S	Fe $^5D_4$	Co $^4F_{9/2}$	Ni $^3F_4$	Cu ^{2}S	Zn ^{1}S

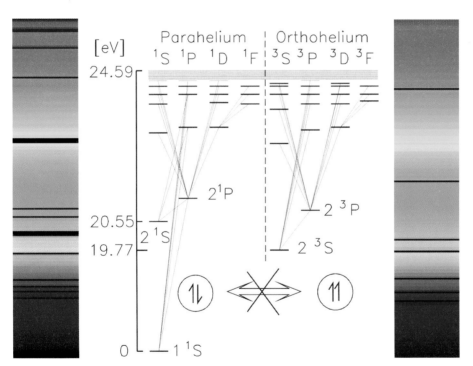

Figure 2.11 Term scheme of para- and orthohelium. Both "species" are in reality one molecule "helium" where the singlet state corresponds to the para- and the triplet state to the orthohelium. The concomitantly forbidden transition between both systems (*intersystem crossing*) led to the early assumption of two independent molecules.

In order to describe the spectral properties of atoms with several electrons, the simple scheme as used for the hydrogen atoms is no longer sufficient but we have to take into account transitions between different terms. However, not all electronic transitions are allowed or observable, only those that fulfill certain *selection rules* according to symmetry considerations of the electronic states are involved. From the viewpoint of wave mechanics, electronic transitions are usually more often forbidden than allowed – allowed transitions are the exception. A quantum mechanical calculation provides the following rules (and in addition to these, information about the polarization properties of absorbed and emitted radiation is given in Section 2.3.3.4):

- The total angular momentum J of an atom does not change at all or only by ± 1 (however, transitions from $J = 0$ to $J = 0$ are forbidden).
- The multiplicity is not allowed to change: relaxation of this rule by various factors is important in molecular spectroscopy.
- The angular momentum l changes by exactly ± 1.
- The magnetic quantum number m_l is not allowed to change, or only by ± 1.

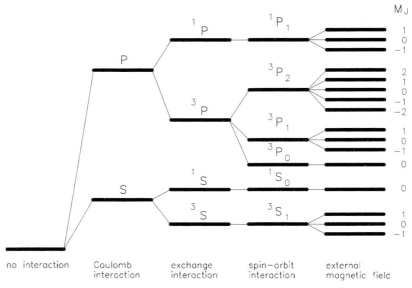

Figure 2.12 Schematic visualization of the complete splitting of an electronic transition of helium (initially totally degenerate) from $n = 1 \rightarrow n = 2$ upon successive consideration of assumed "disturbances" into 16 lines $\{2 \times [2 + (3 \times 2)]\}$.

2.3.2.3 The Helium Atom

On the basis of the discussion of the hydrogen atom as the simplest system, we will now take a look at the helium atom with its two electrons. It actually shows an unexpectedly complex spectrum, which historically was only understood after extensive research efforts. However, the understanding of spectral processes significantly advanced as a result of the helium atom, and led to the study of larger, multi-electron molecules. The spectroscopic data of helium can be subdivided into two distinct groups. Helium exists as *parahelium* and *orthohelium*. As there are two electrons in the helium atom, two spin states, singlet and triplet, are possible (Figure 2.11, see p. 31). Spectral transitions between the two spin states are forbidden and only take place within one system having the same spin state (Section 2.3.2.2, the second rule).

The ground state of helium (parahelium) is designated as $^1S_0$, and in the lowest excited state one of the two electrons is in the quantum number $n = 2$ level with the two possible electron configurations $1s^1 2s^1$ and $1s^1 2p^1$. While for a hydrogen atom the transition from $n = 1$ to $n = 2$ yielded a single spectral line whose wavelength can be calculated accurately with the Rydberg formula, the corresponding helium line splits into 16 lines after removal of all degeneracies as illustrated by Figure 2.12.

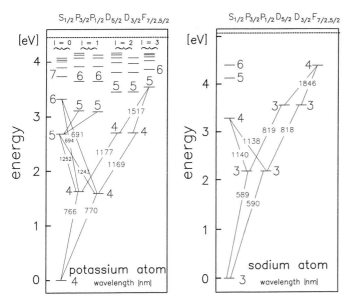

Figure 2.13 Term scheme of the alkaline earth metals potassium (detailed presentation) and sodium with their hydrogen-like spectra. The energy values are referred to the ground state. The red potassium line and the yellow sodium line are double lines corresponding to the $S_{1/2}$ → $P_{1/2}$, $P_{3/2}$ transitions. In particular the bright, yellow sodium D-line (D = double) serves to evaluate the resolution of spectrophotometers for test purposes.

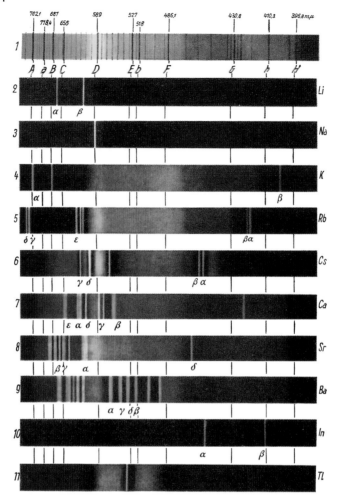

Figure 2.14 Line spectra of various elements in the visible spectral range as obtained with a grating spectrophotometer. The (continuous) sun spectrum shows a series of dark lines that are generated by the absorption of sunlight in the chromosphere of the sun. For example, the Fraunhofer lines **C** and **F** are the first two lines of the Balmer series of hydrogen (cf., Figure 2.6), **D** is the sodium double D-line (double line, Figure 2.13), **E** and **G** are two of many iron lines (cf. Figure 2.10), **A** and **B** are O_2-oxygen lines (cf. Figure 2.19).

S-terms are typically lower lying than P-terms with the same principal quantum number n, while the probability of finding s-electrons is greater in the immediate neighborhood of nuclei than that of p-electrons, and similarly the electrostatic and Coulomb interaction. In the triplet state both electrons have the same spin; they are in different orbitals and the Pauli exclusion principle prohibits the electrons from coming too close. However, in the singlet state the electrons are paired (total spin = 0), no Pauli exclusion principle is in operation, allowing the two electrons to come close. This results in repulsion, which raises the singlet term above the triplet term. The so-called *spin orbit interaction* finally leads – after vector addition – to the quantum numbers of the total angular momentum J, where the terms with higher J are more stable and thus lying at a lower energy. Finally, if all internal interactions are taken into account, the remaining degeneracies with respect to the quantum number M_J are removed by an external magnetic field. M_J and J are analogous to the values m_l and l, as discussed previously for hydrogen atoms.

Figure 2.13 (see p. 33) shows the term schemes for potassium and sodium in the visible and infrared region. The individual lines are doublets, because the total spin $S = 1/2$, as holds true for all alkali metals ($2 \times 1/2 + 1$). The spectra of various elements in the visible wavelength range are shown in Figure 2.14.

The well known "yellow sodium-D-line" (doublet lines) at 589 and 590 nm is due to transitions $^2S_{1/2} \rightarrow {}^2P_{3/2}$ and $^2S_{1/2} \rightarrow {}^2P_{1/2}$ and it often serves as a wavelength calibration.

As pointed out earlier, the spectra of all two-electron systems such as helium, alkaline earth metals or a hydrogen molecule display singlet and triplet series. In optical spectroscopy and in the photochemistry and biology of larger molecules, the differential roles of the triplet and singlet states are of interest.

2.3.3
Simple Molecules and their Spectral Properties

2.3.3.1 The Chemical Bond
The ground state of nearly all molecules is a singlet state, with molecular oxygen (O_2) as the most significant exception. The spectroscopic properties of atoms are determined by their electronic structures, but the spectral properties of molecules are dependent upon the structure and the chemical bonding of their individual components.

Each chemical bond in a molecule is based on the mutual interaction of electrons and nuclei. The electrons of the outer, partially filled shells, known as the "valence electrons", play the predominant role when forming a chemical bond.

In general the "deeper" lying shells are completely filled, i. e., they have a "closed" and totally symmetrical wavefunction. Exceptions can be found in the lanthanide and actinide groups (which are important in fluorescence spectroscopy).

It is a good idea to distinguish between various subclasses of chemical bonds, even if the distinction is not always clear-cut:
• Ionic bonds
• Covalent bonds
• Coordination bonds
• Hydrogen bond
• Van der Waals forces
• Crystal bonds in solids, metal bond

The ionic bond is based on the electrostatic attraction of positively and negatively charged partners (e. g., Na^+ and Cl^- in NaCl) and not on the immediate interaction of electrons. The same argument holds true for van der Waals forces, which depend on the interactions of induced dipole moments, dispersion interactions and dipole–dipole interactions. The hydrogen bond, on the other hand, depends on an electronic interaction and has an, often even indirectly, modifying influence on optical spectra. Solids exhibit special properties that are manifested by co-operation of atoms and molecules in the lattice. Most amenable to optical spectroscopic analysis, however, are covalent bonds and metal coordination bonds, which we will discuss in more detail below.

2.3.3.2 Electronic Spectra of Diatomic Molecules

Again, the basic laws are best understood with the simplest system, the hydrogen molecule, which is built from the covalent bond between two hydrogen atoms. The results are easily transferred to more complex molecules.

Some ground rules will simplify our understanding. Previously we were acquainted with the atomic wavefunction φ. This is a mathematically complex function with real and imaginary parts including a sign, "+ " or "−", also known as the phase. A covalent bond is only established between atomic orbitals with the same phase, i. e., sign: a new orbital is generated, the molecular orbital (*MO*), which again can accept a maximum of two electrons of opposite spins. A bonding orbital requires sufficient spatial overlap of the atomic orbitals. For example, bonding between a 1s- and a 3d-orbital is not really possible; however a 2p- with a 1s-, or a 3p- with a 4d-orbital will easily form molecular orbitals. Finally, we have to take into account an important rule of quantum mechanics which states that the total number of orbitals remains constant.

Let us assume that two distant hydrogen atoms with their 1s-orbitals φ_A and φ_B approach each other, and there are mutual interactions, resulting in a molecular orbital (MO). How is the MO generated from the atomic orbitals? The basic idea, first formulated by R. S. Mulliken, was originally a mathematical solution. According to Mulliken, atomic orbitals combine linearly and additively or "in-phase" to form a molecular orbital ψ, $\psi_b = \varphi_A + \varphi_B$ (with the subscript b representing bonding). This procedure is known as the *linear combination of atomic orbitals* (*LCAO*). Both electrons are mainly confined in time and space to the connecting line between both nuclei. The electrons thus occupy the *bonding orbital*. According to the rule determining the number of orbitals (Table 2.3), we can expect a second *MO*. We can inspect the energetic properties using the potential curves in Figure 2.15. These describe the total energy of the system as a function of the nuclear coordinates. On the one hand the nuclear–nuclear approach leads to the expected energy minimum, i. e., bonding (with the bond length r_0; any further approach is forbidden by the strong electrostatic repulsion, as the potential curves increase steeply). On the other hand we observe an immediate repulsion right from the beginning of the approach, which is described by an *antibonding orbital*. In the first case both electrons fill the bonding orbital. In the second case one electron

Table 2.3 Conservation rule for the number of orbitals. When molecules are formed through covalent bonds between atoms the total number of orbitals is conserved; e. g. 2 × 3 atomic orbitals yield 6 molecular orbitals.

Atomic orbitals	Molecular orbitals
2 *s*-orbitals →	1 σ, 1 σ* (2 orbitals)
6 *p*-orbitals →	1 σ, 1 σ* (6 orbitals)
	2 π, 2 π*
10 *d*-orbitals →	1 σ, 1 σ*
	2 π, 2 π* (10 orbitals)
	2 δ, 2 δ*

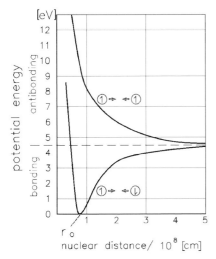

Figure 2.15 Potential curves of two hydrogen atoms approaching each other from large distances. We can distinguish a bonding and an antibonding state. Upon binding, at a distance of $r_0 = 0.74$ Å the covalent binding force and the electrostatic repulsion keep the nuclei in equilibrium, and a potential well builds up. In the case of antibonding, there is no a stable situation and we refer to an impact pair.

fills the bonding, the other the anti-binding orbital. The resulting bonding energy is zero, i. e., no bond is formed (antibonding).

The out of phase linear combination of atomic orbitals results in the formation of an antibonding MO, $\psi_a = \varphi_A - \varphi_B$ (subscript "a" represents antibonding) as shown by the upper curve in Figure 2.15. The MO changes its sign at the nodal plane between the two nuclei, where the electron density and thus the bonding energy is zero (Figure 2.16).

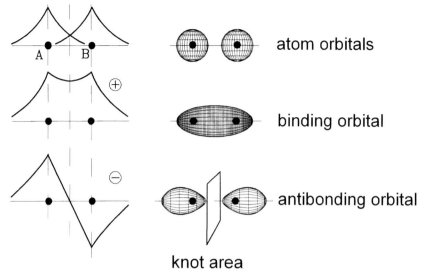

Figure 2.16 Energetic and spatial visualization of the formation of molecular orbitals on the basis of the linear combination of atomic orbitals using hydrogen as an example (*linear combination of atomic orbitals, LCAO*). Top: atomic orbitals, middle: binding molecular orbital, $\psi_b = \varphi_A + \varphi_B$ (electron density is maximum along the line connecting the nuclei), bottom: antibonding molecular orbital, $\psi_a = \varphi_A - \varphi_B$ (nodal plane between the nuclei).

In analogy with the terminology for atoms, the molecular states adopt Greek notations (Table 2.4). Instead of s-, p-, d-orbital electrons we have σ-, π- and δ-electrons in molecules; the total angular momenta are termed Λ rather than L, the individual states with Λ-values of 0, 1, 2 and 3 are Σ, Π, Δ and Φ. Antibonding orbitals are marked with a superscript asterisk (*). As with hydrogen atoms, both MOs of the hydrogen molecule have no angular momentum and are termed σ-orbital (bonding) and σ^*-orbital (antibonding).

Table 2.4 The Arabic characters used for the angular momenta for atoms and Greek characters for molecules.

System	Angular momenta			
	1	**2**	**3**	**4**
Single electron (atom)	s	p	d	f
Multi-electron system (atom)	S	P	D	F
Single electron (molecule)	σ	π	δ	φ
Multi-electron system (molecule)	Σ	Π	Δ	Φ

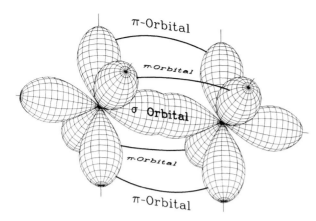

Figure 2.17 Spatial distribution of the triple bond in the nitrogen molecule: Two π-bonds and one σ-bond. For clearer representation the π-orbitals are shown – as usual – in a more elongated form compared with reality (cf. Figure 2.8).

In a molecular system, atoms often form multiple bonds; e. g., a nitrogen molecule even forms a triple bond. In the nitrogen atom there are three p-orbitals with one electron each (4S-term), which can be combined into one σ- and two π-molecular orbitals. A simple geometrical representation can visualize that two of the three p-orbitals (p_x, p_y, p_z) can be placed parallel to each other, where they overlap above and below the two planes containing the two atomic nuclei: The two bonding electrons have the angular momentum $l = 1$ and thus are π-electrons. On the other hand, the two p_x-orbitals aligned along the axis of the two nuclei overlap to form a molecular orbital. The angular momentum associated with this molecular orbital is $l = 0$ which is therefore a σ-orbital (Figure 2.17). In analogy with atomic orbitals, molecular orbitals with the angular momentum $l = 2$ are "δ-orbitals".

Figure 2.18 demonstrates, with a so-called correlation diagram, the electronic configuration for the diatomic molecules H_2, He_2, N_2 and O_2. Two s-orbitals become one σ- and one σ^*-molecular orbital, two from each of the three p-orbitals (p_x, p_y, p_z) become one σ- and one σ^*-molecular orbital and two bonding π- and antibonding π^*-orbitals. Lower lying atomic orbitals where the energy is not or only slightly influenced by the binding process are termed "non-binding", the electrons are "n-electrons". To construct the correlation diagram, we fill the MOs in order of increasing energy using the existing total number of electrons from "bottom to top". The electronic properties of the O_2 molecule are most important and are shown in detail in the molecular orbital diagram in Figure 2.19. As mentioned previously the ground state of oxygen is a triplet.

In order to visualize the energetic properties of a molecule in a straightforward manner – corresponding to the measurement of a current – instead of the molecular orbital diagram we usually choose a method of representation invented by Jablonski (Figure 2.20). It is not the energetic molecular states themselves that are shown but rather the *energetic differences* between individual transitions with the appropriate terminology. For example, $S_1(n,\pi^*)$ describes the energetically lowest excited state of a molecule with singlet character after an electronic transition $n \rightarrow \pi^*$. Then if the first triplet state is reached by a $\pi \rightarrow \pi^*$ transition we call

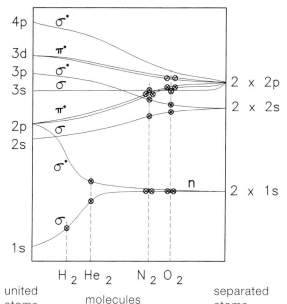

Figure 2.18 Homonuclear, diatomic correlation diagram. Schematic representation of the change in energy of the molecular orbitals as a function of nuclear distance. For some molecules the electron distribution is given, from which the spectroscopic properties can be derived immediately. The *n*-orbitals of a molecule (non-binding) correspond to pure atomic orbitals and do not contribute to molecular spectra. Antibonding orbitals are indicated by asterisks.

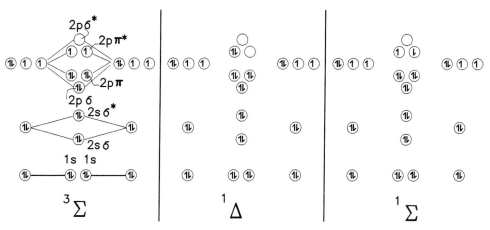

atom molecule atom

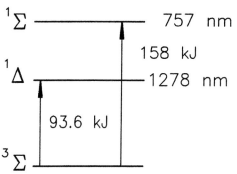

Figure 2.19 Term scheme and energies of the three most important states of the oxygen molecule. The important first singlet state $^1\Delta$ (singlet oxygen) with 1278 nm is located in the near-infrared spectral range.

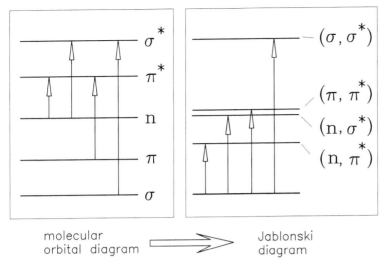

molecular orbital diagram $\Longrightarrow$ Jablonski diagram

Figure 2.20 Conversion of a molecular orbital diagram in a practical, energetically oriented Jablonski diagram, in which all observed transitions are drawn on the basis of a common energy level. Each transition in a Jablonski diagram reflects a difference in energy and the types of source and target states are indicated (if known).

it $T_1(\pi, \pi^*)$. Such terminology is not only of theoretical value but immediately reflects the spectroscopic/analytical properties, as we will see later in more detail.

Most spectra have their origin in small regions of a molecule only. Thus the electron of a hydrogen atom in a methyl group (–CH₃) is confined within just a few Ångströms (10^{-10} m). On the other hand, there are molecules where the electrons are delocalized across large areas and more orbitals in the order of 10 Å or even multiples of these. Conjugate double bonds are the reason for the intensive color of molecules ("pigments") where electronic transitions not only take place in the ultraviolet but predominantly in the visible wavelength range. These pigments are of great importance in optical spectroscopy, which we will discuss in more detail later.

2.3.3.3 Molecular Vibrations

In addition to the properties of atoms that have exclusive electronic energy levels and display spectral *lines*, diatomic and polyatomic molecules can be excited to higher vibrational and rotational levels by the absorption of infrared and microwave frequencies. While the absorption and emission properties of individual atoms with respect to "lower lying" electrons (non-valence electrons, X-ray range) in a bond in a molecule remain unchanged, the molecules display quite different spectral properties. There are individual spectral lines, spectral bands and whole systems of more or less overlapping spectral bands of great complexity,

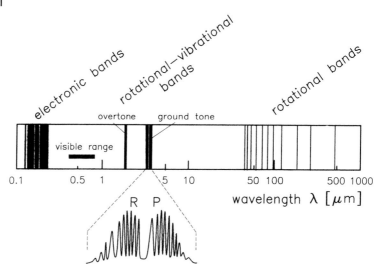

Figure 2.21 Complete absorption spectrum, which extends over orders of magnitude, of a polar, diatomic molecular using hydrochloric acid (HCl) as an example. Corresponding to the three transition types we can distinguish electronic, vibrational and rotational transitions, the energies of which are successively lowered by a factor of approximately 15 (a rule of thumb). The electronic bands in the ultra-violet cannot be resolved into individual lines. In contrast, the rotational–vibrational bands R and P in the infrared can be further resolved into individual lines as shown.

which are obviously related to the outer shell, i. e., valence electrons. It required the work of a whole generation of scientists to explain these spectra and to make optical spectroscopy into a powerful analytical tool in science and then later also in industry and technology. Again, we will abbreviate this long and tiresome path through a type of bird's eye overview. For example, Figure 2.21 shows the spectroscopic properties for a simple molecule (HCl) over the whole spectral range from 100 nm (vacuum-UV) up to 1 mm (far-infrared). We can recognize three accumulations of lines and bands. In the first spectral range we see three types of transitions taking place simultaneously: electronic, vibrational and rotational. The second spectral range encompasses the so-called rotational–vibrational bands and in the third (far-infrared) there are pure rotational transitions.

The quantization of rotational and vibrational energy of molecules (even more difficult to visualize than the electronic quantization) was first convincingly suggested by the step-wise change of specific heat C_V with temperature, which is a macroscopically defined entity (Figure 2.22). When the temperature is lowered, certain degrees of motion freeze and, consequently, are no longer capable of absorbing energy. The simplest models for the investigation of molecular vibrations are diatomic molecules: they are capable of vibrating against each other, similar to two masses that are connected by a spring. According to Hook's law (energy being proportional to the square of the displacement) the vibration can be described by a

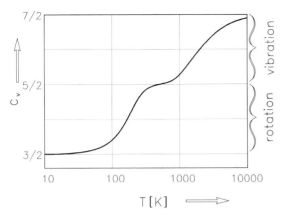

Figure 2.22 Change of heat capacity C_V (at constant volume) with the absolute temperature T for a diatomic molecule (H_2), expressed as $\varkappa = R/L$, where Boltzmann constant $\varkappa$ is the gas constant for a single molecule and L is the Loschmidt number. The degrees of freedom of oscillation are already frozen at room temperature (300 K), however not those of rotation: the molecules are in their vibrational ground state.

parabolic potential curve of the harmonic oscillator (Figure 2.23). Schrödinger's equation for a harmonic oscillator yields discrete energy values $E_v = h\nu\,(v + 1/2)$, where v is the vibrational quantum number and ν the vibrational frequency. The energy scheme shows equidistant lines of distance $h \times \nu$, where the lowest possible value – as forced by Heisenberg's uncertainty law – is not zero but $h\nu/2$. This is called the *zero point energy*. Again, theory delivers the important selection rule of $\Delta v = \pm 1$: the vibrational quantum number v must change by exactly 1. Thus, the harmonic oscillator "possesses" only one single spectral line of energy $h\nu$, because the line distance is constant over all. For small vibrational quantum numbers the harmonic oscillator is a realistic model. With greater deflections the deviations from the ideal harmonic oscillator become stronger, the energy levels approach a certain limit where the "spring breaks" and the molecules dissociate. A realistic potential curve that describes the real properties fairly well is named after the inventor, the *Morse curve*. It is asymmetric and leads to inharmonic vibrations as shown schematically in Figure 2.23. The stringent selection rule for the harmonic oscillator is weakened in such a way that, with decreasing probability, i. e., intensity, transitions with $\Delta v = \pm 2, \pm 3,...$ are also possible.

Theory and practice show more precisely that with the Morse potential curve the *decrease* in distance for subsequent levels is constant.

If two masses m_1 and m_2 vibrate against a common center of mass S, the so-called *reduced mass* $m = m_1 \times m_2/(m_1 + m_2)$ is the decisive entity which determines the vibrational frequency. Figure 2.24 shows a collection of vibrational frequencies for the most important chemical groups as a function of the reduced mass m. It is obvious that the number of possible vibrations (*normal vibrations*), increases

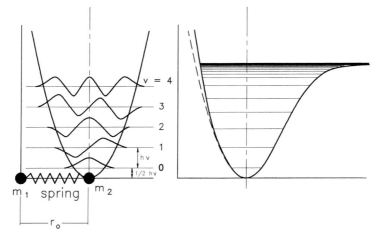

Figure 2.23 Left: The ideal harmonic oscillator with its quantum mechanically defined vibrational modes corresponding to Hook's law (two masses m_1 and m_2 vibrating against each other, connected by an elastic spring) for a parabolic potential well. The wavefunctions with the quantum numbers of vibration $v = 1$, 2..., n are shown. In its ground state a harmonic oscillator possesses a finite, so-called zero-point energy $1/2h\nu$. Right: Comparison of the potential curves of a harmonic oscillator (dashed line) with the more realistic potential curve of Morse.

with the number of individual atoms N, corresponding to the three space coordinates. These are $3N - 6$, respectively, $3N - 5$ (3 degrees of freedom of translation plus 3, respectively, for linear molecules plus 2 for the rotation of the whole molecule have to be subtracted). Thus, for a water molecule we obtain $3N - 6 = 3$ normal vibrations, i. e., two *valence vibrations* and one *deformation vibration* (Figure 2.25). With larger molecules in particular we have to distinguish between localized and delocalized vibrations. As expected, when localized valence vibrations are excited too much (according to the Morse curve, Figure 2.23) individual bonds will break and the corresponding molecule or molecular group dissociates as a whole. Indeed, when using a laser with the "correct" monochromatic energy certain photochemical reactions can be induced. The first example of the removal of a fluorine atom out of a molecule of sulfur hexafluoride (SF_6) by an infrared beam of an intense carbon dioxide laser became known as the possible start of industrially used laser chemistry.

We will now discuss the concomitant change to the electronic and vibrational states of a molecule. Here the *Frank-Condon principle* comes into action. This states that the nuclear distance during the comparatively fast electronic transition remains unchanged and will be adjusted to a new value much later. Following the theory, transitions take place predominantly between states with a large overlap of the corresponding vibrational wavefunctions (Figure 2.26). The so-called *0–0 transition*, i. e., the electronic transition without any change to the vibration quan-

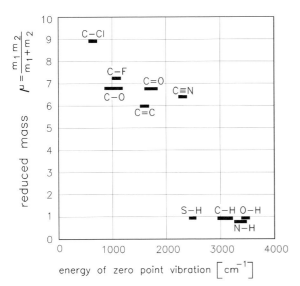

Figure 2.24 Vibrational energies of the important covalently bound chemical groups as a function of their reduced masses.

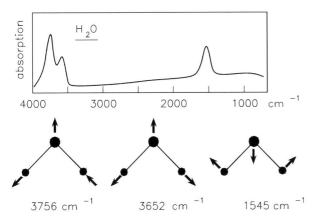

Figure 2.25 The three vibrational bands of a water molecule as a result of the corresponding modes of vibration.

tum number, as is the case for atoms, is only fully realized in the absorption spectra of molecules because the distance between nuclei typically changes on transition. The structures of molecular spectra are dominated by the Frank–Condon principle.

2.3.3.4 Transition Moments

The probability, P, that a model molecular system passes from state φ_1 to state φ_2 by the absorption of light, depends on the scalar product of the electrical light vector E and the transition moment M:

$$P \propto |\, E \cdot M\,|^2 \tag{2.18}$$

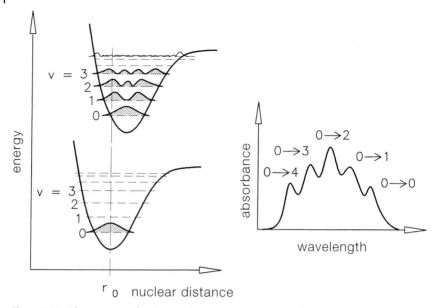

Figure 2.26 The structure of a vibration band is essentially defined by the *Frank–Condon principle*: the larger the overlap of electron densities of the ground and excited states the more intensive is the corresponding spectral band (here having a maximum absorption for the transition $v = 0 \rightarrow v = 2$).

The transition moment is defined by the equation

$$M_{12} = e \int \varphi_1 \, \mathbf{r} \, \varphi_2 \, dV \qquad (2.19)$$

where e is the electric charge, $\mathbf{r}$ the position vector, dV the infinitesimal volume element and $\mathbf{M}$ has the dimensions of a dipole moment. The (mathematical scalar) product P in Eq. (2.18) and the individual symmetry properties of φ_1 and φ_2 have two consequences, namely the selection rule and polarization effects observed in absorption and emission of molecules. Light is only being emitted or absorbed when $M_{12} \neq 0$, i. e., the transition dipole moment has to change on transition. A two-dimensional example in the x–y plane in Figure 2.27 shows the transition between two hypothetical wavefunctions φ_1 and φ_2 with certain symmetry properties, where the space vector is decomposed into its x (top) and y (bottom) components. The integrand (in the right-hand picture) reveals a nodal line in the direction of the y-component and M_{12} vanishes; however, in direction of the x-component M_{12} has a finite value. Thus, by absorption and emission experiments with polarized light we can study the structure and spatial arrangement of molecules.

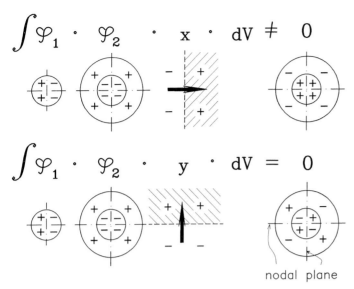

Figure 2.27 Visualization of the transition (dipole) moment on the basis of two hypothetical molecular orbitals. The symmetry properties of the contributing orbitals as well as the direction of the so-called *position vector* are important. In this example the transition shows total polarization in the x-direction, in the y-direction the transition moment disappears completely.

2.3.3.5 The Rotational Spectrum

While a diatomic molecule has only two rotational degrees of freedom around the center of mass S (the rotation around the inter-nuclear axis is inapplicable), larger molecules posses three independent possibilities of rotation and thus energy absorption (Figure 2.28). The harmonic oscillator that applies to the vibration is replaced by a "free rotator" for the rotational motion. A diatomic molecule with the atomic masses m_1 and m_2, which rotate with a radius r_0 around a common center of mass S, has the *reduced mass m*, which we mentioned previously for a harmonic oscillator. If we discuss rotational motions, it is not the mass but the distribution of mass that is defined by the moment of inertia $\theta = \mu \times r^2$ that determines the rotational energy. Assuming that a diatomic molecule rotates around an axis perpendicular to the inter-nuclear axis of both atoms (plane rotator), we obtain $P_R = \theta \omega$ for the angular momentum, where ω is the angular velocity. For the angular momentum we obtain, from Eq. (2.8), the quantum condition

$$P_R = J \times \hbar \text{ with } J = 1,2,\ldots \tag{2.20}$$

Using the relationship for the rotational energy, known from classical mechanics, $E_{rot} = \frac{1}{2}\theta\omega^2$, we obtain

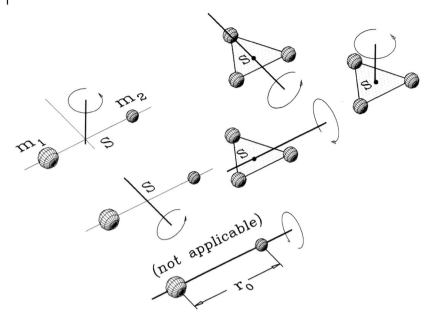

Figure 2.28 Freedom of rotation of a diatomic (2, left) and a triatomic molecule (3, right). Rotations take place around a common center of gravity.

$$E_{\text{rot}} = \frac{\hbar^2}{2\,\Theta} \times J^2 \tag{2.21}$$

where J is the rotational quantum number. If we take into account that in principle a molecule has three degrees of freedom of rotation (a free rotator), we have – following the theory – J^2 again by the quantum mechanical square $J(J + 1)$ and we obtain

$$E_{\text{rot}} = \frac{\hbar^2}{2\,\Theta} \times J\,(J + 1) \tag{2.22}$$

If we replace, as is typical in infrared spectroscopy, E_{rot} by the wavenumber $n = 1/\lambda$ [identical to Eq. (2.3)] and introduce the rotational constant $B = \hbar/(2c\theta)$, for the energy we obtain the simple expression

$$v = B\,J(J + 1) \tag{2.23}$$

Here v is usually used instead of k. On the other hand, $v = c/\lambda$ is also used for the frequency of light [e. g., in Eq. (2.1)]. Such confusing double usage is typical in all science and we are frequently confronted with such confusing notations.

A selection rule also exists for rotational transitions, namely $\Delta J = \pm 1$, i. e., each rotational state can only combine with its immediate neighbor states under emission or absorption of radiation. Thus the energy gap between subsequent rotatio-

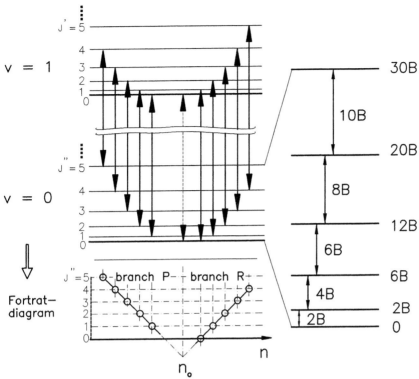

Figure 2.29 Schematic *Fortrat diagram*. According to the selection rule of rotational–vibrational transitions of diatomic molecules there are energetically different branches R ($\Delta J = +1$) and P ($\Delta J = -1$). The frequencies (energies) are plotted at the bottom (reflected by the lengths of arrows). On the right-hand side magnified energy levels are reproduced; each distance increases by $2B$ with increasing rotational quantum number.

nal states is $2B$ (Figure 2.29). Therefore, rotational spectra allow the immediate determination of the moments of inertia of molecules. However, this is true only for such molecules that change the electrical dipole moment on transition. These are molecules for which the center of charge does not coincide with the mechanical center of mass (cf. Figure 2.21, HCl). Therefore, O_2, H_2, N_2, etc., do not possess rotational spectra. However, Raman spectroscopy (Section 8.3) provides a method for treating the corresponding molecular data.

2.3.3.6 **Rotational–Vibrational Bands**

A rotational–vibrational band is caused when each vibrational transition is super-imposed by several rotational bands. In Figure 2.29 the series for two rotational terms which belong to the vibrational levels $v = 0$ and $v = 1$ are simplified schematically. According to the selection rule $\Delta J = \pm 1$ there are two branches of the resulting rotational–vibrational bands corresponding to the transitions $\Delta J = +1$ and $\Delta J = -1$. The line series which corresponds to $\Delta J = +1$ and which decreases in wavelength with increasing J is termed the *R-branch*, in analogy the line series for $\Delta J = -1$ in which there is an increase in wavelength with increasing J is called the *P-branch*. According to the selection rule the pure rotational transition (0–0 transition) at first glance is forbidden and is shown in Figure 2.29 as a dashed line and termed n_0. However, the spectroscopic findings do not meet this expectation exactly, and therefore our model has to be refined. Firstly, the moment of inertia changes on transition; secondly, owing to the interaction the rotational and vibrational energies should not simply be additive. This property is taken into account by additional mixed terms that lead to satisfactory results. In the lower part of Figure 2.29 the energetic course (the length of the arrows correspond to the energy) versus the rotational quantum number of the lower state (J'') is depicted. This representation is known as a *Fortrat diagram* and provides the best overview of this model.

2.3.3.7 **Vibrational, Rotational and Electronic Transitions**

In the most complex case, which in optical spectroscopy is in fact the real situation, electronic, vibrational and rotational transitions take place simultaneously. The electronic transitions, which are energetically dominant, are found in the ultraviolet–visible wavelength range, hence optical spectroscopy. On electron excitation the electron arrangement is changed – and therefore the electric dipole moment of a molecule is also changed. This has two consequences. Firstly, electronic band spectra are observable even with symmetrical diatomic molecules such as H_2, O_2, etc.; secondly, transitions without any change in the rotational quantum number J also take place. Thus the selection rule for J is extended: $\Delta J = 0, \pm 1$. In addition to the P- and R-branches which belong to $\Delta J = +1$ and $\Delta J = -1$ transitions, respectively, another Q-branch belonging to $\Delta J = 0$ is observed. Finally, it should be mentioned that in all electronic bands, transition between states with a total angular momentum of zero is missing, because no change in the electrical dipole moment would be possible.

A more precise inspection shows that the line distances in the P- and R-branches with higher rotational excitations are different. A more realistic presentation of a Fortrat diagram (which is characteristic for each individual molecule), is shown in Figure 2.30. According to Eq. (2.2) the single branches approximately reflect the shape of a parabola. If, for example, the *R*-branch provides the edge of the spectrum, we call this situation red-shadowed if the P-branch provides the edge, we refer to it being violet-shadowed. In principle, such a spectral edge is *not* the border line to the continuum as observed for electronic line systems and discussed for the hydrogen atom (Figure 2.7).

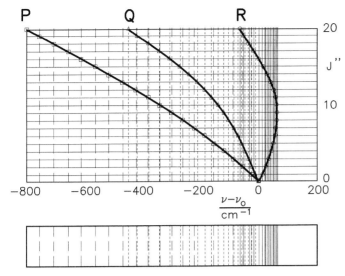

Figure 2.30 A more realistic Fortrat diagram with R-, Q- and P-branches for a molecule with non-zero angular momentum and the rotational constant $B' = 0.9B''$ (G. Wedler, Lehrbuch der Physikalischen Chemie, 1985, Verlag Chemie, Weinheim, Germany, modified).

Band spectra, particularly their "shadowing" provide information about the combining molecular states. Quantitative analysis of the absolute values of the rotational constants yields moments of inertia and bond distances.

2.3.3.8 Spectra of Atoms and Small and Large Molecules: A Comparison

While the spectral properties of atoms are largely independent of external influences such as temperature, solvent or aggregation state, molecular spectra are influenced greatly by these factors. For example, hydrogen bonding inhibits rotations and vibrations, which is reflected immediately in the rotational and vibrational structure of the spectra. In contrast to room temperature spectra, spectra obtained at low temperatures (e.g., at $-196\,°C$, the boiling point of nitrogen) often exhibit a significant, highly specific vibrational structure, which is caused by selective freezing of the degrees of rotational freedom. This effect is often used in biological/chemical analysis (*fingerprints*).

Figure 2.31 shows an example of how strongly the vibrational structure of the n, π^*-transition of 1,2,4,5-tetrazine 1 (a small molecule) depends on the relationship between the solvent and temperature. It is interesting that a limited number of normal vibrations typically dominate the spectra, even if theoretically there are a large number of normal vibrations (cf. Chapter 10 on near-infrared spectroscopy). Even the spectra of large molecules, particularly with chromophoric groups exhibiting spectra in the visible region (chromoproteins, see below),

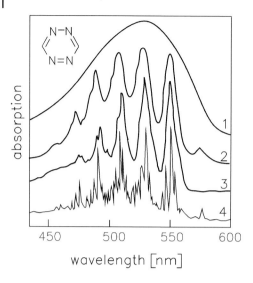

Figure 2.31 Vibrational structure of the *n,π**-transition of 1,2,4,5-tetrazine 1 (Mason, S. F., *J. Chem. Soc.*, 1959, modified). 1, Spectrum in water at room temperature. 2, Spectrum in cyclohexane at room temperature (spectrum shifted by 6.3 nm towards longer wavelengths). 3, Spectrum at −196 °C (liquid air) in an isopentane–methylcyclohexane matrix. 4, Vapor spectrum at room temperature.

show only a few, marked vibrational peaks or shoulders at low temperature that can serve for analysis.

In spite of their complexity, the theory of spectra of smaller, inorganic molecules is understood. However, to date the spectra of macromolecules can not be analyzed in detail. Nevertheless, spectrophotometric analysis is very useful in all scientific disciplines. It is not the precise molecular analysis and the molecular structure of macromolecules that is the focus of spectroscopy but rather identification, concentration, reaction mechanisms, kinetics and embedding (e. g., in plasma membranes) and the roles they play in complex systems, such as living cells. The phenomena observed are often more important even if a theoretical explanation is sometimes required, as we will see in subsequent chapters.

As a summary, spectral properties will be illustrated by one of the most important pigment molecules in biology, namely chlorophyll with its extended delocalized π-electron system (Figure 2.32). The measured spectra are shown on the left-hand side, and the Jablonski diagram derived from these is on the right.

This representation serves to give a better visualization, even if it is not drawn quite correctly: on the *y*-axis the wavelength is shown, rather than the energy.

We can see that there is a complex, not particularly structured system of bands and shoulders. The electronic transitions are not resolved distinctly, but the composition of the spectrum by electronic, vibrational and rotational transitions is indicated. A special feature is that the strict distinction between singlet- and triplet-states is shown, where the prohibition of *intersystem crossing* is not very stringent. The electronic transition $S_1 \rightarrow S_0$ is not forbidden, thus the lifetime of the S_1-state is short and is in the nanosecond range.

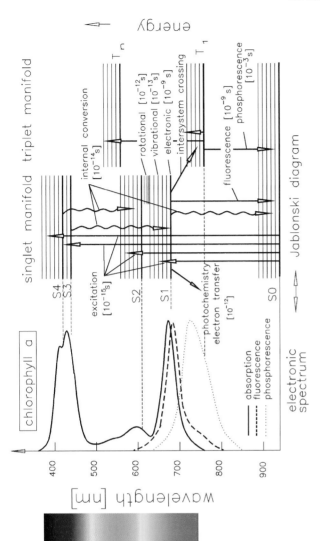

Figure 2.32 Comparison of the measured absorption, fluorescence and emissions spectra and the corresponding Jablonski diagram of the complex molecule shown, i. e., for chlorophyll. This is the green pigment (found for example in plant leaves) that is responsible for the conversion of light into chemical energy in all photosynthetic organisms. The superposition of electronic, vibrational and rotational states results in a specific, individual structure for a spectra (*fingerprint*). In complex systems, as here, only the distinction between triplet ($T_1, \dots T_n$) and singlet states ($S_1, \dots S_n$) is significant. The important features and characteristic time constants are given.

Different systems of nomenclature for electronic transitions are used: by Mulliken (e. g., V ← N), by Platt (e. g., B ← A), by Kasha (e. g., $n \leftarrow \pi^*$) or on the basis of the symmetry classes of group theory (e. g., $^1B_{2u} \leftarrow {}^1A_{1g}$). In this book we will use the so-called *numeral notation* exclusively, which simply counts the states according to the Jablonski diagram and taking the multiplicity into account (e. g., $S_0 \rightarrow S_3$).

On the other hand the transition $T_1 \rightarrow S_0$ is "forbidden", and thus improbable and the lifetime of T_1 is about 1000 times longer than that of a typical excited singlet state, which has important consequences in photochemistry, photophysics and photobiology. $S_1 \rightarrow S_0$ and $T_1 \rightarrow S_0$ transitions, which take place concomitantly with the emission of light, fluorescence and phosphorescence, will be discussed in more detail in Chapter 6 on luminescence spectroscopy.

In the next chapter we will focus on the theory and practice of various optical aspects of optical spectroscopy. The basic understanding of spectroscopy based on quantum mechanics as discussed in the present chapter is helpful for successful applications of optical spectroscopy.

2.4
Selected Further Reading

Atkins, P. W. *Molecular Quantum Mechanics: An Introduction to Quantum Chemistry*, Oxford University Press, Oxford, **1970**.

Blinder, S., Blinder, S. M. *Introduction to Quantum Mechanics, Material Sciences and Biology*, Academic Press, New York, **2004**.

Brand, J. C. D. *Lines of Light: The Sources of Dispersive Spectroscopy, 1800–1930*, Gordon & Breach Sciences, New York, **1995**.

Esposito, G. *From Classical to Quantum Mechanics: An Introduction to the Formalism, Foundations and Applications*, Cambridge University Press, Cambridge, **1994**.

French, A. P., Taylor, E. F. *An Introduction to Quantum Physics*, CRC Press, Boca Raton, **2001**.

Griffiths, D. J. *Introduction to Quantum Mechanics*, Prentice Hall College Division, Englewood Cliffs, **2005**.

Hanna, M. W. *Quantum Mechanics in Chemistry*, 3rd edn., Addison Wesley, London, **1981** (a short introduction into quantum mechanics for non-physicists).

Lindon, J. C., et al., eds. *Encyclopedia of Spectroscopy and Spectrometry*, Academic Press, London, Sydney, Tokyo, **1999**.

Modino, A. *Quantum Theory of Matter: A Novel Introduction*, John Wiley & Sons, Chichester, **1996**.

Phillips, A. C. *Introduction to Quantum Mechanics*, John Wiley & Sons, Chichester, **2003**.

Ratner, M. A., Schatz, G. C. *Introduction to Quantum Mechanics in Chemistry*, Prentice Hall, Englewood Hills, **2000**.

Workman J., A. W. Springsteen, eds. *Applied Spectroscopy: A Compact Reference for Practitioners*, Academic Press, New York, **1998**.

3
Optics for Spectroscopy

Cum deus calculat, fit mundus
(Leibniz)

With God calculating, the world was created

3.1
Introduction

Optical spectroscopy consists of all analytical methods that rely on the interaction between light and matter. According to Figure 1.1, we can distinguish between absorption, reflection, scattering and emission spectroscopy. In all cases a "monochromator" selects monochromatic light of specific wavelength from the white spectrum of an appropriater light source/lamp directed onto the sample to be analyzed. It is then possible to determine experimentally the portion of light that is either reflected (reflection spectroscopy), transmitted (transmission or absorption spectroscopy) or somehow scattered (scattering spectroscopy) (Figure 3.1). As discussed in Chapter 2, light of wavelength λ_1 excites the sample which, after a time, relaxes and emits light of another wavelength λ_2 on transition back to the ground state (emission spectroscopy).

Considering the terminology, historically, we call the measuring device a *spectroscope* when we observe the spectrum with the bare eye, and in general a *spectrometer* when we utilize a calibrated scale for the determination of wavelengths, a *spectrograph* when the spectrum is registered using a photochemical film or a photoelectric device and finally a *spectrophotometer* when a combination of a monochromator and photoelectric detectors is used for the determination of the optical parameters.

The basic components of all spectrophotometers are a light source, a monochromator, the sample to be analyzed and the photodetector that transforms the received light signal into an electrical signal, which is then processed by a computer-based system. The particular selection and set-up determines the specification and working mode of a spectrophotometer. Prior to going into detail, we will first introduce the radiometric and photometric light units, these being essential for all optical measurements. The symbols for photometric quantities are the same as

Optical Spectroscopy in Chemistry and Life Sciences. W. Schmidt
Copyright © 2005 WILEY-VCH Verlag GmbH & Co. KGaA, Weinheim
ISBN 3-527-29911-4

absorption, transmission

reflection, scattering

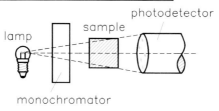

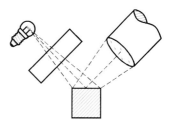

fluorescence, phosphorescence

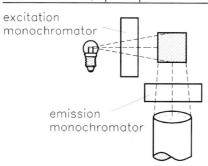

Figure 3.1 The three principal optical arrangements in optical spectroscopy according to measurements of *transmission, reflection* and *luminescence.*

those for the corresponding radiometric quantities. When it is necessary to differentiate between them, the subscripts v and e, respectively, should be used.

3.2
Physical Light Units

With respect to terminology, the physical light units as used by an optical spectroscopist are termed "radiometric", and light units as used by an illumination technician and related to the physiological recognition by the human eye "photometric" (see Chapter 4). Light as a form of energy is measured in units of ergs, Joules, Watt seconds, calories, electron volts, etc. As light is a quantized entity, it is also feasible simply to count quanta, assuming we know the relevant wavelengths. The "intensity" of irradiation [*radiant flux* Φ_e, measured in Watts (W)] is defined by the quotient energy/time. Next to the radiant flux we are particularly interested in its spatial and spectral distributions. A basic prerequisite to the use of irradiation is its reception on a specified area of some type of light sensitive receiver, A_2. In this case the decisive entity is the so called *radiant flux density* and is defined by

$$E_e = \frac{\Phi_e}{A_2} \quad (\text{W m}^{-2}) \qquad (3.1)$$

The same radiant flux leads to a higher radiant intensity I_e, when it is confined to a smaller spatial volume. In order to define the spatial volume around a point-like light source we introduce a solid angle with the unit steradian (sr). This corresponds to the definition for a radian (rad) of angle in two dimensions and is defined by the area R^2 cut out of the surface of a unit sphere with radius R (Figure 3.2). Because the surface of the sphere is $4\pi R^2$, the whole solid angle Ω is 4π sr ($R = 1$), and the radiant intensity is

$$I_e = \frac{\Phi_e}{\Omega} \quad (\text{W sr}^{-1}) \qquad (3.2)$$

A plane light source of area A_1 with radiant intensity I_e appears to be brighter when it is emitting in a smaller area. If a plane light source is observed not at right angles to the surface but rather at an angle ε to the normal, we find experimentally that the radiant intensity is proportional to $\cos \varepsilon$. To visualize this law, only the projection of the area $A_1 \times \cos \varepsilon$ is active and the radiance is given by (Lambert law, 1760):

$$L_e = \frac{I_e}{A_1 \times \cos \varepsilon_1} \quad (\text{W sr}^{-1} \text{ m}^{-2}) \qquad (3.3)$$

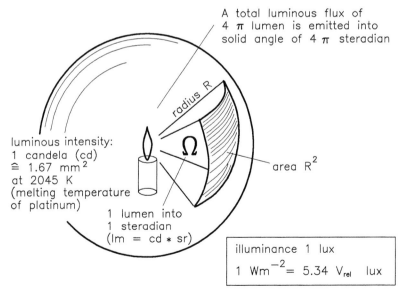

A total luminous flux of 4 π lumen is emitted into solid angle of 4 π steradian

radius R

luminous intensity:
1 candela (cd)
≙ 1.67 mm^2
at 2045 K
(melting temperature of platinum)

area R^2

1 lumen into 1 steradian (lm = cd ∗ sr)

illuminance 1 lux
1 Wm^{-2} = 5.34 V$_{rel}$ lux

Figure 3.2 Visualization of photometric light units on the basis of a unit sphere with radius R. The luminous intensity (*candela*, cd) generates a luminous flux (*lumen*), which leads to the luminous flux density (*lux*) inside the sphere.

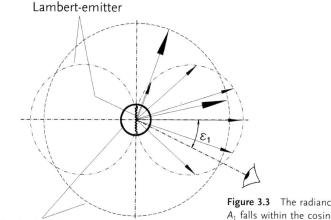

Lambert-emitter

spherical emitter

Figure 3.3 The radiance L_e of an area A_1 falls within the cosine of the observation angle ε_1.

Replacing the plane, light-emitting area A_1 by a glowing cylinder located perpendicular to the plane of the paper, the radiance will be independent of ε (a sphere radiator, Figure 3.3). From this it is clear that the filament of a bulb is a sphere radiator in the plane perpendicular to the axis of the filament coil; however in a plane that contains the axis of the filament coil it is a Lambert radiator.

In order to define the spectral composition of light, the spectrum is divided into "infinitesimal" sections $\Delta\lambda$ (e. g., 1 nm intervals). On this basis the spectral radiant flux is:

$$\Phi_e\,(\lambda) = \frac{\mathrm{d}\Phi_e}{\mathrm{d}\lambda}\ (\mathrm{W\ nm^{-1}}) \tag{3.4}$$

The total irradiant flux Φ_e is calculated by integrating over the total spectral range λ_1 to λ_2 emitted by the light source.

$$\Phi_e = \int_{\lambda_1}^{\lambda_2} \Phi_e\,(\lambda)\ \mathrm{d}\lambda \tag{3.5}$$

The following formulae allow the simple conversion of various light units without performing the physically correct calculation, for example, for a light quantum of 600 nm (orange):

$E = hc/\lambda$

$= (6.6 \times 10^{-27}\ \mathrm{erg\ s})(3 \times 10^{12}\ \mathrm{cm\ s^{-1}})/(6 \times 10^{-7}\ \mathrm{cm})$

$= 3.3 \times 10^{-12}\ \mathrm{erg}$

When λ is expressed in nanometers (nm), the energy of one quantum is:

$$
\begin{aligned}
&2 \times 10^{-9}/\lambda \text{ erg} \\
&2 \times 10^{-16}/\lambda \text{ J} \\
&0.48 \times 10^{-16}/\lambda \text{ cal} \\
&\text{or } 1240/\lambda \text{ eV}
\end{aligned}
\tag{3.6}
$$

It is often advantageous to know the energy of one mole of light quanta (6.022×10^{23}), known as one *Einstein* (E), which is:

$$
\begin{aligned}
E &= 12 \times 10^{14}/\lambda \text{ erg} \\
&= 285 \times 10^{5}/\lambda \text{ cal}
\end{aligned}
\tag{3.7}
$$

In some cases, for theoretical reasons, we would like to know the number of quanta (quantum density) rather than the energy. Following simple formulae allows the mutual conversion where the wavelength λ is again expressed in nanometers (nm):

$$
\begin{aligned}
1 \text{ E m}^{-2} \text{ s}^{-1} &= 12 \times 10^{14} \, \lambda^{-1} \text{ erg m}^{-2} \text{ s}^{-1} \\
&= 12 \times 10^{7} \, \lambda^{-1} \text{ W m}^{-2}
\end{aligned}
\tag{3.8}
$$

and

$$
\begin{aligned}
1 \text{ W m}^{-2} &= 10^{7} \text{ erg m}^{-2} \text{ s}^{-1} \\
&= 8.3 \times 10^{-9} \times \lambda \text{ E m}^{-2} \text{ s}^{-1}
\end{aligned}
\tag{3.9}
$$

In this context a few interesting data referring to the natural irradiation of the sun onto the earth will be given, as a huge number of photo-physiological responses of animals, plants and men depend on them, as do "alternative energy sources", such as photovoltaic cells and solar panels on roofs of houses for water heating. The total irradiance is composed of direct sun irradiation and indirect radiation in the atmosphere generated by various scattering processes. The total irradiance of the sun on the earth's surface amounts to 1353 W m^{-2}, when the sun is located in the zenith and without any air cover (*solar constant*). However, under natural, clear sky conditions a maximum of 1120 W m^{-2} is measured. On cloudy days this value is diminished and thus we obtain a yearly total average of approximately 100 W m^{-2} (50% direct, 50% diffuse light). Surprisingly, and so far unexplained, is the finding that a completely overcast sky has the expected color temperature of between 4000 and 6000 K (surface temperature of the sun), however, a blue, clear sky shows much higher values between 15000 and 30000 K (cf. Section 3.4.1). It could possibly be that this is somehow caused by the temperature of the sun's corona, which is also much higher than the temperature of the sun's surface.

3.3
Photometric Light Units

These are subjective and depend on the spectral sensitivity of the human eye. Typically they are utilized for general illumination purposes in the human environment (room illumination, photography, etc.). Even though the optical spectroscopist and photobiologist/chemist should avoid physiological units and use physical units instead, i.e., objective light units, even today scientific publications often deal with physiological units (e.g., "luxmeter"). A large number of radiometric and photometric light units are confusing, and their possible relationships are given in Table 3.1. However, knowing the spectral sensitivity of the human eye, physiological and physical units can be approximately converted into each other.

A candela corresponds – for historical reasons – to the "irradiation power" of a common house-hold candle. Today a candela (cd) is a black body light source with an area of 1.67 mm^2 at the melting temperature of platinum (2045 K). A (point-like) candela generates a luminous flux into the total surrounding space [4π steradian (sr)] of 4π lumen (lm = cd $\times$ sr). The resulting luminous flux density at the surface of the sphere with a radius of 1 m around the light source is defined as 1 lux. This unit simply relates to "what we really see", i.e., it depends on the

Table 3.1 Summary of the important radiometric units (photometric units in parentheses relating to the human eye).

Quantity	Symbol	Definition	Unit	Description
Radiant flux Fluence rate [Luminous flux (lm)]	Φ_e (Φ_v)	dQ_e/dt	W	Power dQ_e that is emitted by a light source into the whole space per time unit dt
Radiant intensity [Luminous intensity (cd = lm sr^{-1})]	I_e	$d\Phi_e/d\Omega$	W sr^{-1}	Power $d\Phi_e$ that is emitted by a light source into a solid angle $d\Omega$
Radiant flux density at a surface Fluence [Luminous flux density (lux = lm m^{-2})]	E_e	$d\Phi_e/dA_2$	W m^{-2}	Power $d\Phi_e$ per detector area dA_2
Radiance [Luminance (cd m^{-2})]	L_e	$dI_e/d(A_1 \cos \varepsilon_1)$	W sr^{-1} m^{-2}	Power $d\Phi_e$ that is emitted by a Lambert irradiator of surface $d(A_1)$ into a solid angle $d\Omega$
Spectral radiant energy [Spectral luminous flux (lm nm^{-1})]	$\Phi_e(\lambda)$	$d\Phi_e/d\lambda$	W nm^{-1}	Power $d\Phi_e$ emitted within a defined wavelength interval $d\lambda$

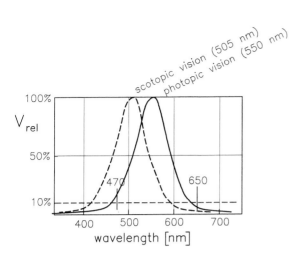

Figure 3.4 Spectral sensitivity of the eye V_{rel} (l) for daylight vision (photopic vision = color vision; *cones*, solid line) and night vision (scotopic vision = black and white vision; *rods*, dashed line).

spectral sensitivity curve of the human eye (when adapted to day light, i. e., the color sensitive cells called cones, are active; for a dark adapted eye, the sensitivity curve looks similar, however it is shifted by 50 nm to shorter wavelengths, Figure 3.4). Thus, for the correct conversion of lux into physical, objective units we have to take into account the sensitivity curve. Its maximum is located in the green–yellow spectral range. For a given monochromatic light V_{rel} can be simply determined and the conversion performed according to:

$$1 \text{ lux} = 1.61 \times 10^{-7} \times V_{rel} \text{ (W cm}^{-2}) \tag{3.10}$$

However, the conversion is difficult if we use broad band light or even "white light". Then we have to integrate over the various wavelengths and the spectrum has to be precisely known. Different spectral energy distributions can lead to identical energy values or identical luminous fluxes. On the other hand, different spectral distributions typically lead to different photo-physiological and photochemical responses/reactions. Clearly, the precise definition of spectral distribution, or at least the conditions under which a well defined light source is operated, is definitely required for scientific measurements and investigations.

3.4
Light Sources

Light sources convert energy – most often introduced in an electrical form – with a given efficiency into visible light (Figure 3.5). We generally distinguish between two types of light sources. The first type consists of the so-called thermal radiators, which are heated until they begin to glow and thus – according to Planck's law – emit light. Incandescent lamps are typical thermal radiators. The second type of light source, the so-called luminescence radiators, utilize the electronic excitation of atoms which – upon relaxation – emit light. Important luminescence radiators in optical spectroscopy are gas discharge lamps, light emitting diodes and lasers. The most important feature of light sources is their spectral radiant energy Φ_e. We can distinguish between a continuums radiator in which energy is distributed over a wide spectral range and a line radiator where the energy is confined to only one or to a set of distinct spectral lines. The quantum efficiency of a light source is defined as a quotient of the radiant flux divided by the introduced power P: $\eta = \Phi_e/P$.

3.4.1
Black Body Radiation

According to Planck's radiation law every ideal black body (i. e., one that absorbs all incident light quanta) emits electromagnetic radiation in the range $d\lambda$:

$$L_e \, d\lambda = \left[\frac{hc^2}{\lambda^5 \left(e^{\frac{hc}{\lambda kT}} - 1 \right)} \right] \times \Omega_0 \, d\lambda \tag{3.11}$$

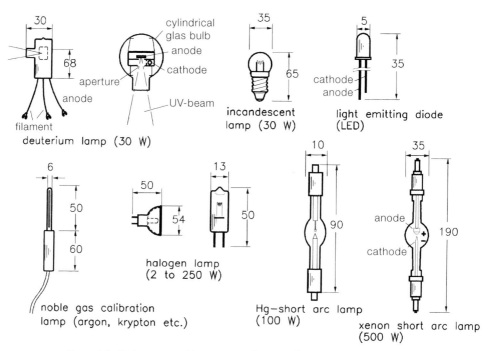

Figure 3.5 Various types of lamps as used in optical spectroscopy including their geometric dimensions in millimeters (note the different scales).

with h = Planck's constant, $L_e(\lambda)$ = spectral radiance, c = velocity of light, $\varkappa$ = Boltzmann's constant (sometimes represented by k), T = absolute temperature (K) and unit solid angle Ω_0 = 1 sr. All non-black thermal radiators, so-called "gray radiators", show similar energy distributions of the emitted light, even if their total emitted energy is smaller. Figure 3.6 shows the calculated emission curves of an ideal black body radiator for temperatures from 3000 to 7000 K according Eq. (3.11). With increasing temperature there is a rapid increase in emitted power and the irradiation maximum is shifted to shorter wavelengths as described by Wien's law [first derivative of Eq. (3.11)]:

$$\lambda_{max} = 2.8978 \times 10^6 \times T^{-1} \ (nm) \tag{3.12}$$

The energy maximum of the sun's spectrum is at 500 nm, with the greatest sensitivity for the human eye for photopic vision (daylight vision) being at 555 nm. The argument is often put forward that this obvious conformity reflects the phylogenetic adaptation of the visual pigments to the sun's spectrum and that in this way daylight is utilized most effectively. However, this conclusion is wrong. For the photochemical efficiency of the photoreceptors it is not the energy, but the

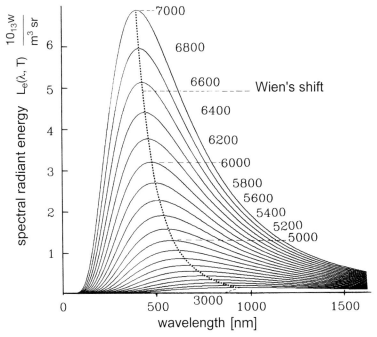

Figure 3.6 Radiant flux density of a *black body* as a function of absolute temperature (K), calculated according to *Planck's equation* of radiation [Eq. (3.11)]. *Wien's* law describes the red shift of the maxima, and the Stefan–Boltzmann law describes the decrease in areas enveloped by the Planck curves with decreasing temperature.

Ludwig Boltzmann (1844–1906)

number of quanta that is the critical factor. The photoreceptors of the eye are quantum counters, not energy detectors. Therefore, it is only the absorption of a quantum that counts, whereas it is the energy/wavelength that determines the probability of absorption. According to this premise the maximum of the sun's spectrum is 640 nm. Another argument could explain this discrepancy. During evolution, when the pigments were selected, the sun was significantly colder, i.e., redder with a maximum of quantum emission at 640 nm. However, this remains speculation.

The total emitted power of any thermal radiator at a given temperature is described by the Stefan–Boltzmann law, which is obtained by integrating Planck's equation [Eq. (3.11)] *over the entire wavelength range:*

$$L_e = \frac{\sigma}{\pi} \times T^4 \times \frac{1}{\Omega_0} \quad (\text{W m}^{-2} \text{ sr}^{-1}) \tag{3.13}$$

Intense radiance L_e and efficiency of visible radiation is obtained by raising the incandescent temperature as high as possible. Each specific temperature of a black body leads to only one specific spectral distribution as described by Planck's law – and vice versa, Thus, the term "color temperature" was introduced to define a spectral distribution. It describes quantitatively the spectral distributions of incandescent and luminescent radiators.

3.4.2
Incandescent Lamps

Incandescent lamps are still being used in optical spectroscopy because they are simple, cheap and do not require dedicated power supplies. A double coiled tungsten wire (melting point 3653 K) in an evacuated glass bulb is heated by an electric current (Figure 3.5). Traces of tungsten always evaporate which leads to progressive blackening of the bulb. Therefore larger incandescent bulbs are filled with a noble gas to slow down the blackening (Ar, Kr, Xe; filament temperature approximately 2500 K). A real break-through in light techniques came about with the addition of a halogen gas (typically bromide). In a cyclic process the eva-

porated tungsten atoms combine with bromide, which settles on the glowing filament and dissociate again to give the tungsten element. Using a small, short and
compact filament the filament temperature can be increased up to 3450 K. In
order to accelerate the cyclic process, the dimensions of the bulb must be kept
small, which is advantageous for many optical applications. Radiance (physiological: luminance) and high efficiency in the visible wavelength range (spectrum in
Figure 3.7) of halogen lamps fulfill the requirements of optical spectroscopy for
an efficient point light source. Particularly useful are lamps with an integrated
dichroic reflector that is transparent in the (near) infrared, thus resulting in
less photolytic destruction of the sample (Figure 3.5).

3.4.3
Gas Discharge Lamps

Electrical discharges play a key role in many types of lamp (Figure 3.5). If a volume of an inert gas such as neon containing traces of mercury is exposed to a
high voltage, stray electrons will be accelerated to high kinetic energies that are
sufficient to ionize the gas. The subsequent recombination leads to excited
atoms of the noble gas. These finally transfer their energy to the mercury
atoms which in turn are stimulated to emit light. This is the principle of a mercury discharge lamp. The spectral characteristics strongly depend on the pressure,
i. e., on the various recombination processes. Low pressure lamps are categorized
as those with an Hg partial pressure of approximately 10^2 Pa, which primarily
emit the 253.7 nm line, medium pressure lamps with a pressure of approximately
10^5 Pa require higher electric currents and show a much greater number of lines.
High pressure lamps with a partial pressure of more than 3×10^6 Pa and a small
electrode distance exhibit broad, overlapping bands ranging from the ultraviolet to
the red wavelength range (Figure 3.7). However, the radiant flux in the red and
blue–green spectral range is relatively small. Therefore, in this wavelength
range high pressure xenon lamps are preferable (Figure 3.5), with a color temperature of nearly 6000 K (similar to the daylight spectrum). These lamps do
not reach the spectral radiant energy of Hg lamps in the UV range, but their spectrum in the visible range is much "smoother", which is better suited to spectral
analysis, when continuous spectra have to be recorded. In addition, their spectral
radiant energy in the visible spectral range is higher than that of an Hg lamp –
even if there are some strong (and thus interfering) spectral lines in the blue
and near-infrared. Therefore, xenon lamps are used universally in fluorimetry
(Chapter 6).

The best, generally available "continuum-light source" in the ultraviolet is a hydrogen lamp (Figure 3.5), where the light emission depends on the recombination of $2H \rightarrow H_2$ at cold surfaces (Figures 3.5 and 3.7, with respect to synchrotron
irradiation cf. Section 3.4.6). In a quartz bulb there is a tungsten anode and a
cathode which can be heated. The lamp is preheated to approximately 100 °C
and finally "ignited" at 220 V. The loss of energy by heat conduction is relatively
high, therefore the spectral radiant energy is small. Heat conduction is signifi-

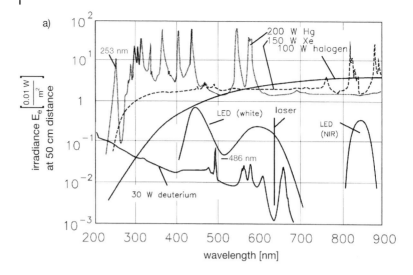

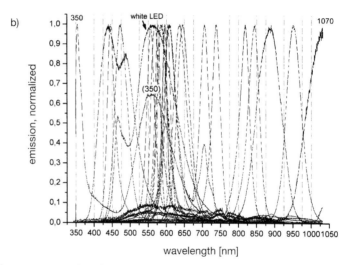

Figure 3.7 (a) Radiant flux density E_e of some important light sources on a logarithmic scale, as a function of wavelength (top). (b) Collection of various LEDs (emission spectra normalized) covering the whole visible, and in addition, the ultraviolet and near-infrared wavelength range below (source of LEDs: www.roithner-laser.com, Vienna, Austria).

cantly diminished if we use deuterium with its double atomic mass compared with hydrogen and the radiant energy increases by approximately 30%. The emission maximum of a deuterium lamp is at 220 and of a hydrogen lamp at 290 nm. The single peak (H_β of the Balmer series, cf. Figure 2.6) is at precisely 486.12 for hydrogen and at 485.99 nm for deuterium, and thus can be used for wavelength calibration.

Small, penlight discharge lamps filled with highly pure gas (Ar, He, Kr, Ne, Xe or Hg), together with suitable power supplies and reference spectra are offered by various manufacturers as calibration lamps for the whole ultraviolet–visible (UV–VIS) wavelength range. In addition, Hg calibration lamps can be utilized as small but excellent low cost light sources in the UV (254 nm) for fluorescence excitation. Figure 3.8 shows the idealized emission line spectrum of a mercury–argon (low pressure) calibration lamp (No. 6035, ORIEL). Figure 3.9 shows the spectrum of a krypton calibration lamp scanned with a 2 nm slitwidth (logarithmic ordinate scale).

In laboratory spectrophotometers, two types of lamps are generally used: deuterium lamps for the UV range and halogen lamps for the visible–near-infrared spectral range (320–2500 nm). Automatic switching between these lamps requires a significant opto-mechanical expenditure. This is avoided with a *see-through lamp* (from Hamamatsu): light from the halogen lamp is guided through the shield and aperture of the deuterium lamp thus resulting in a continuous emission spectrum over the whole wavelength range from UV to the near-IR.
    A flash lamp (such as those used in photographic flashlights) is also a gas discharge lamp, and is operated in a pulsed mode. It was invented in the 1930s by Harold E. ("Doc") Edgerton. Usually it is filled with xenon, because this gas converts electrical energy most efficiently into light energy (for the near-infrared, krypton is preferred). Two opposite electrodes are connected to a low voltage condenser of high capacity. In the unionized state the electrical resistance of the lamp is a multiple of 10 MΩ. An ignition electrode on the outer surface of the lamp bulb generates the required (*breakdown*) voltage of approximately 10 kV: The gas is ionized, the resistance drops to a few milli-Ω and generates – depending on the particular design of the electronics – a flash between 1 μs and 50 ms. The peak power ranges from kilo- to megawatts. The lamp's envelope is made from glass or quartz, and the spectrum for a xenon (flash) lamp ranges from 200 to more than 1100 nm (cf. Figure 3.7). There are two bulb types: linear

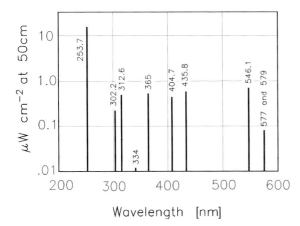

Figure 3.8 Idealized spectrum of a mercury calibration lamp.

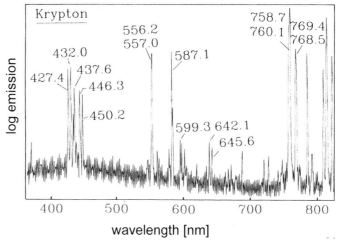

Figure 3.9 Spectrum of a krypton calibration lamp with important wavelengths as monitored with a resolution of 2 nm (logarithmic ordinate scale).

and the hemispherical lamps. The latter is preferred for opto-spectroscopic applications (point like, short flashes, a high repetition rate and long lifetime), e. g., for the detection of triplet states and free radicals.

3.4.4
Light Emitting Diodes

Light emitting diodes (semiconductor luminescence diodes, LEDs) are solid state devices. Recombination of electrons and "holes" generate the emission of radiation. The wavelength depends on the semiconductor material used and ranges in the semiconductor band from UVA (350 nm) to the near-infrared spectral range [1550 nm, Figure 3.7(b)]. Even if the radiance and efficiency of LEDs are relatively weak – e. g., compared with simple incandescent lamps of the same power – their advantages in optical spectroscopy are a long lifetime (10^5 h), small spectral emission bands, low price and the possibility of modulation frequencies up to the nanosecond range. On increasing the temperature, the spectral emission curve is shifted bathochromically by 0.1 to 0.3 nm K^{-1}. With green LEDs the change in luminous intensity is barely recognized by the human eye as the peak wavelength is close to the maximum of the $V–\lambda$ curve (Figure 3.4). However, with red and blue LEDs the peak wavelengths are located just on the flanks of the $V–\lambda$ curve, which results in strong color variations on temperature change. A "white LED" with comparably high luminosity is available, with some blueish tinge: a blue emitting LED was placed in a typical LED case that was covered by a white fluorescing material similar to ordinary white fluorescent tubes. However, the spectrum reaches only up to 700 nm (Figure 3.7). It is important to note that some LEDs show spurious light emission.

3.4.5
Lasers

Since their first realization in 1960 by Maimann, lasers (**l**ight **a**mplification by **s**timulated **e**mission of **r**adiation) have been used as light sources in various fields because of their unique properties and the possibilities for applications cannot be overestimated:
- There is high temporal (cf. coherence length, Section 2.2) and spatial coherence at the same time (collimated light beam of small cross section, point light source).
- The wavelengths are exactly defined (monochromatic) and can be easily tuned.
- The laser beam can be focused down to the sub-micrometer range. This allows micro-spectrophotometric experiments of high resolution and makes coupling into very thin (monomode) glass optical fibers easier.
- By special acousto–optical methods, ultra short laser pulses of high repetition rate down to the femto second time range can be generated.

Solid, liquid and gaseous substances can be excited to emit laser radiation. Laser radiation covers the whole range from the ultraviolet to the infrared. It is emitted continuously ("cw", that is a "continuous wave" for most gas lasers such as helium, neon, argon or carbon monoxide lasers) or in the form of short pulses. The laser process is based both on photophysical as well as optical principles. The first pre-requisite is a long lifetime, an electronic state of the atom or molecule which leads to population reversion and more particles residing in an excited (W_2) rather than the ground state (W_1) (Figure 3.10). Thus, in addition to the "normal" absorption transition, a reverse transition from the excited to the ground state can be induced with light of energy $h\nu$ (within the natural "long" lifetime of, e. g., 10^{-3} s in contrast to the 10^{-9} s of spontaneous emission), yielding two quanta of the energy $h\nu$.

In the three-level laser operation, the laser active state W_2 is reached via *internal conversion* from a higher energy state W_3 (Figure 3.11); the induced transition W_1 → W_2 is termed "pumping". A ruby laser is an important example of this and it carries Cr^+ ions embedded into Al_2O_3 as the active material and emits red light at 694.2 nm. The well known neodymium–YAG laser (ground wavelength at 1060 nm), is based on neodymium ions embedded into $Y_3Al_5O_2$ (yttrium–aluminum garnet) and is a leading example of a so-called four-level laser. It is characterized by the fact that the ground state W_1 is *not* also the lower laser state W_2. The latter is almost vacant and thus allows population reversion, W_3 versus W_2, even with a low pumping power. Neodymium–YAG lasers can also be operated in a conti-nuous mode (cw), whereby an iodide–tungsten lamp serves as the pumping

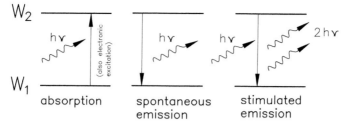

Figure 3.10 Comparison of absorption/excitation, spontaneous and induced (stimulated) emission between two
electronic states W_1 and W_2.

light source. With an efficiency of 2 %, a maximum laser light output of 100 W
can be achieved.

In order that the two photons which appear as a result of the induced transition
"from top to bottom", initiate a "chain reaction", the lasing (laser light emitting)
material is enclosed in an optical resonator that carries a mirror on both ends
(Figure 3.12b: "confocal resonator"). These provide multiple reflections through
the lasing material. One of the windows is a half mirror through which the
laser beam leaves the system. Typically this window is tilted by the "Brewster
angle" φ_P (see later, Figure 3.16, Section 3.5.2), which allows the light vibrating
in the plane of incidence to leave the resonator without any loss, thus supporting
the lasing (laser light emitting) process. The other perpendicularly vibrating component sustains high losses. Along the length L of the resonator, standing waves
in various axial modes are generated [$L = m\lambda/(2n)$, scattering order m, refraction
index n, wavelength λ], as the last requirement is fulfilled for different wavelengths λ. Various transversal modes are built up (TEM: transversal electro-

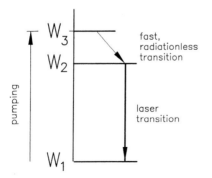
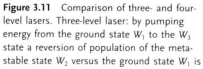
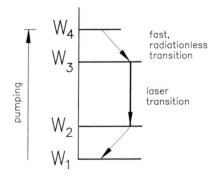

Figure 3.11 Comparison of three- and four-
level lasers. Three-level laser: by pumping
energy from the ground state W_1 to the W_3
state a reversion of population of the metastable state W_2 versus the ground state W_1 is
accomplished; this is the prerequisite to an
inducible laser transition $W_2 \rightarrow W_1$. Four-level
laser: the reversion of population is much
easier to obtain as prior to induction the W_2
state is completely empty.

magnetic modes, Figure 3.13) with indices indicating the nodal planes in x- and y-directions.

An very sensitive method for measuring extremely small absorbancies (Chapter 5), such as measurement of specific overtones in the infrared range (cf. Chapter 10), is achieved by insertion of the sample (solid, liquid or even gas) into a specific cuvette with Brewster windows into the laser-resonator itself. Even the slightest light losses by absorption result in huge changes in the output beam. Special care has to be taken to protect the photodetector from the highly intense laser.

An overview of various lasers and their wavelengths that are suitable for optical spectroscopy is given in Figure 3.14. This list is completed by the nitrogen laser (λ = 337 nm) as well as by tunable dye lasers in different wavelength ranges (350 nm $<\lambda<$ 1100 nm). Most solid state lasers work in the near-infrared range. A frequency doubled, tripled or quadrupled Nd:YAG laser is commonly utilized in optical spectroscopy. The medium powers of typical lasers vary between approximately 1 mW and several watts. Excimer and nitrogen lasers are pulsed lasers,

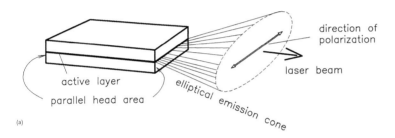

(a)

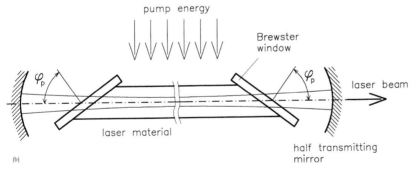

(b)

Figure 3.12 (a) Scheme for a solid state diode laser which in principle works like a luminescence diode. The parallel front faces act as resonator mirrors. Owing to the very thin emitting rectangular surface, for physical reasons the emitted light is strongly divergent and elliptically polarized. The direction of polarization is indicated. (b) Typical laser. The volume filled with active laser material is closed by windows arranged under the Brewster angle φ_p. Thus the light vector vibrating within the entrance plane is transmitted without loss, while the perpendicularly vibrating component is blocked: laser light is strongly polarized. The confocal resonator with convex mirrors as shown here is particularly simple, stable and easily adjusted.

Figure 3.13 Various transversal electromagnetic vibrational modes (i. e., field distributions, *TEM*) of lasers. The indices represent the number of nodal planes in *x*- and *y*-directions.

with pulses between 1 and 20 ns and a repetition rate of up to 500 Hz. Pulsed lasers allow much higher radiant fluxes than continuous lasers and are modulated by resonator variation ("Q-switched"). All other lasers can be operated in the continuous mode.

The semiconductor or diode laser is replacing the classical laser tube in many areas of optical spectroscopy. The first prototype was realized in 1970 [Figure 3.12(a)]. It is only a few millimeters in size, it is low priced and can be modulated up to the gigahertz range. The emission principle is similar to that of a light emitting diode (LED, cf. Section 3.5.3). The laser diode is in principle a four-state laser. The wavelengths that are currently available are around 680, 800, 1300 and 1500 nm, which seriously limits their application range. Through frequency doubling an ultraviolet wavelength as low as 315 nm can be obtained, but the quantum flux is moderate. As the emitting layer is only a few micrometers thick, the emanating bundle is strongly divergent (action of a slit), which is partly corrected by collimating optics [Figure 3.12(a)].

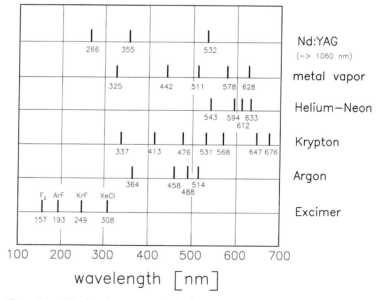

Figure 3.14 Wavelengths for most of the lasers suitable for optical spectroscopy. The ground vibration of the well known *Nd:YAG* laser is 1060 nm (after Schmidt and Schneckenburger, 1990, modified, Carl Hanser Verlag, Munich, Germany).

The most important liquid lasers are dye lasers. Low concentrations of specific organic pigments (e. g., 10^{-4} M rhodamine in ethanol or fluorescein in water) serve as lasing (laser light emitting) material, emitting over a wide spectral range (> 50 nm). The dense molecular vibrational levels serve as lasing (laser light emitting) states (cf. Figure 2.22). Using set-ups with dispersing elements such as gratings or prisms in the light paths between the mirrors the laser is adjusted continuously. Thus, the whole visible range is covered with only five different dyes.

Similarly to tunable dye lasers, laser diodes can be made to be wavelength adjustable. The laser diode is part of an external resonator. With the help of a grating (so-called Littman or Metcalf configurations) and perhaps an additional etalon (Littrow configuration) the adjustment over a wavelength range of $\Delta\lambda \geq 20$ nm is feasible. Tunable dye lasers are already utilized in **l**aser **a**tomic **a**bsorption spectroscopy (LAA, Chapter 4).

3.4.6
Synchrotron (Cherenkov) Radiation

An "exotic light source" for optical spectroscopy is synchrotron radiation, which has become very important. It includes a wide range of electromagnetic radiation from infrared to γ-rays (cf. Figure 2.2); it is the strongest light source for X-rays and UV radiation with a continuous "smooth" emission spectrum (in contrast to an Hg lamp, Figure 3.7). The working principle is seemingly simple: each charge, which is somehow accelerated, emits electromagnetic energy, also in the form of light. In a synchrotron, electrons are moving in a circular path thereby experiencing a continuous acceleration angle. Synchrotron radiation is polarized and is emitted with a scatter angle of only $0.006°$, it allows pulse widths narrower than 10^{-10} s and a repetition rate of up to 5×10^8 s^{-1}.

3.5
Geometric Optics and Wave Optics

3.5.1
Refraction and Reflection

A light beam is a straight, imaginary line along which light energy propagates. The properties of light beams can be described without taking into account the dual nature of light, being both a wave and a particle: that is the principle of reversibility, the reflection and refraction laws. The principle of reversibility states that on reversal of its propagation direction a light beam follows exactly the same path. However, this property only applies to the light path, not the light energy.

The reflection law says that for a light beam that is reflected from the boundary layer (e. g., air versus glass), the incident angle φ is the same as the exit angle ψ. Angles are always measured against the normal to the boundary layer. In addition to reflection, refraction into a second medium is observed when the entrance angle is changed.

This law is described quantitatively by Snell's law:

$$\frac{\sin \varphi}{\sin X} = \frac{n_2}{n_1} \tag{3.14}$$

where φ is the incident angle (= exit angle ψ), X the angle of the refracted beam, n_1 and n_2 are the refractive indices of the two media.

The refractive index, n, is a function of the wavelength and describes the properties of the medium through which the light beam passes. It is defined as the ratio of light velocity in a vacuum (c_0) to that in a medium (c_n):

$$n (\lambda) = \frac{c_0}{c_n} \tag{3.15}$$

For example, the velocity of light in a crown glass of n (600 nm) = 1.51 yields a velocity of $c_n = c_0/n = 198\,537$ km s^{-1}.

The refractive index n_0 of a vacuum is "1" by definition. Figure 3.15 demonstrates for a given incident angle φ the refraction angle X for the transition from air to various translucent materials. On transition into a medium of higher refractive index a light beam is always refracted towards the normal to the boundary layer, and vice versa. Consequently a specific angle θ_g exists for the transition between media of lower to higher refractive indices. If this angle θ_g is exceeded ($\theta > \theta_g$), total reflection occurs; this is the basis of light fiber technology (Section

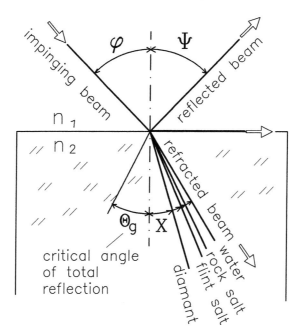

Figure 3.15 Visualization of the refractive law for the transition from air (n_1) into some selected materials (n_2). Angles are always measured from the normal.

3.5.5) and also of ATR reflection spectroscopy (ATR, attenuated total reflection, Section 8.6).

3.5.2
The Fresnel Formulae

The relationships known by this name were discovered by the French physicist Fresnel (1788–1827) and are of great importance in optical spectroscopy. They describe the relationships between intensity on refraction and reflection, which are not provided by the laws discussed so far. The so-called Fresnel reflection (*specular reflection*) is a reflection at the surface of a sample which is primarily defined by the different refractive indices of the two media that are separated by a boundary layer.

The light energy of an electromagnetic wave is stored in electrical (E) and magnetic fields (H) and equals the product of $\varepsilon \times E^2$, where ε is the dielectric constant of the medium in which the light propagates. The reflectivity R at a surface is defined by:

$$R = \text{reflected power/incident power}$$
$$= E_r^2 / E_e^2 \tag{3.16}$$

There are significant differences in the angle-dependent reflectivity for non- or strongly-absorbing reflectors. Without discussing the mathematically complex Fresnel formulae, we will describe the essential features with the example given in Figure 3.16. Depending on the direction of the polarization, light beams are reflected and refracted to different extents (cf. Section 9.7 on ellipsometry). In particular, there is an incident angle at which light polarized parallel to the entrance plane is no longer reflected. This effect is utilized to produce highly efficient plane polarized light. The angle φ_p at which this happens is called the Brewster angle after its discoverer (Sir David Brewster, 1781–1868). In this case, the refracted and reflected light beams are perpendicular to each other according to the equation:

Augustin Jean Fresnel (1788–1827)

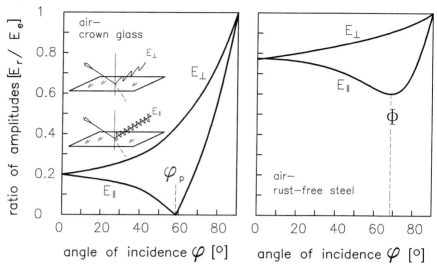

Figure 3.16 Visualization of Fresnel's formulae for reflection and refraction at the interface air → glass (not absorbing), respectively, air → steel (strongly absorbing; for details see text).

$$\frac{\sin \varphi_p}{\sin X} = \frac{n}{1} = n$$

$$\sin X = n \sin \varphi_p = n \sin (90 - \varphi_p) = n \cos \varphi_p \tag{3.17}$$

$$n = \tan \varphi_p$$

Thus, the refractive index n is measurable by determining the Brewster angle φ_p. With $\varphi = 0$, i.e., at perpendicular incidence, for the reflectivity against air, the Fresnel formulae yield a very useful relationship which only contains the refractive index n of the reflecting medium:

$$R = \left(\frac{n-1}{n+1}\right)^2 \tag{3.18}$$

For example, window glass with a refractive index of approximately $n = 1.5$, has a reflectivity of $R = 4\%$; for diamond with $n = 2.4$ we obtain a value $R = 17\%$, and for germanium with $n = 4$, R is as high as 36%.

According to the above formula, producing reflection-free surfaces, as is required for spectrophotometric lens systems, appears to be impossible. However, with the help of interference effects (cf. Sections 3.6.3, 3.5.5) surfaces can be made "reflection free", at least for a given wavelength range. This is accomplished by coating them with crystalline layers of KBr, CaF_2 and other modern materials of specified thickness.

With strongly absorbing and diffusely reflecting surfaces, particularly of biological samples, specimens *in vivo* or in suspensions, the difference in polarization between the parallel and perpendicular components of the electrical light vectors is less pronounced. Thus, the term "Brewster angle" (φ_p) is replaced by the less rigid term "mean angle of incidence" (Φ).

Summarizing: all refraction and reflection processes are concomitant with changes in the polarization, which cannot be ignored when polarization experiments are performed (Section 6.4). In particular, when spectrally resolved polarization is measured, real and artificial polarization effects have to be rigorously differentiated (Chapter 6.4).

3.5.3
Lenses and Mirrors

The most important imaging, direction-influencing components in geometric optics are plane mirrors, converging (convex, "+") and diverging lenses (concave, "−"), and correspondingly converging and diverging mirrors. Here we only discuss the most important relationships between so-called "thin lenses". Thick lenses, i. e., real lens systems, require a more sophisticated treatment and are largely omitted here. Lenses and non-planar mirrors are primarily defined by their focal lengths and apertures. Three parameters are used to distinguish them: the lens midpoint O, the focal length f and double the focal length $2f$, which is identical for thin lenses with a spherical radius R (Figure 3.17). With the object distance as a and the image distance as b, for an ideal lens and a spherical mirror the following equation holds:

$$\frac{1}{a} + \frac{1}{b} = \frac{1}{f} \tag{3.19}$$

An object at infinite distance ($a = \infty$) in the object area is projected onto the focal point in the image area (focal length f), an object at the spherical midpoint ($R = 2f$) is projected without change in size onto the second spherical midpoint, however as an inverted image. For spherical mirrors this relationship is valid only for rays that are close to the axis (Figure 3.18). For more distant rays the focal point is closer than $R/2$ which results in "spherical aberration". Because of this problem, in optical spectrophotometers parabolic mirrors are primarily used. The parabolic mirror sphere generated by rotation of a parabola $y^2 = 2px$ is largely aberration free, with $f = p/2$ (Figure 3.19).

In optically well defined systems, such as in spectrophotometers, only surface mirrors are acceptable. These are generated by evaporation of a thin aluminum layer onto a glass matrix that is protected by another thin layer of quartz (SiO_2). Low cost back surface mirrors ("razor mirrors") are generated by coating the backside of the mirror and then a lacquer protection. Additional reflection at the front glass layer is not acceptable for optical instruments.

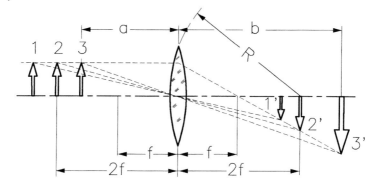

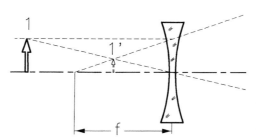

Figure 3.17 Visualization of the imaging equation of thin convex (top) and concave lenses (bottom). Parallel rays become focus rays, and vice versa, midpoint rays pass through the lenses without changing the direction. Concave lenses only produce virtual images.

An optical system is only capable of generating a well defined image of a subject when the object point is projected only onto one image point. This statement is quantified by the famous *sine-condition* of E. Abbe (1873) and H. v. Helmholtz (1874):

$$y \times \sin u = y' \times \sin u' = \text{const.} \tag{3.20}$$

Image and object sizes (y, y') as well as the numerical apertures $(\sin u, \sin u')$ are explained in Figure 3.20. The products of object or image sizes with the corresponding aperture always yields the same constant.

Ernst Abbe (1840–1905)   **Hermann von Helmholtz** (1821–1894)

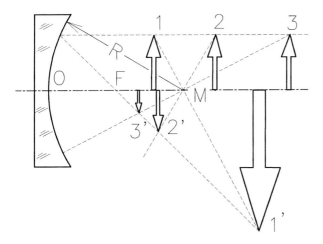

Figure 3.18 Visualization of the imaging equation for ideal, spherical concave mirrors.

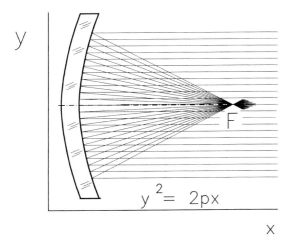

$$y^2 = 2px$$

Figure 3.19 While spherical convex mirrors exhibit strong imaging errors, parabolic mirrors have much better imaging properties, in particular the same focal point for rays off the central axis.

The term "aperture" is used to describe a variety of different situations, which can be rather disturbing. In one case it is the active diameter of a lens, mirror, etc., that is completely filled with light. "Numerical aperture" terms refer to the sine of the half opening angle, i.e., the angle between the outer most ray and the optical axis (Figure 3.20, angle *u*). Finally, "aperture" can simply mean a physical stop that limits the light path.

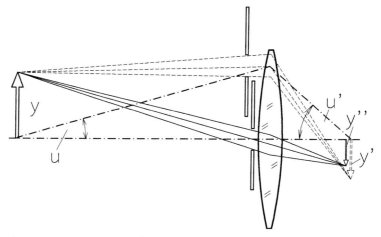

Figure 3.20 Representation of the sine condition. This is fulfilled if each object point is uniquely connected to an image point – in contrast to the two images y′ and y″ in this figure.

If the conversion of size depends on the distance from the central ray of an optical element such as a lens or mirror, the sine condition is no longer obeyed, and the image is distorted and blurred.

For the various spectral regions different materials are applied as optical components, Table 3.2.

Most applications, such as high quality imaging of the entrance slit onto the exit slit of a monochromator, require two or more lenses. If we compare various configurations as illustrated in Figure 3.21 (unit conjugate ratio), this will help our understanding of relay lenses. (a) In the single lens application at large apertures there is significant spherical aberration; i. e., different rays are not focused at the same point. (b) To improve the system, a single lens can be replaced by two individual lenses, each having twice the focal length of the lens in (a). This yields

Table 3.2 Summary of the materials used for various optical components over different wavelength ranges.

Spectral range	Mirror	Lens	Windows
X-rays	Not possible	Not possible	Beryllium
Ultraviolet	Aluminum	Synthetic quartz (*fused silica*)	Synthetic quartz (*fused silica*)
Visible range	Aluminum	Glass	Glass
Near-infrared	Gold	Glass	Glass
Infrared	Copper, gold	ZnSe	ZnSe, NaCl, BaF$_2$

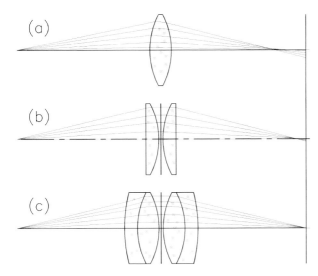

Figure 3.21 Visualization of the improvement to the imaging when using "thick lenses" as realized by combinations: (a) in the single lens application; (b) improvement by a combination of two lenses, each having twice the focal length as the one in (a); and (c) further improvement by achromatic lenses.

a higher quality image and decreases the amount of spherical aberration at lower *f*-numbers (*vide infra*). The surfaces are nearly in contact, as in a typical condenser set-up. (c) The next step in improving the system is to use achromates. Achromatic lenses are superior to single lenses for infinite conjugate distances and large apertures. Spherical aberration is negligible at large apertures, with the addition of color correction.

3.5.4
Light Paths

Figure 3.22 shows a typical optical set-up of a monochromator with largely independent, but intersecting light paths (after Köhler). We can identify: (a) the object and the illumination beams, (b) the view parallel to the slit height and (c) the view at right angles to the slit height.

(a) *Illumination beam*: The lamp's filament (or light arc) is imaged through the condenser onto the entrance slit (aperture stop); behind this collimator 1 generates a wide, parallel light beam (telecentric light path), which evenly illuminates the total dispersing element (prism or grating, cf. Section 3.6). After the angular dispersion collimator 2 finally images the entrance slit with the picture of the filament onto the exit slit in a "dispersed" form. The dotted line at the exit slit indicates the dispersed spectrum.

(b) *Object beam*: The aperture is brightly illuminated and by means of field lens 1 and collimator 1 is fully imaged into the aperture of the dispersing element. The field lens at the exit slit is required for an evenly illuminated cuvette or sample, or to couple the monochromatic light into a light fiber (Section 3.5.5). Köhler's basic idea of interlaced light paths is to maximize light throughput and to obtain maximum wavelength resolution simultaneously. Optimal spectral resolution requires

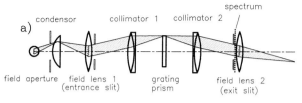

object beam

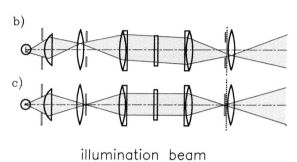

illumination beam

Figure 3.22 The two interconnected light paths of a monochromator provide: (a) for the best light throughput and concomitantly optimal spectral resolution; (b) sits in vertical cross section; (c) slits in horizontal cross section; the nomenclature is adopted from the "imaging optics".

faultless imaging of the entrance slit onto the exit slit, i. e., the fulfillment of the sine condition (cf. Section 3.5.3).

A numerical measure for the light throughput of any optical system such as a monochromator is the so-called *f*-number. This is defined by the ratio of the focal length to the aperture *A* (here in the sense of the diameter of the circular illumination of the entrance aperture):

$$f\text{-number} = f/A \tag{3.21}$$

The light throughput is inversely proportional to the square of the *f*-number. Typical *f*-numbers for monochromators are for example 3 or 4.2 (written as *F*/3, *F*/4.2) corresponding to a light throughput of 0.111 and 0.057; they differ by a factor of two.

3.5.5
Optical Fibers

In Section 3.5.1 we showed that, upon entering a medium of higher refractive index at an angle larger than a specific angle θ_g, a light beam is totally reflected back into the original medium, virtually without any loss in energy (cf. Section 8.5). This effect is utilized in so-called light guides. The simplest light guide is a glass rod with flat ends, in which total reflection takes place at the boundary layers between the surface of the rod and air. However, in real light fibers the core fiber is covered with a cladding made of material that has a lower refractive index in order to have well defined conditions: each individual fiber is optically isolated from its neighbors. In addition, without any mechanical protection and because of small scratches and irregularities, the individual fibers can easily break. For this reason another layer, the so-called buffer, is required. This defines the mechanical, thermal and chemical properties of light fibers, which depend on the specific application. The great advantage of light fibers is the possibility of bending them to certain radii, thus allowing light to be guided "around corners". Therefore they are built up from thin, individual, flexible fiber elements.

The construction within a single fiber is shown in Figure 3.23. It consists of a so-called core of quartz, glass or a polymer with a refractive index n_1, and is covered with a fluorine doped fused silica cladding of refractive index n_2. The maximum entrance angle u at the front end corresponds to half the cone angle and results from the condition $\theta > \theta_g$. For the numerical aperture we obtain (cf. Section 3.6.3):

$$\sin u = \sqrt{n_1^2 - n_2^2} \tag{3.22}$$

For straight fibers the cone angle $2u$ is just the emission angle with a characteristic energy profile (Figure 3.24). For a typical numerical aperture of $NA = 0.22$ we calculate $u = 12.7°$. Because the entrance aperture of a spectrophotometer is usually smaller than this value (e. g., 0.1), coupling of light into or out of the spectrophotometer mediated by light fibers requires adaptation by special tele-optics.

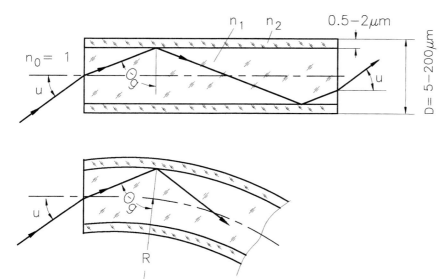

Figure 3.23 Single light fiber including light guide: top, straight and bottom, bent fiber (bending radius R).

With bent cylindrical fibers the above formula is generalized to

$$\sin u = \sqrt{n_1^2 - n_2^2 \times \left(1 + \frac{D}{2R}\right)^2} \qquad (3.23)$$

where D is the core diameter of the fiber [for $R \rightarrow \infty$ we obtain Eq. (3.22)]. Depending on the diameter, bend radii of a few centimeters down to a few millimeters are feasible. Quartz light fibers allow transmission ranging from the near-infrared down to the far-ultraviolet (UV-C, 200 nm) (Figure 3.25). The core material differs in OH content: quartz fibers with high OH content (600–1000 ppm) are used in the UV–VIS range between 230 and 800 nm. For applications below 230 nm special "solarization-resistant" fibers are used, which, however, have a limited lifetime. The term "solarization" describes the opaqueness of the fibers caused by UV-C light, which is absorbed by the "color centers" at 214 nm.

Modern fibers exhibit excellent transmission T, i. e., low losses: $T = 10^{-LK/10}$, where L is the length of the fiber and K the loss in dB m^{-1}. Thus, e. g., $K = 0.5$ dB m^{-1} (a good value obtainable today) is equivalent to a transmission of 0.89 m^{-1}. The Bel (B) is similar to an *Einstein* or *Dalton,* in that it is not a real unit but a relative value. The decibel (dB) is 0.1 B. Primarily it allows energetic values I_0 and I to be compared on a logarithmic basis. For example, if we couple light with an energy I_0 into a light fiber and after 1 m we only measure $I = 0.89 \times I_0$, then the loss is $K = \log[I_0/(0.89 \times I_0)] = 0.0506$ B m^{-1} = 0.506 dB m^{-1}. A fiber of this quality and of 10 m length has a transmission of $T = 10^{-LK} = 10^{-(10 \times 0.0506)} = 0.32$ or 32 %.

Branched light fibers with several "arms" find their applications, for example, in reflection spectroscopy (Section 8.5), low temperature fluorescence spectroscopy (cf. Section 6.3.4) or in dual wavelength spectroscopy (Section 5.5). In these cases, the excitation light is guided through one light fiber arm to the sample, and the emitted or reflected light is guided back from the sample to the detector (cf. Section 5.5.2). Light fibers with special connectors (e. g., SMA 905, ST, FC-PC)

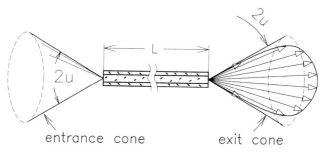

entrance cone   exit cone

Figure 3.24 Only light beams which come into the entrance cone with an aperture of 2u (aperture angle u) are guided. With straight glass fibers entrance and exit angles are identical, within the emission cone light intensity varies corresponding to the arrow lengths.

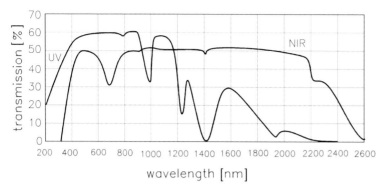

Figure 3.25 Spectral properties of light fibers, manufactured for the visible (UV–VIS) and near-infrared range (NIR). Typical are the marked breaks in transmission curves at certain wavelengths.

are easily and flexibly handled and their applications in optical spectroscopy are growing steadily. Nevertheless, for each individual application one has to determine whether a "light-fiber-free" light path is preferable as it saves money, avoids problems of light-coupling out of or into the light fibers and does not limit the spectral and energetic throughput.

3.5.6
Ulbricht or Integrating Sphere

Most light sources are not Lambert radiators, i.e., perfectly diffuse radiation into space. Rather they exhibit a more (lasers) or less pronounced directional characteristic, which renders the measurement of the total radiant flux Φ_e with a spatially limited photodetector nearly impossible. At the change of the last century (1900) Ulbricht solved this problem with a sphere, which he called an integrating photometer (first described by Sumpner in 1892, later studied more intensively by Taylor, 1920). The Ulbricht sphere is a white empty sphere, which is whitened inside and has the light source to be measured in its center. This light source either emits actively itself, or it is excited to do so (reflection, fluorescence, phosphorescence) through a small hole in its shell (Figure 3.26). After multiple reflections inside the sphere an equilibrium of radiation is established and the inner volume of the sphere is completely, evenly and homogeneously illuminated. With only a small photodetector of area ΔS, which is located somewhere inside the sphere, a signal $\Delta \Phi$ is measured, which is exactly proportional to the radiant flux Φ_e. Care must be taken that the photodetector is completely shielded from direct light of the light source in the center of the sphere.

Originally, the white layer of the inner Ulbricht sphere was made from a paste of zinc oxide, alcohol with camphor and a colorless celluloid. This was a time consuming process and the layer was not very durable. Nowadays we have a series of

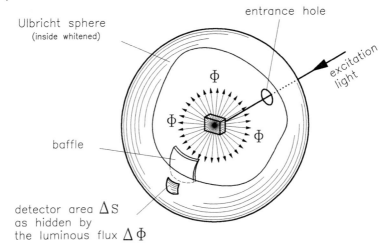

Figure 3.26 The integrating sphere (Ulbricht sphere) allows a nearly perfect homogenization of the radiant flux Φ emanating from a light source and its measurement with a small light detector of area ΔS. The detector is protected by a baffle from the direct light of the light source. Samples such as fluorescing algae can be induced to emit light by an external light. Spherical radii of between 10 and 50 cm are typical. However, for special applications there are spheres which are measured in meters.

excellent layer materials available based on barium sulfate, and several patented compounds such as Spectralon, Spectraflect, Duraflect or Infragold, which display good reflection properties of between 80 and 99 %, depending on the wavelength (Labsphere).

If the reflectivity of the sphere's inside is ϱ and if the small surface of the sphere ΔS is hit by the radiant flux $\Delta\Phi$, then it becomes a radiator itself. Each element of the sphere contributes to the first reflection of the total radiant flux density E_1:

$$E_1 = \frac{\varrho}{4\pi R^2} \sum \Delta\Phi \tag{3.24}$$

After the second reflection we obtain analogously:

$$E_2 = \frac{\varrho^2}{4\pi R^2} \sum \Delta\Phi \quad \text{etc.}$$

The total radiant flux E_e is then calculated by

$$E_e = E_1 + E_2 + E_3...$$
$$E_e = \frac{\varrho^2\Phi}{4\pi R^2(1 - \varrho)} \tag{3.25}$$

The ratio

$$k = \frac{E_e}{\Phi} \tag{3.26}$$

is a constant which specifies the sphere.

Another important entity when the sphere is used as an ideal Lambert radiator is throughput, τ. This is defined as the ratio of total flux out Φ_e to the total flux in Φ_i. It depends on the reflectance properties of the sphere, the number of holes and their diameters:

$$\tau = \frac{\Phi_e}{\Phi_i} = \frac{\varrho A_e/A_s}{\left[1 - \varrho\left(1 - A_p/A_s\right)\right]}$$

where Φ_e is the total flux at the entrance hole, Φ_i the total incident flux, A_e the area of the exit hole, A_s the surface area of the sphere, A_p the sum of all hole areas and ϱ the sphere wall reflectance.

Deviations from an ideal Ulbricht sphere are caused by various factors. The reflectivity ϱ at the inner wall of the sphere is wavelength dependent, light is not perfectly scattered but partly absorbed. The baffle in front of the detector and various holes in the sphere's mantle disturb the geometry and homogeneity, and the detector does not follow the Lambert law perfectly (cf. Figure 3.3): tilted incoming radiation is less weighted. In particular, homogeneity is improved with larger spheres. Therefore larger spheres are preferable to smaller ones.

Ulbricht spheres do not only serve to homogenize and integrate evenly distributed light. Integrating spheres are used for different applications and purposes. They might contain various holes in their mantle to set up samples, measuring devices and special optical components in appropriate relationships. Thus, on the basis of a calibration standard, the transmission or reflection properties of differently scattering samples are measurable, or actively radiating or reflecting color sources can be measured in order to establish their locus in a normalized CIE-color triangle. Last but not least, an Ulbricht sphere yields completely isotropic and diffuse light to the outside: if we place a typical incandescent lamp in the center of the sphere, a small hole in the mantle represents an ideal Lambert radiator (cf. Section 3.2).

3.5.7
Modulators

For many decades, there has been a need for methods to control the intensity or amplitude of a light beam. In optical spectrophotometers such as double- or dual-beam spectrophotometers, fluorimeters, CD spectrometers or phosphoroscopes, mechanical rotating or vibrating shutters, choppers or modulators (mirror polygons) are used, which work down to the kilohertz frequency at best. Mechanical shutters as used in photographic cameras are often sufficient, even if they are far from being absolutely lightproof, as verified with single photon counting.

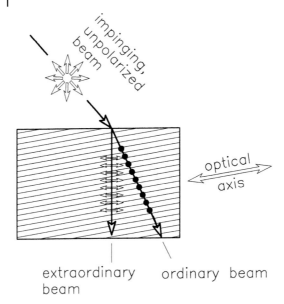

Figure 3.27 A birefringent crystal with an optical axis as indicated splits an incident, unpolarized light beam into an ordinary and an extraordinary beam (cf. Figures 3.16 and 3.27). Both beams are plane polarized and their *E*-vectors are perpendicular to each other.

An electrooptical modulator is based on induced electrical birefringence changes when an electric field is applied. Various types of electrooptical modulators can modulate the radiant intensity I_e spatially and time-resolved, as they allow shutter speeds down to nanoseconds and modulation frequencies in the gigahertz range. They depend on the Kerr effect of birefringence (Figure 3.27).

In optical spectroscopy the term birefringence is very important both theoretically and practically. It is closely related to other terms such as optical activity, optical rotatory dispersion and circular dichroism. Most chemical groups in an atomic system or a crystal, as for example the asymmetric C atom with its four different ligands, are also asymmetric with respect to polarizability by light, i. e., the motility of electrons along the covalent bond that defines the optical axis. In this direction the polarizability α is greatest, the velocity of light smallest and consequently the refractive index is also at a maximum [Eq. (3.15)], and perpendicular to the direction when it is at a minimum. The refractive index varies with the direction of the extraordinary beam. When the light vector is orthogonal to the optical axis of the crystal, then the propagation direction is unimportant and we have normal, ordinary refraction: ordinary beam (cf. Figures 3.16 and 3.27). The extraordinary component undergoes retardation with respect to the ordinary component (the retardation is expressed in Φ as number of wavelengths; the *half-wave voltage* is that which results in $\Phi/2$ (the *electrooptic constant* of a modulator is expressed in picometers per volt for a given wavelength). If the incoming light beam is plane polarized, composition of the two beams emerging from the modulator results in elliptically polarized light (cf. Section 9.7).

In addition to the permanently birefringent crystals some materials such as nitrobenzene exhibit birefringence. We term them "electrooptic materials". The dipole moments that are induced by an electric field E of the order 10000 V cm^{-1} are proportional to the field strength E. Moreover, the already polarized molecules become more and more aligned with increasing fields, contrary to the motility caused by heat. For this reason the birefringence increases with E^2. A shutter is set up as follows (Figure 3.28). Between two crossed polarization filters (polarizer, analyzer) is a cuvette with a Kerr-active liquid, in which the Kerr effect is induced by two electrodes. Without an applied induction voltage the light beam is blocked by the analyzer. After induction of the Kerr effect the light beam is split into ordinary and extraordinary beams, and only one beam can pass through the analyzer.

Because Kerr cells require high switching voltages and, in addition, nitrobenzene of high purity, Pockels cells become more important. The birefringence of ammonium dihydrogen phosphate (ADP, $NH_4H_2PO_4$) and potassium dihydrogen phosphate (PD_2PO_4, the deuterated form only requires half the switching voltage) in the Pockel cell increases linearly with the applied voltage.

Refractive indices for potassium dideuterium phosphate as a function of wavelength are shown in Table 3.3.

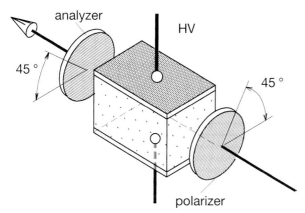

Figure 3.28 A Kerr cell utilizes the bifringence that can be induced in liquids by high voltages. Two crossed polarizing filters (polarizor and analyzer), the transmission vectors for which are tilted by 45° each in relation to the electrical field vector E, are consequently "crossed" (90° to each other). On this basis a "fast" optoelectronic shutter is realized. Without a high voltage being applied the shutter is "closed", but opens with increasing voltage.

Table 3.3 Refractive indices for potassium dideuterium phosphate as a function of wavelength.

Wavelength (nm)	Potassium dideuterium phosphate	
	Ordinary ray	Extraordinary ray
200	1.621	1.563
400	1.524	1.480
600	1.509	1.468
800	1.520	1.463
1000	1.496	1.461
1200	1.490	1.458
1400	1.483	1.456
1600	1.467	1.454
1800	1.468	1.452
2000	1.460	1.450

3.6
Monochromators

3.6.1
Filters

Light filters are round or square optical elements that diminish the intensity of an incident light beam or change it in some other way (e. g., retardation, polarization) in response to wavelength, incident angle or polarization. Neutral density filters ("grey filters") diminish the incident light energy by a given factor (e. g., 75 %) and are mainly independent of wavelength.

However, for theoretical reasons transmission curves of neutral density filters always show some variations in density over the wide spectral range, in practice they are not as "neutral" as is required. Black, metal or fabric webs with sufficiently fine texture, as shown in Figure 3.29, have much better neutral filter properties, particularly in the UV region. Thus, in order to minimize interference effects due to their fine structure, in spectrophotometers the filter(s) should be placed – if access is possible – between the field aperture and the field lens (cf. Figure 3.22).

Color filters only allow specific wavelengths or bands of wavelengths to pass. They are based either on interference effects and reflection, or on the specific absorption of filter materials (absorption filters).

For physical reasons the transmission wavelengths of interference filters depend on the angle of incidence of the light beam. The interference filter is a modified Fabry–Perot interferometer. It is based on two transparent plane-parallel, partly reflective plates a small distance apart, which –

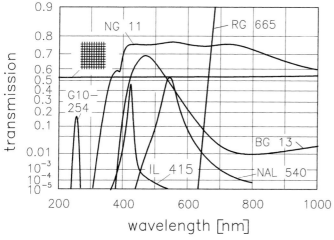

Figure 3.29 Typical transmission curves for some important optical filters as used in optical spectroscopy (for ordinate scale, see the text). G10-254, narrow "mercury-line filter" (Corion); BG 13, blue filter glass (Schott, Mainz, Germany); NAL 540, broad-band interference filter (Schott); IL 415, interference-line filter (Schott); RG 665, long path filter (Oriel); NG 11, grey filter (Schott). Fine meshed nets are preferable to grey filters made from glass, as they attenuate light practically independently of wavelength (first introduced by W. L. Butler).

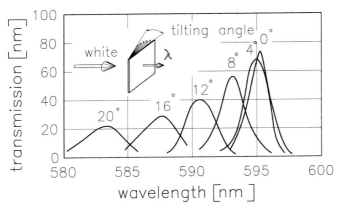

Figure 3.30 Owing to their physical working principles, the transmission properties of interference filters depend on the tilt angle relative to the incident (collimated) light. For example, for this line filter with a maximum transmission wavelength at 596 nm, a simple turning device allows a wavelength range of $\Delta\lambda$ = 10 to 20 nm to be covered (decreasing transmission!).

owing to interference effects – only allow a small spectral range to pass. Interference filters, as compact assemblies, consist of several layers. The transmission wavelength is specified for perpendicular incidence (cf. Section 3.7.2). Even small deviations of a few degrees lead to a significant spectral shift of the maximal transmittance to shorter wavelengths – with concomitant broadening and some reduction in the transmitted band. This effect can be utilized by a small rotational device, covering up to 20 nm with a single interference filter (Figure 3.30). Note, the often strongly reflecting side of interference filters should point towards the light source to reduce the heat generated by light absorption.

Optical filter properties are defined by the degree of transmission $T(\lambda)$ and are illustrated in the form of filter curves (Figure 3.29). In order to visualize transmission T, typically the quantity $[1\text{-}\log(\log T^{-1})]$ is plotted against wavelength. Cut-on filters show a steep increase in transmission T, and cut-off filters a sharp decrease with increasing wavelength. Recently cut-on/cut-of filters have become available (Coherent Auburn), which allow a continuous change in the cut-on/cut-off wavelength. Band filters transmit a wider spectral range while line-filters transmit only a narrow wavelength band in the form of a bell-shaped curve. The half-width (HW) of line filters is defined as the width of the spectral band, in nanometers, where the intensity of the peak is reduced to 50 % of the maximum ($\Delta\lambda_{0.5}$). Another definition of the half-width (HW) of a filter refers to the transmission wavelength λ_m and is given in percent:

$$HW = \frac{\Delta\lambda_{0.5}}{\lambda_m} \times 100 \quad (\%)$$

e. g., 1 % at $\lambda_m = 500$ nm corresponds to a half-width of 5 nm. The steepness of the transmission spectrum at the cut-off point γ of the cut-off filters is defined by

$$\gamma = 100 \times \frac{\lambda_{0.8} - \lambda_{0.05}}{\lambda_{0.5}} \quad (\%) \tag{3.27}$$

For example, for a high pass filter at $\lambda_{0.5} = 520$ nm with $\lambda_{0.8} = 530$ nm and $l_{0.05} = 505$ nm we obtain $\gamma = +4.8$ (low pass filters result in negative values).

Continuously variable interference filters come in the form of round or rectangular plates. The transmission wavelength changes continuously with the rotation angle or the direction along the long axis of the plate, e. g., between 400 and 750 nm.

Polarization filters are based on the effect of birefringence of anisotropic crystals such as tourmaline or herapathite (quinine sulfate periodide), where the refractive index is anisotropic and depends on the direction of light propagation. This results in two differently refracted beams which are polarized perpendicularly (Figure 3.27), where one of the two beams is strongly absorbed. Polarization filters with a large area are manufactured from stretched films of poly(vinyl alcohol) with embedded pigments. Good polarizers work largely independently of wavelength (Figure 3.31).

A specific type of interference filters are so-called *rejection-band*-filters. These filters suppress a narrow wavelength band (typically $\Delta\lambda = 15$ to 40 nm at $T = 10\%$)

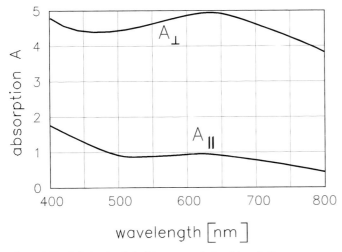

Figure 3.31 Polarization foil of low transmission ($\Delta_{||} = 1$ A) and maximum suppression ($\Delta_{\perp} = 5$ A).

with an efficiency of between 10^{-3} and 10^{-4} ($A = 3-4$); however they transmit approximately 75 % of the light outside the suppressed band (except at the higher and lower harmonics of the suppression wavelength). These so-called Notch-filters are utilized in Raman spectroscopy in particular. They have an extremely narrow blocking range of a few nanometers with more than 7–8 absorbance units at the suppression wavelength; however outside this band they allow more than 90 % transmission.

3.6.2
Dispersion Prisms

The advantage of optical filters is their high transmittance and they are relatively cheap, small and light. However, their transmittance is fixed or is only variable over a small wavelength range (cf. Figure 3.30). Dispersion prisms utilize the wavelength dependency of the refractive index n of all transparent materials such as glass or quartz in order to disperse light, i.e., to separate a light beam spatially according to wavelength (Figure 3.32). The intersection of the two prism planes is known as the breaking edge, the corresponding angle the vertex angle. A beam that runs in the drawing plane is displaced by the angle $\delta = \alpha + \varepsilon - \beta$ (which is negative in a clockwise direction). The exit angle is

$$\varepsilon = \arcsin\,[\cos\alpha\,\cos\beta - \sin\,\alpha\,(n^2\,\sin^2\beta)] \tag{3.28}$$

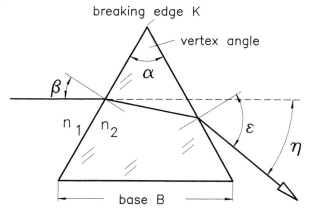

Figure 3.32 Beam deflection η of a prism with an angle of incidence β (for parameters see the text).

If a beam enters the prism at an angle β

$$\beta = \arcsin\ n \times \sin\ \frac{\alpha}{2} \tag{3.29}$$

transverses it symmetrically ($\beta = \varepsilon$) which results in the minimum of deviation (Figure 3.33)

$$\eta_{min} = \alpha - 2\beta \tag{3.30}$$

With Eqs. (3.29) and (3.30) we obtain

$$\sin\ \frac{\alpha - \eta_{min}}{2} = n\ \sin\ \frac{\alpha}{2} \tag{3.31}$$

This equation allows a precise determination of the refractive index n, because α and η_{min} can be determined very precisely.

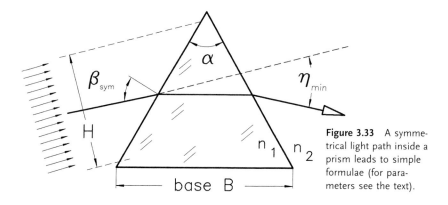

Figure 3.33 A symmetrical light path inside a prism leads to simple formulae (for parameters see the text).

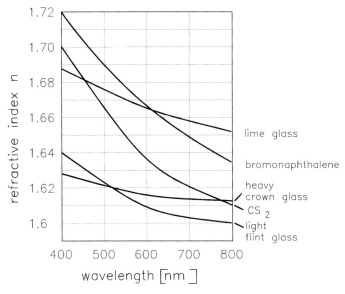

Figure 3.34 Refractive index *n* of various optical materials as a function of wavelength; CS$_2$, carbon disulfide.

As the deviation η depends on n and therefore on λ, a beam of white light will be dispersed into its monochromatic components. Using Eqs. (3.29) and (3.30) and the abbreviations B and H as given in Figure 3.33 we obtain:

$$\frac{d\delta}{d\lambda} = \frac{d\delta}{dn} \times \frac{dn}{d\lambda} = -\frac{2 \sin\frac{\alpha}{2}}{\sqrt{(1 - n^2 \sin^2\frac{\alpha}{2})}} \frac{dn}{d\lambda} \qquad (3.32)$$

$$\frac{d\delta}{d\lambda} = -\frac{B}{H} \frac{dn}{d\lambda}$$

The differential quotient $dn/d\lambda$ depends on the type of glass (Figure 3.34). Thus the angle of dispersion is only proportional to the length of the base B and inversely proportional to the height of the beam H of the incident light beam, however not on the height of the prism or the vertex angle.

3.6.3
Dispersion Gratings

If a parallel light beam hits two slits of comparable distance and widths with respect to the wavelength of the light, we observe interference phenomena. According to Huygens, both slits are the source of elementary or spherical waves that interact with each other in space, i.e., they show interference (for the requirements of coherence see Section 2.2): the wave maxima enhance and the opposite amplitudes chancel each other. If we draw wave maxima as circles

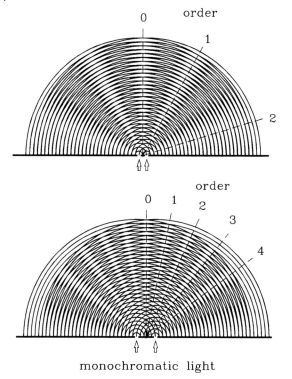

monochromatic light

Figure 3.35 Overlapping of only two spherical waves of monochromatic light generates – depending on their relative distance (grating constant *a*) – typical interference patterns with marked maxima (lines are crossing) and minima in various directions (spacings are crossing). Relative to the normal we define maxima of 0, 1st, 2nd, etc., scattering order. The further apart the slits, the more orders are covered (Echelle grating, cf. Figure 4.8)

then the two slits result – depending on their relative distance *a* (gratings constant) – in a specific interference pattern as shown in Figure 3.35. In some directions the elementary waves enhance each other, in others they cancel each other. This simple example serves as a visualization of the action of real optical gratings with several hundred grooves ("slits") per millimeter. The direction of maxima are determined by the fact that according to Figure 3.36 homologue rays just differ by one wavelength – or a multiple of it (grating equation):

$$\sin \beta - \sin \varepsilon = \pm m \frac{\lambda}{a}$$

$$(3.33)$$

$$\beta = \arcsin \left(\pm \frac{m\lambda}{a} + \sin \varepsilon \right)$$

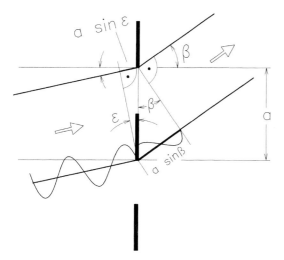

Figure 3.36 Mathematical realization of the results shown in Figure 3.35: interference patterns by light scattering with two slits at a distance a. The crests of the spherical waves which emanate from the two slits are additive when the phase shift $a \times (\sin \beta - \sin \varepsilon)$ corresponds to a multiple m of the wavelength λ.

Christiaan Huygens (1629–1695)

The plus sign is used for transmission, the minus sign for reflection gratings.

The resolution of a grating $R = \lambda/d\lambda$ is a measure of its ability to separate two narrow spectral lines. According to Rayleigh this is the case when the maximum of one falls into the minimum of the second. If N is the total number of grooves on the grating, then

$$R = \frac{\lambda}{d\lambda} = m \times N \tag{3.34}$$

The greater the order m, the larger the number of grooves and the higher the resolution.

Thus a grating with 600 lines mm^{-1} and a width of 50 mm yields for the first-order scattering a numerical resolution of $R = 600 \times 50 = 30000$. If we substitute Eq. (3.33) into (3.34) and write for the total width of the grating $W = N \times a$ then we obtain

$$R = W \times \frac{\sin \varepsilon - \sin \beta}{10^{-6} \lambda} \tag{3.35}$$

Thus, the resolution of a grating depends on its width W, the corresponding "mid-wavelength" λ and the geometry used.

The resolution should not be confused with the bandwidth. This is calculated, for example, for 500 nm to give $d\lambda = 500/30000 = 0.0166$ nm. However, because the effective bandwidth of a spectrometer also depends on the slitwidth and on various optical bias such as aberration, astigmatism or coma, in practice the actual bandwidth will always be worse.

Angle dispersion is defined as the angle difference, $d\beta$, that is obtained for two wavelengths which differ by $d\lambda$ ($d\beta/d\lambda$). Differentiation of Eq. (3.33) yields

$$\frac{d\beta}{d\lambda} = \frac{\sin \varepsilon + \sin \beta}{\lambda \cos \beta} \tag{3.36}$$

The linear dispersion $d\lambda/dx$ of a grating system (dispersing element plus optics which images the entrance slit onto the exit slit) equals the reciprocal product of angle dispersion and focal length f of the exit collimator [cf. Eq. (3.22)]:

$$\frac{d\lambda}{dx} = \frac{a \cos \beta}{m f} \tag{3.37}$$

With a = line distance (grating constant) and $n = 1/a$, the number of lines per millimeter, we obtain

$$\frac{d\lambda}{dx} = \frac{10^6 \cos \beta}{m \, n \, f} \left(\frac{nm}{mm}\right) \tag{3.38}$$

Thus a grating with 2 nm mm^{-1} (at the exit slit) has a higher linear dispersion than one with 4 nm mm^{-1}.

Of practical importance is the energy distribution I of a grating across the whole spectral range of interest. With the given grating constant of a, the energy distribution is given by:

$$I(\varepsilon, \beta) = \frac{I_0 \sin^2 \left[W \pi \lambda^{-1} (\sin \varepsilon + \sin \beta) \right]}{\sin^2 \left[a \pi \lambda^{-1} (\sin \varepsilon + \sin \beta) \right]} \tag{3.39}$$

For a = 6000 nm and N = 2 to 100 lines we obtain the energy distribution as shown in Figure 3.37. Obviously the scattering maxima sharpen with the number of grooves, and the resolution increases.

Figure 3.38 illustrates a general limiting problem with gratings, in contrast to dispersion prisms and continuous filters. As a demonstration, the deviation (dispersion) of monochromatic light of discrete wavelengths ranging from 200 to 1000 nm for a given grating is illustrated for the first to the fourth orders. The linear dispersion $d\lambda/dx$ increases proportionally with the scattering order [Eq. (3.37)]. Thus, the dispersion for the first order (m = 1) is unequivocal. The same is true for 300–600 or 400–800 nm, i.e., generally for one octave (doubling of the wavelength). If the incident light beam is white, it is clear that we obtain

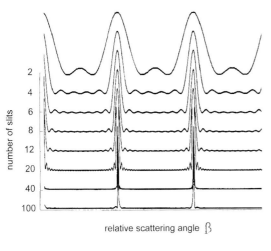

Figure 3.37 Relative scattering intensity of a grating with 2–100 slits at a fixed distance a with a given angle of incidence $\varepsilon = \theta$ as a function of the scattering angle β.

number of slits

relative scattering angle β

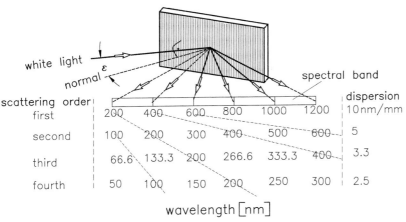

wavelength [nm]

Figure 3.38 Deviation (dispersion) of monochromatic light with discrete wavelengths from 200 to 1000 nm of a given grating in the first up to the fourth order. The linear dispersion $d\lambda/dx$ increases with the order [Eq. (3.37)].

Thus, within one octave (doubling of the frequency) the dispersion is always unequivocal. For higher orders the linear dispersion increases and we obtain significant overlap.

overlapping orders. "Unwanted" light of the second order of 200 nm falls into the range between 400 and 800 nm of the first order. This problem is solved in practice by blocking the unwanted orders by a longpass filter (in our example the 200 nm light). Therefore we define a new important parameter, the "effective wavelength range":

$$\Delta\lambda = \frac{\lambda}{m} \tag{3.40}$$

In the most favorable case we obtain for $m = 1$ and 200 nm a range $\Delta\lambda = \lambda$, i.e., for 200 nm from 200 to 400 nm. For 600 nm and $m = 4$ spectra are unequivocal only in the interval 600–750 nm (however with four-fold resolution!).

Dispersion gratings also have the disadvantage of unfavorable energy distributions, as in general only one single scattering order (the first) is utilized [except for Echelle gratings in inductively coupled plasma (ICP) spectroscopy, Chapter 4]. With a simple trick, however, the energy can be concentrated within one scattering order, e.g., the first one. The grooves are formed having triangular cross sections with the steps of the slope being ε, where the scattering angle β is identical to the "ordinary" reflection angle ε for specular reflection (Figure 3.39); this angle is termed the "blaze angle" and correspondingly the wavelength of maximum scattering the blaze wavelength. In the example shown the grating is "blazed in the first order". The dashed line indicates the resulting energy distribution. In Chapters 4 and 5 we will discuss further details of the practical usage of dispersion gratings.

The energy profiles of blazed gratings are determined in a so-called Littrow configuration, where incoming/scattered light beams are pointing in the same direction, they are auto-collimated while the grating rotates ($\beta = \varepsilon$). In this case we obtain for the blaze wavelength λ_B

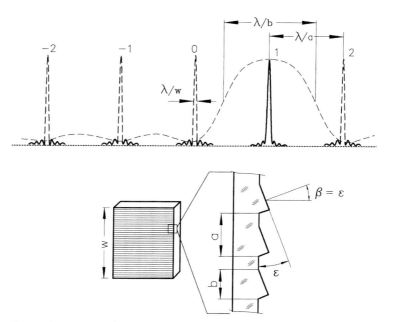

Figure 3.39 By giving the grooves in the gratings a specific shape (Echelle reflection grating, etalon) the scattered energy of a grating can be concentrated largely into one order; e.g., into the +1st order corresponding to the envelope in this example (dotted bell-shaped curve; after Jobin-Yvon, modified).

$$\sin \beta \; - \; \sin \varepsilon = \; - m \times \frac{\lambda}{a}$$

$$\omega \; = \; \varepsilon \; = \; -\beta \, , \text{ with } \; \omega \; = \; \text{blaze angle} \qquad\qquad (3.41)$$

$$2 \times \sin \omega \; = \; \frac{m \times \lambda_B}{a} \times 10^6 \quad (\lambda \text{ in nm})$$

For example, for a grating of 600 lines mm^{-1} at 500 nm we obtain a blaze angle of $\omega = 8.627°$.

Physically, a dispersion grating is an arrangement of evenly distributed, parallel optical "disturbances". In the gratings we have discussed so far the lines have been engraved mechanically or by light techniques in a (glass) matrix, and have reflection or transmission properties different to the matrix itself. In *acoustooptical gratings* (known as *acoustooptical modulators*, AM) mechanical compression waves are generated by the piezoelectric effect in crystals (tellurium dioxide TeO$_2$, lead molybdate PbMoO$_4$, quartz glass or germanium) on which piezoelectric plates are attached yielding a corresponding pattern of the refractive index (ultrasound grating). In the so-called "Bragg regime" only a single scattering order is generated (Figure 3.39). By varying the power of the high frequency generator the light intensity is modulated, its frequency determines the grating constant *a*. An AM mainly serves as an intensity and directional modulator (cf. Section 3.5.7), but it can also be utilized as a wavelength scanner, particularly in near-infrared spectrometers (Chapter 10).

3.7
Photodetectors

Photodetectors are generally classified on the basis of the outer or inner photoeffect. In addition, there are detectors that depend on the detection of heat or on photochemical reactions. The sensitivity of the various photodetectors differs for individual ranges of the electromagnetic spectrum and is utilized accordingly (Figures 3.40 and 3.41). Table 3.4 shows some semiconductor materials with their spectral sensitivities.

Table 3.4 Semiconductor detector materials as used for various spectral ranges.

Detector material	Wavelength range (μm)
Si	0.2–1.1
Ge	0.8–1.8
In(Ga)As	1.0–3.8
InSb	1.0–7.0
InSb (77 K)	1.0–5.6
HgCdTe (77 K)	1.0–25

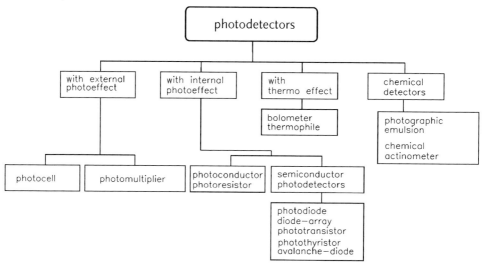

Figure 3.40 Classification of photodetectors.

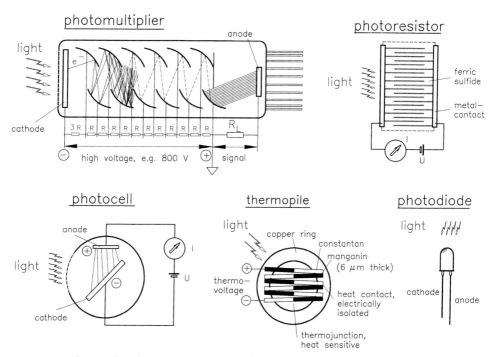

Figure 3.41 Schematic representations of the most important photodetectors in optical spectroscopy.

3.7.1
Photodetectors with an Outer Photoeffect

These photodetectors utilize the so-called Einstein effect or *outer photoeffect*. Photons are capable of removing electrons from a photocathode which consists of alloys of alkali metals, such as Sb–Cs, Na_2–K–Sb, Cs–Te and others, whereby the cathode thus becomes positively charged. If electrons are fed into the cathode and removed them through the anode by applying an appropriate voltage, then we observe a light driven electrical current, the "photocurrent". In order to avoid interactions with air molecules the photocells are largely evacuated. The photocurrent can be strongly amplified by a "photomultiplier" (PM). A series of so-called dynodes with increasing electrical potential is arranged in series. A cascade of electrons is removed sequentially from the dynodes, known as secondary, tertiary and subsequent electrons until the electron current its amplified 10^6- to 10^7-fold. Photomultipliers (PM) are among the most sensitive photodetectors; however, they are comparatively expensive, large-sized, difficult to handle, mechanically sensitive and require technical maintenance, for example, a stabilized high voltage. Figure 3.42 shows the linear relationship between the photocurrent of a high quality photomultiplier and the incident light stream. Six to seven decades of amplification are easily obtained with a photomultiplier.

Depending on the cathode and window materials PMs are only utilized in practice in the UV–VIS range (Figure 3.43). However, a photomultiplier with a novel cathode type (InGaAs) has been introduced by Hamamatsu, which covers the near-infrared range up to 1700 nm; note that cooling is essential. We will distinguish between two types of photomultipliers that differ in their mechanical and geometric arrangement of their photocathodes, namely the *side-on* and the *head-on* PM (Table 3.5).

Owing to thermal processes in the presence of light, a photomultipliers generate, as with all other photodetectors, an unwanted background signal, i. e., dark

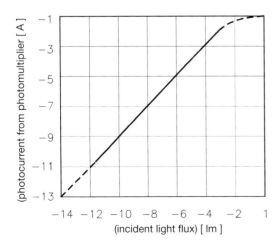

Figure 3.42 Dependency of the photocurrent of a high quality photomultiplier as a function of the incident light flux (after RCA, modified).

Table 3.5 Comparison between *side-on* and *head-on* photomultipliers.

Side-on	Head-on (Figure 3.41)
Reflecting photocathode (sensitivity is strongly dependent on spatial light distribution)	Transmitting photocathode (more homogeneous signal reception)
Moderately high voltage	Greater high voltage
"Low cost"	ca. 5 times more expensive than a side-on
Only small solid angle coverage	Large solid angle covered (close to 2π)
High amplification, high sensitivity (circular dynodes)	

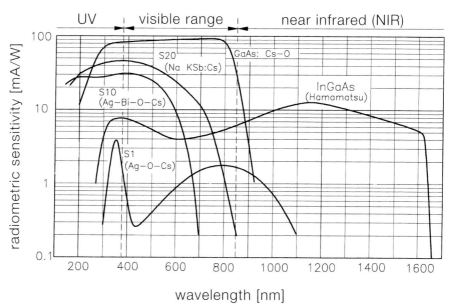

Figure 3.43 Spectral light sensitivities of photomultipliers as a function of the cathode material. Each alkali alloy has its own terminology such as S1, S10, etc. In addition, in the UV range the window material plays an important role. The application of photomultipliers is largely limited to the UV–VIS wavelength range (after RCA, modified).

current. This decreases the sensitivity and can be somewhat improved by cooling. Sensitivity can be defined as the measure of radiant flux Φ_e at 100 times the dark current. Figure 3.41 shows a typical voltage cascade for continuous reception of spectra (cw mode, pulse counting PMs show a slightly different electronic circuit; in the latter the negative potential is to ground). The values of the resistors R of the cascade are between 20 kΩ and 5 MΩ (the cascade current should have a multiple value of the anode current in order to guarantee a significant dynode voltage). A high sensitivity of the photomultiplier is obtained through a high voltage (i.e., with a large resistor = $3R$) between the photocathode and the first dynode. The negative potential of the cathode is "high", while the positive potential of the anode is set to ground. The "load-resistor" R_L, which yields the electrical voltage signal, is typically chosen to be 51 Ω, the impedance of the coaxial cable that allows optimum signal processing. For special applications (e.g., space technology) there are photomultipliers as small as a match box (Hamamatsu), in which the PM including high voltage generator are integrated. Here, only a single external dc voltage supply is required.

A modern version of the photomultiplier is the *multi-channel plate*, Figure 3.44. This technique was originally used in electronic image amplification. Micro-channel plates are made from molten glass fibers with a glass core and lead silicate glass cladding. After etching the core we have small tubes with an inner diameter of less than 15 μm and a length of approximately 500 μm. Subsequently the inner surfaces of these tubes are covered with a semiconducting layer and the plates are covered on both sides with metal electrodes, through evaporation. Thus, each individual tube acts as a photomultiplier. With an applied voltage of, e.g., 1000 V, an amplification factor of approximately 10^4 is obtained. To further increase the amplification (and concomitant loss of resolution) it is possible to use two micro-channel plates in series. These image amplifiers are utilized successfully in diode array or CCD spectrometers.

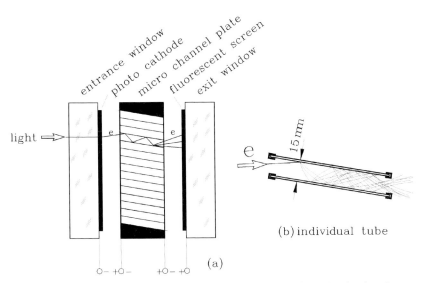

Figure 3.44 Operating mode of the micro-channel plate (MCP). Light falls through the entrance window onto the photocathode. Owing to the *outer photoeffect* electrons are ejected from the cathode and accelerated in the individual tubes of the MCP by the applied high voltage, and finally hit the phosphorescing light screen. Here the amplified electron stream is reconverted into light, which can be detected in a conventional manner, e.g., by a diode array or a CCD detector.

3.7.2
Photodetectors with an Inner Photoeffect

In semiconductor detectors the incident light quanta increase the number of free charge carriers within the solid state (*inner photoeffect*). Typical sensitivity curves of a photoresistor (Cd–Se) and two photodiodes (Si, Ge) are shown in Figure 3.45, compared with the sensitivity of the human eye, along with the emission spectrum of an incandescent lamp. Photoconductors and -resistors are made from iron sulfide or calcium compounds and change the resistance independently of the current direction. These photodetectors are very slow (<10 Hz). Semiconductor detectors with a pn-boundary interface are termed – depending on the electric circuit – *photoelements* (operation without applied voltage) or *photodiodes* (operation with the voltage in the reverse direction). In photoelements the voltage increases logarithmically with the radiant flux density (this effect is utilized in absorption spectroscopy, Chapter 5), in photodiodes the reverse current is proportional to the radiant flux density. Semiconductor materials used are germanium (high sensitivity) and silicon (low temperature dependency of the dark current). The frequency limit is in the MHz range. Special diode arrays with 512 and 1024 densely packed individual diodes are used as the detector in diode array spectrometers, which we will discuss in more detail in Chapter 5. Phototransistors have two pn-boundary layers and allow additional amplification; however, the "dark noise" will be amplified as well, with no improvement in the limit of detection. Avalanche diodes are based – similar to the photomultiplier – on the mechanism of an internal amplification where small signals can rise above the noise level. Pairs of charge carriers are generated by light, which are accelerated by a reverse voltage in such a manner that the kinetic energy suffices to generate further pairs of charge carriers by impact ionization. The amplification factor of avalanche diodes is approximately 100 (i. e., still some orders below that of photomultipliers). Their strong temperature dependence is a serious disadvantage for certain applications.

3.7.3
Photodetectors with a Thermoelectric Effect

Radiation receivers that depend on heating the detector are pure energy or thermal detectors; their sensitivity is primarily wavelength independent. Thus, as many photons as possible have to be absorbed – independent of the wavelength. To achieve this it is best to blacken the receptor, or better still, radiation is absorbed by a hollow that is within a blackened hollow body with a small window (black body). In addition, the heat capacity of the detector has to be small in order to give a large temperature change with a small energy transfer. Bolometers depend on the temperature dependence of the electrical resistance of certain materials. If a constant voltage is applied the current through the bolometer increases approximately with the absorbed energy. The thermoelement consists of a pair of soldered bimetallic (or semiconductor) connections. One connection is

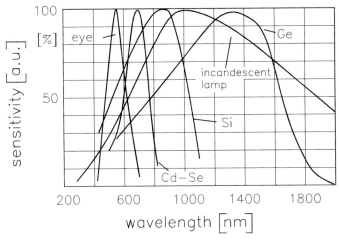

Figure 3.45 Various semiconductor detectors in comparison with the spectral sensitivity of the eye and the emission spectrum of a typical tungsten lamp. The germanium detector is particularly useful for the near-infrared spectral range (cf. absorbency of singlet oxygen, Figure 2.18).

heated by light while the temperature of the reference connection is kept constant by thermal connection to a larger mass (the copper ring in Figure 3.41). Typical sensitivities are some tenths of a microvolt per degree Celsius. A *thermopile* consists of a series of thermoelements that are connected in series with linearly increasing sensitivity.

3.7.4
Photochemical Detectors

Photoelectric and thermoelectric light measurements are simple, relatively precise and fast. However, absolute light measurements as required for the determination of quantum efficiency in photosynthesis, for example, are difficult and problematic, as the geometry of the detectors can not really be standardized and the linearity range varies with detector age. In terms of absolute values, from radiometer readings we can expect measurement errors as high as $\pm 50\,\%$; the possible absolute precision is generally overestimated. Therefore, applying the more laborious methods of photochemical light measurement (actinometry) is sometimes unavoidable. The measurement is integrated in space and time and the measuring cuvette can be adapted to the light flux to be measured. A photochemical actinometer is essentially a solution in which a photochemical reaction of high and precisely known quantum efficiency takes place (similar to the definition of quantum efficiency in fluorescence; see Section 6.2.6). The concentration of the reactants at the beginning is known. After a certain illumination time we measure the concentration of a particular product and for the radiant flux of the light source to be measured we obtain:

$$photons \ (s^{-1}) = \frac{number \ of \ product \ molecules \ generated}{quantum \ efficiency \times irradiation \ time}$$

A classical actinometer, after *Hatchard* and *Parker*, measures the photodissociation of ferri-oxalate ions into ferro-oxalate:

$$2 \ Fe \ (C_2 \ O_4)_3^{3-} \xrightarrow[\text{to be measured}]{\text{light}} 2Fe(C_2 \ O_4) + 3C_2O_4^{2-} + 2CO_2$$

After irradiation a solution of phenanthroline is added yielding a colored complex with the reduced iron. Finally, its concentration is determined by absorbance at 510 nm.

3.8
Cuvettes

Liquid samples are measured in special glass containers, so-called *cuvettes*. Depending on the particular application and type of measurement there are various styles and qualities of cuvettes available, which differ greatly in price. Which to choose depends on sample thickness, sample volume, spectral range to be measured and the measuring temperature. Measurements in the UV range below 320 nm, for example, requires quartz glass; for measurements at temperatures other than room temperature special cuvette holders can be thermostatically controlled. Low temperature measurements, e. g., at the temperature of liquid nitrogen or helium, are performed in so-called Dewar flasks. For biochemical applications in particular, samples sometimes have to be anaerobic, i. e., measurements are performed with atmospheric oxygen excluded. This is accomplished by gas flushing the sample with a gas such as argon (Ar), xenon (Xe) or nitrogen (N_2). A standard cuvette has outer dimensions of $12.5 \times 12.5 \times 45$ mm, inner dimensions of 10×10 mm and it is made from plastic, glass or quartz glass (Figure 3.46, left); a small volume cuvette as offered by one manufacturer has the same outer dimensions, however, the inner cross section is only 10×1.25 mm. A continuous-flow microcuvette has a volume of only a few microliters and allows – e. g., in column chromatography – spectral measurements during elution. Triangular cuvettes allow the measurement of fluorescence with the excitation light incident on the same cuvette face as that where the measurement takes place – in contrast to conventional fluorescence measurement which is performed perpendicular to the exciting light with fluorescence light traversing a larger volume. Thus in the former, "self-absorption" of the fluorescing molecule is diminished (see Section 6.3.2).

It is somewhat surprising that spectrophotometric standard cuvettes (as well as most spectrophotometers) are typically positioned for horizontal light paths. The essential disadvantage of a typical cuvette is that only a small fraction of the sample is illuminated by the measuring light (with standard cuvettes less than 10 %).

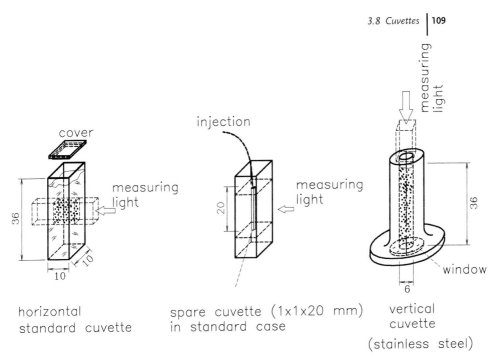

cover

injection

measuring
light

measuring
light

measuring
light

36

10

10

20

36

6

window

horizontal
standard cuvette

spare cuvette (1x1x20 mm)
in standard case

vertical
cuvette
(stainless steel)

Figure 3.46 Comparison of the light paths in a conventional cuvette with defined thickness, an ultra-microcuvette and a steel cuvette with vertical light paths.

If samples are precious and only very small volumes are available, microcuvettes or even ultra-microcuvettes with small volumes of only 50 µL or even 2.5 µL can be used. Clearly, the small volumes lead to capillary effects and problems with air bubbles, possibly requiring degassing. Finally, recovering the sample material is critical.

A vertical light path, on the other hand (as proposed by W. L. Butler, Figure 3.46, right) offers 100 % usage of the available sample volume, or in other words, an approximately 20 times smaller sample volume can be used compared with a standard cuvette. However, in this case the poorly defined sample thickness is limiting, if quantitative rather than qualitative determinations are required. For the measurement of small volumes the sample thickness – i. e., the precise concentration – is often sacrificed, relative to the spectral shape resolution. In addition, for these measurements reflection spectroscopy (e. g., ATR, Chapter 8) or the light fiber arrangement as utilized in *optrodes* are possibilities (Figure 6.34).

According to the Lambert–Beer law, the absorption is proportional to the path length of the radiation through the sample. Thus multiple traversing of the measuring beam through the sample volume is required. This is accomplished by multiple reflection within the cuvette. This procedure is used particularly in the absorption spectroscopy of weakly absorbing gases, similarly in ATR-infrared spectroscopy (*attenuated total reflection*, Chapter 8.6 and Figure 8.27b); however it is less important in UV–VIS spectroscopy.

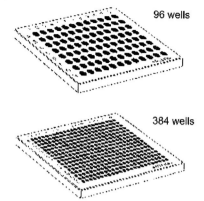

96 wells

384 wells

Figure 3.47 The two common types of microplates with 96 and 384 wells.

In protein and DNA research in particular, the *simultaneous* measurement of as many as 96 or 384 individual small cuvettes (wells) allows highly efficient analyses to be made. The wells are arranged in *microplates* (Figure 3.47), which are automatically measured by special spectrophotometers called microplate readers. Absorption as well as fluorescence measurements are feasible.

3.9
Selected Further Reading

Albani, J. R. *Structure and Dynamics of Marcomolecules: Absorption and Fluorescence Studies*, Elsevier Science, Amsterdam, **2005**.

Bass, M., Styland, E. W. *Handbook of Optics: Fundamentals, Techniques, and Design*, McGraw Hill, New York, **1994**.

Bass, M., Styland, E. W. *Handbook of Optics: Devices, Measurements, and Properties*, McGraw Hill, New York, **1994**.

Dereniak, E. L., Crowe, D. G. *Optical Radiation Detectors*, John Wiley & Sons, New York, **1984**.

Denney, R. C. *Visible and Ultraviolet Spectroscopy (Analytical Chemistry by Open Learning)*, John Wiley & Sons, Chichester, **1987**.

Engstrom, R. W. *Photomultiplier Handbook*, RCA/Electrooptics and Devices, Lancaster, PA, **1980**.

Hutley, M. C. *Diffraction Gratings*, March, N. H., Daglish H. N., eds., Academic Press, New York, **1982**.

O'Shea, D. C. *Elements of Modern Optical Design. Wiley Series in Pure and Applied Optics*, Goodman, J. W., advisory ed., John Wiley & Sons, New York, **1985**.

Petruzzellis, T. *Optoelectronics, Fiber Optics and Laser Cookbook*, McGraw Hill, New York, **1997**.

Robinson, J. W. ed., *Handbook of Spectroscopy*, CRC Press, Boca Raton, **1981**.

Slayter, M. E. *Optical Methods in Biology*, Wiley-Interscience, John Wiley & Sons, New York, **1970**.

Smith, W. J. *Modern Optical Engineering. The Design of Optical Systems*, McGraw Hill, New York, **1966**.

Wilcox, C. H. *Scattering Theory for Diffraction Gratings (Applied Mathematical Sciences Series Vol. 46)*, Springer-Verlag, Berlin, Heidelberg, New York, **1984**.

Yen, W. M., Levenson, M. D. eds., *Lasers, Spectroscopy and New Ideas: A Tribute to Arthur L. Schawlow (Springer Series in Optical Sciences, Vol. 54)*, Springer-Verlag, Berlin, Heidelberg, New York, **1987**.

4

Spectroscopy of Atoms

Natura in minimis maxima
(From the Pythagorasera)

In nature, the smallest is the greatest

Felix qui potuit rerum cognoscere causas
(Virgil)

Blessed are those who succeed at understanding the principles of the world

In accordance with Kirchhoff's rule, atomic spectra can be recorded under absorption as well as under emission conditions. All elements exhibit specific and often intense spectral lines that are available for analysis, e. g., Al 308.215, 396.152; Ca 315.887, 317.933; Cd 214.438; Mg 279.079, 279.553; Pb 220.353; K 404.721, 766.491; Na 330.237, 588.995, 589.592; or Zn 206.095, 213.856 nm. For various reasons, which will become clear in the subsequent chapters on absorption and emission spectrometry of molecules, atomic emission spectroscopy (AES and ICP-OES) is far more sensitive than atomic absorption spectroscopy (AAS). Therefore, it is not surprising that more than 5000 ICP spectrometers have been purchased worldwide since 1979 and that the market is still growing rapidly.

Nevertheless, AA is one of the most widely used trace element analytical techniques. In terms of absolute sensitivity, from a theoretical perspective, emission will provide bigger signals, however, in analytical terms, such as detection limits, then both flame AA and ICP, for many elements show similar detection limits. In virtually all cases graphite furnace AA shows better detection limits than ICP. These differences are due to sample uptake rates. Flame AA nebulizers present much more sample into the flame than ICP nebulizers, where approximately 98 % of the sample goes down the waste tube! On the other hand, the biggest advantage of ICP is the multi-element capability and the large dynamic range, in many cases the sample throughput rates far outweigh the extra expense of the argon consumption in an ICP.

Optical Spectroscopy in Chemistry and Life Sciences. W. Schmidt
Copyright © 2005 WILEY-VCH Verlag GmbH & Co. KGaA, Weinheim
ISBN 3-527-29911-4

4.1
Atomic Absorption Spectroscopy

In chemical/biological systems, particularly in enzymatic reactions, metal ions play a key role. Therefore, their qualitative and quantitative determination is the basic prerequisite for an efficient analysis such as in environmental monitoring. AAS has been proven to be a fast, reliable, cheap and effective method of sufficient specificity and sensitivity, and is a useful supplement to mass spectrometry. As pointed out in Section 1.2 on the history of spectroscopy, the so-called Fraunhofer-lines (Table 1.1) reflect the atomic absorption of various elements in the atmosphere of the earth and the sun.

Typically, an atomic absorption spectrometer is set up for the determination of the concentration of only a single element [Figure 4.1(b)], particularly in complex mixtures. This is because each element exhibits its own individual absorption lines, well defined in wavelength (and intensity, see Chapter 2 and introduction to this chapter). The most widely used light source is the so-called hollow cathode lamp. Its hollow cylindrical cathode contains only the metal to be determined – or an alloy of it. The lamp is filled with a noble gas such as argon or neon at low

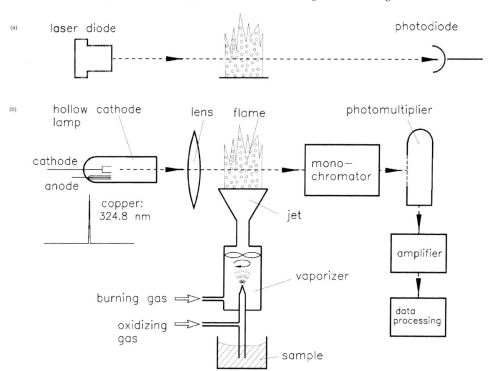

Figure 4.1 (a) Principle schematic diagram of an atomic absorption spectrometer with a tunable laser diode as the light source and photodiode as detector or (b) with a hollow cathode lamp as the light source.

pressure. When the current is passed the gas is ionized at the anode and its anions are accelerated towards the cathode. The resulting bombardment of the cathode material generates an atomic vapor, which is raised to the first excited state, finally emitting a single atomic line that can be used as a strong specific light source characteristic of the element (essentially a metal) to be determined.

Multi-element lamps are also available in which an alloy of several metals is used as the cathode material. In addition to hollow cathode lamps there are "electrodeless" atomic lamps available which contain a small amount of the metal to be determined and are filled with an inert gas at low pressure. Similar to ICP spectroscopy (see below) this mixture is excited to the plasma state by microwaves, finally emitting light several hundred times more intense than that attainable with a hollow cathode lamp.

A new alternative to a hollow cathode lamp is a diode laser [Figure 4.1(a)]. Because of the extraordinary spectral sharpness, a diode laser allows the highly selective determination of various elements and their isotopes to be made, by scanning the laser wavelength ($\Delta\lambda \leq 20$ nm, Section 3.4.5). In addition, the absorption line profile serves to eliminate the unspecified absorption of the sample matrix. Simple modules for the extension of conventional AA spectrometers to laser diode atomic absorption spectroscopy (LAAS) are available today. Unfortunately, so far sufficiently intense laser diodes are only available for wavelengths longer than 630 nm, which seriously limits the usage of LAAS.

AAS requires transfer of the atoms to be determined into the gas phase – in contrast to molecular optical spectroscopy which allows measurement of solid, liquid and gaseous samples. This is accomplished either by simply vaporizing the sample in a flame, or by flameless evaporation in a graphite furnace (non-flame cell). In flame-AAS only liquid samples can be measured. However, in a graphite (electrothermal atomizer) furnace the measurement of liquids, slurries and even solid samples is feasible. In flame-AAS an aerosol of the sample containing the element to be determined is mixed with a fuel gas, such as acetylene and air, and transported to the flame. The graphite furnace method is about 100 to 1000 times more sensitive than the flame method. The reason is that all the sample is deposited in the atomizer and there are no losses due to nebulizer efficiency and dilution by the fuel gases as in flame-AAS. On the other hand, a typical analysis time by graphite furnace-AAS is 2 to 3 min, and with flame-AAS it is less than 10 s.

The linewidth of atoms is essential to the understanding of AAS. Because of the principle by which a hollow cathode lamp works, it emits many individual atomic lines characteristic of the particular cathode metal, e. g., copper or iron. The linewidth is determined by two factors: (1) the lifetime of the excited atom and (2) the Doppler broadening. Lifetime and linewidth are correlated by Heisenberg's uncertainty principle:

$$\Delta\nu_{LT}\,\Delta\tau_{LT} \geqslant 1$$

For atoms this corresponds to the average time interval between collisions. In air of 1000 hPa (1 bar) and 2500 K we obtain $\Delta\nu_{LT} = 10^{11}$ Hz. Choosing a wavelength of 300 nm, corresponding to a frequency of 10^{15} Hz, we get $\Delta\nu/\nu \approx \Delta\lambda/\lambda = 10^{-4}$ and thus $\Delta\lambda \approx 0.03$ nm = 0.3 Å.

Doppler broadening is defined by:

$$\Delta\lambda/\lambda = v/c$$

where v equals the relative velocity of the light source and the observer, and c the velocity of light. For 2500 K and a medium velocity of 1500 m s^{-1} we obtain $\Delta\lambda/\lambda = 5 \times 10^{-6}$, a significantly lower value than that obtained by collision broadening.

The spectral situation in atomic absorption spectroscopy is shown in Figure 4.2. In principle, a dominant line emerging from the specific (low pressure) hollow cathode lamp of approximately 10^{-4} Å spectral width and the intensity I_0 is transmitted through the flame or heated graphite tube where the apparent spectral absorption width of the same line of the analyte is approximately 1000 time larger. After the corresponding attenuation, light of the intensity I is detected by the sub-

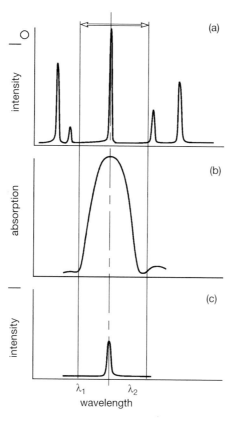

Figure 4.2 (a) Owing to the low pressure (1 hPa), a hollow cathode lamp (HCL) emits atomic lines with a spectral width of only 10^{-4} Å and an intensity of I_0. (b) Depending on the AAS method used, this light traverses either the flame or graphite tube containing the analyte at normal pressure (1000 hPa) and thus produces a 1000 times larger linewidth of 10^{-1} Å. (c) A monochromator with a slitwidth ($\lambda_2 - \lambda_1$) of from 1 to 10 Å selects a single individual atomic line. After the corresponding attenuation light of the intensity I is detected by the subsequent photomultiplier and converted into absorbance ($A = \log I_0/I$). For further details on background correction see Section 4.1.3.

sequent photomultiplier. For additional information on background correction see Section 4.1.3.

4.1.1
Flame Method

The actual flame temperature is not crucial. What is more important is the length of the flame traversed by the measuring light beam, as the amount of absorption depends on the volume of atomic vapor traversed, as well as on the concentration of atoms in the flame in their ground state. Therefore the light is passed through a long narrow flame produced by a slit burner, (e. g., x = 10 cm path length, see the Lambert–Beer law, Chapter 5).

According to Boltzmann's distribution law

$$\frac{N_1}{N_0} = e^{\frac{E_0 - E_1}{\varkappa \times T}}$$
(4.1)

with N_0 and N the number of atoms in the ground and excited states, respectively, E_0 and E_1 the corresponding energies, $\varkappa$ the Boltzmann constant and T the absolute temperature. At a flame temperature of 2000 K only 2 % of all atoms to be analyzed are in the excited state, i. e., the majority is still in the ground state. This allows the absorbance to be determined and finally, under defined conditions with baseline correction and after calibration, measurement of the concentration of the element. However, a serious shortcoming needs to be mentioned: with a flame nebulizer more than 90 % of the sample material is wasted, and is not actually being analyzed (Section 3.8).

According to the Lambert–Beer law the concentration c of the gaseous analyte is measured by $A = \varepsilon \times c \times l$ [cf. Eq. (5.5)]. However, because of many uncertainties such as flame temperature, light path, light scattering and molecular absorption, there are significant deviations from the Lambert–Beer law, and AAS requires calibration curves for each individual measurement of the type given in Figure 4.3.

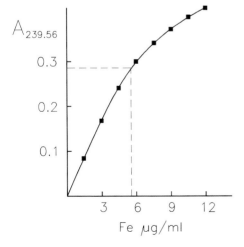

Figure 4.3 Calibration curve for AAS measurement of iron in a specific configuration. A defined concentration in µg mL^{-1} on the x-axis exhibits the absorbance given on the y-axis. The various reasons for deviations from a straight line (Lambert–Beer law) are discussed in Chapter 5.

4.1.2
Graphite Furnace Method

A graphite furnace has many advantages compared with an open flame. It is a more efficient atomizer and allows the measurement of considerably smaller volumes of samples and lower concentrations as a higher concentration of atoms is achieved in a smaller confined volume. In addition, it also provides a "reducing" environment for easily oxidizable samples through the flushing by an inert gas such as argon or nitrogen. Its temperature can be increased precisely and stepwise. Thus the sample material is gently dried, the organic material incinerated and then atomized.

A graphite tube of approximately 30 mm length and an inner diameter of approximately 6 mm is placed in the light path of the atomic absorption spectrometer (Figure 4.4). It contains a small so-called "L'vov platform" of pyrolytic graphite. It is essential that both the tube and platform are manufactured from very high quality graphite with extremely low trace element concentrations. Applying variable voltages at the ends of the graphite tube causes electric currents of up to 400 A, and due to the electrical resistance the tube heats up to 3000 °C. Samples of about 0.1 to 5 mg are typically inserted.

4.1.3
Background Absorption

The light of a hollow cathode lamp is not only absorbed by the particular gaseous analyte atoms, but also attenuated by non-specific absorption and scattering by the matrix. For example, when measuring lead (Pb) in the ppm range in sea water, each individual lead atom is surrounded by 10 million salt particles. Even if the interesting element is largely atomized, it is possible that there are many molecules or even solid particles which can cause a significant but non-specific attenuation of the measuring light. Correction for this non-specific attenuation is indispensable in AAS, particularly in the graphite furnace method.

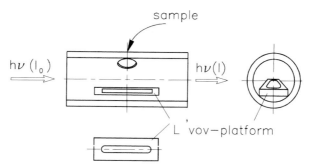

Figure 4.4 The heart of the graphite furnace analysis is the graphite tube of about 30 mm length and an inner diameter of 6 mm. The so-called "L'vov platform" is manufactured of high quality pyrolytic carbon with very low contamination allowing ultra trace analysis.

In 1965 Koirtyohann introduced a method of correction by a continuum light source, which is still widely used today. Firstly, the light beam of a deuterium lamp is passed through the analyte-free matrix flame which subsequently fills the whole entrance slit of the monochromator. This non-specific reference signal is called I_0. Secondly, the sample is injected into the flame and the specific attenuation of the selected line of the hollow cathode lamp (HCL) by the analyte is measured (I). According to the Lambert–Beer law, the concentration c of the analyte is then

$$c = \log\left(I_0/I\right)/(l\varepsilon)$$

with l = path length and ε = extinction coefficient. As in molecular spectroscopy, measurements are performed either in single (Figure 4.5) or in double beam mode (not shown). In modern AA spectrometers switching between observation of the HCL and deuterium lamp light beam is performed mechanically at a frequency of about 50–100 Hz (not shown in Figure 4.5).

Another most powerful background correction procedure is based on the so-called Zeeman-effect. This was discovered in 1896 by the Dutch physicist Pieter Zeeman who was honored for this work, together with H. A. Lorentz, in 1902 with the Nobel prize. In 1969 the German company Zeiss, in Oberkochem,

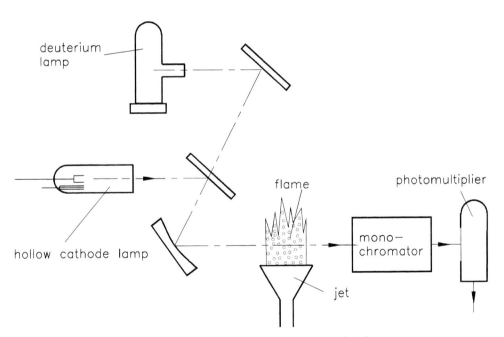

Figure 4.5 Background correction in AAS (flame or graphite tube method) by a continuous wavelength emitter such as a deuterium lamp. The sample is illuminated either by the line or the continuous wavelength source which allows correction for non-specific absorption.

was granted a patent for this correction method, but it was only taken up in 1975 by Hitachi with their first Zeeman-effect atomic absorption spectrometer (ZAAS). The principle is that without an external magnetic field only atomic levels with different magnetic quantum numbers m_l are degenerate, i.e., they possess the same energy (see Figures 2.7 and 2.9). However, when a magnetic field is applied, states with different magnetic quantum numbers m_l are split (Figure 2.10). Then, on transition between electronic states the selection rule for magnetic quantum numbers requires that $\Delta m_l = 0, \pm 1$ (Section 3.2.2.2). Consequently, a transition which shows up as a single line without an external magnetic field, will split into three lines, known as a triplet (Figure 4.6). The middle π-line appears in the original field-free position and is plane polarized parallel to the magnetic field, both σ-lines are polarized perpendicular to the magnetic field. The energy, i.e., difference in the wavelength, between the three lines is proportional to the strength of the magnetic field and is in the order of only 10^{-3} nm. This so-called normal Zeeman-effect takes place with elements in the second main group of the Periodic Table and transition elements, such as Be, Mg, Cd, Pb or Zn. In contrast to the normal Zeeman-effect, the anomalous Zeeman-effect is much more common, but is complex and will not be discussed any further here.

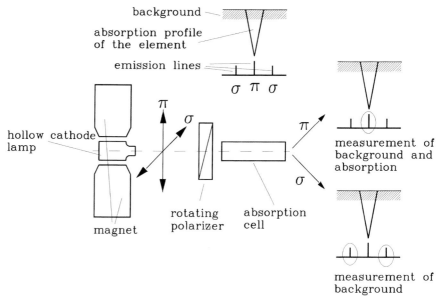

Figure 4.6 Principle of background correction of AAS spectra using the Zeeman-effect. The magnet splits the single emission line (without a magnetic field) into a triplet. The middle line, which is not changed in energy, is termed the "π" and the other two the "σ" lines. The lines are plane polarized perpendicular to each other, as indicated. Using a rotating polarizer and a dual wavelength procedure the background-signal (σ) is easily subtracted from the complete signal (π) to obtain a corrected signal.

The triplet lines emitted are selected sequentially by a rotating polarizer, absorbed by the sample and background (π-line) or just the background (σ-lines), respectively, and finally processed by a dual wavelength method using a lock-in amplifier (Section 5.5). Various Zeeman-effect AAS instruments based on different modifications of the principle described are on the market, but they will not be discussed here.

4.2
Atomic Emission Spectroscopy

Theoretically atomic emission spectroscopy (AES) is the counterpart of AAS and is particularly useful for quantitative multi-element analysis; no selective excitation is required as with AAS. However, the design of the spectrometers, sample preparation, measuring methods and the limits of detection differ substantially. Below we discuss the widely used ICP-OES instrument (inductively coupled plasma-optical emission spectroscopy).

Similar to AAS, ICP-OES involves three sequential steps (Figure 4.7):

1. Utilizing an auxiliary gas the sample is first "nebulized", i.e., dispersed into tiny aerosol droplets of approximately 10 μm in diameter. This process is one of the most critical steps in ICP-OES. In association with a so-called spray chamber, in most cases a pneumatic method is used, sometimes ultrasonic nebulization using oscillating piezoelectric transducers. Utilizing two auxiliary streams of inert gas a fast stream of aerosol is directed into the intense high frequency field of a toroidal load coil.

2. Within the toroidal load coil is a stable "hot" plasma (*torch*) of approximately 10 000 K. The energy for the production of the plasma torch, compared with the actual flame in flame-AAS, see above, is not the result of an oxidation process but stems from a strong electromagnetic high frequency field. The consumption of argon amounts to approximately 15 to 20 L min^{-1} and is a most crucial point in the use of ICP-OES (see the introductory comments to this chapter).

3. Finally, an optical system couples the light of the torch into the monochromator, which precisely selects and detects the individual atomic emission lines for further analysis by on-line computer. Owing to the flowing streams of the nebulizer and auxiliary gases the plasma torch is not homogeneous and shows a spatial structure and temperature distribution containing different contributions of atoms, molecules and ions. The ICP can be observed either from *side-on* or *end-on*.

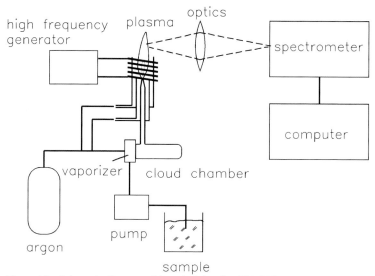

Figure 4.7 Schematic diagram of the principle of an ICP-OES instrument.

A typical high frequency power of 750 to 1500 W is required in order to reach the required plasma temperature of approximately 10 000 K, at which point a signifi- cant portion of all sample atoms is transferred to the excited state [approximately a third; Eq. (4.1)]. For reasons of safety in the high frequency short wave commu- nication bands, the frequency used in ICP-OES is generally 40.68 MHz.

In AAS and AES only a small usable spectral range ($\Delta\lambda = \lambda/m$) needs to be cov- ered, however with high wavelength resolution [$R = m \times N$, Eq. (3.34)], the detect- ing grating monochromator is operated in a high scattering-order (e. g., 100th order, in contrast to the first order utilized in typical UV–VIS spectrophotometers applied in molecular spectroscopy; see Section 3.6.3). On the other hand, a wide spectral coverage from 160 to 782 nm is required. This inevitably leads to prob- lems of interference and overlapping of spectral lines of various elements in many different scattering orders (Figure 3.38). Workers who have gone a long way to solving these problems include Becker-Roß and Florek (see Selected Further Reading). Crucial to the development of ICP-OES instruments was the development of the Echelle spectrograph. We will explain its principle using a modern ICP spectrometer, the Optima 5300 (PerkinElmer):

The basic operating principle of this ICP spectrometer is based on two "crossed" dispersing elements, namely an Echelle grating (originally all ruled gratings with trianglar-shaped cross-sectional grooves were called Echelle grat- ings; today the term is restricted only to gratings with small groove densities allowing dispersion into high scattering orders, see Figures 3.35 and 4.8) and a prism. Thus we obtain a "two-dimensional" dispersion, where the one axis dis- plays the scattering order, and the other the wavelengths.

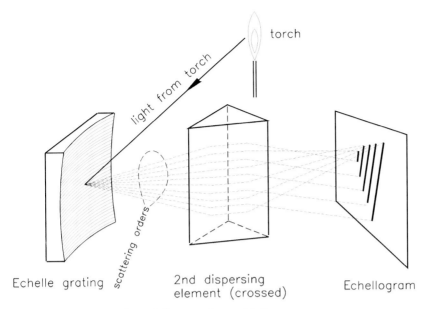

Figure 4.8 Generation of an Echellogram using an Echelle grating as the first and a crossed prism as the second dispersing element.

Such an "Echellogram" (Figure 4.9) contains, for example, for iron hundreds of individual reflections (lines or "spots"). As the position of all these spots in terms of scattering order and wavelength is precisely known (calibration standards!), we only need a suitable two-dimensional detector array for simultaneous analysis. This is now available in the form of the *segmented-array charge-coupled device detector* (SCD). It includes, in contrast to a conventional CCD detector as used in television cameras or video recorders, several hundred subarrays, well separated from each other, with approximately 20 to 80 pixels (ca. 12 × 100 μm) each, which are all localized on the SCD at the 236 positions of the most important spectral lines in ICP spectroscopy. Each subarray has its own electronic components such as register, interface, memory and output electronics, as required for further signal processing.

With the help of the first "Echellogram" Newton could have proven his then revolutionary statement – which later was vehemently contradicted by Goethe and which today appears trivial to us – that white light as dispersed by the first element into the color spectrum can not be further dispersed by the second crossed element. The colors of the rainbow are monochromatic.

Today prisms are mainly utilized in the visible wavelength range, and they often absorb strongly in the UV-B and UV-C wavelength range, so the light paths of two separate wavelength ranges are partitioned. In the visible range the principal scheme shown in Figure 4.8 is used. In the UV range an additional grating (with a small opening) is introduced between the Echelle grating and the prism

Sir Isaac Newton (1642–1727)

Segmented array charge-coupled device detector (SCD)

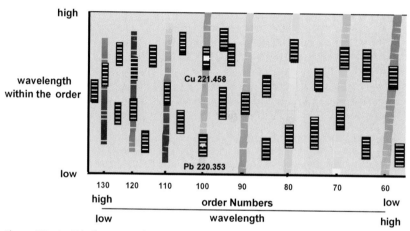

Figure 4.9 An Echellogram is the two-dimensional presentation of several scattering orders. All atomic emission lines are localized at many well defined loci where they are detected by means of small subarrays of 20 to 80 pixels each and further processed by individual electronic circuits and fed to the computer. All arrays and the required solid state components are integrated on a common chip (reproduced with kind permission of PerkinElmer Inc. All rights reserved).

as a "cross-dispersing" element, which allows the visible light to pass; however, the UV portion of the light beam is reflected and directed onto its own detector (not shown in Figure 4.8). In addition, this arrangement has the advantage of delivering undistorted images leading to the best possible resolution in both spectral ranges. In the lower UV range this is about 6 pm and approaches the order of the natural linewidth of 1.5 to 5 pm. Figure 4.10 demonstrates the resolution of the two lines of arsenic and cadmium that are separated from each other by barely 1/100 nm, in spite of their strongly differing absorption cross-sections.

So far the main use of ICP-OES has been for the determination and measurement of metals. However, the latest spectrometers allow precise measurements of

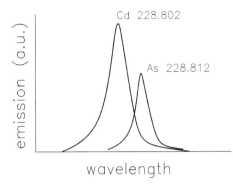

Cd 228.802

As 228.812

emission (a.u.)

wavelength

Figure 4.10 Even very close lying atomic lines of different elements and intensities are easily resolved by ICP spectroscopy, and detected and quantified by measurement of the whole peak structure.

Table 4.1 Detection limits (DL) of some non-metals in aqueous solution.

Element (line nm^{-1})	DL (3σ, μg L^{-1})	Tolerance level, drinking water (μg L^{-1})
Br (163.340)	450	
Cl (134.724)	200	25000
I (161.761)	55	
P (178.290)	3.9	400
S (182.037)	6.5	8300 (total)
S (180.734)	3.5	8300

non-metals as well. The non-metallic contaminants of water and their minimum tolerance levels are shown in Table 4.1 (Data of Analytical Instruments GmbH, Kleve, Germany).

A novel application of ICP-OES is the determination of non-metals in organic samples such as waste oils, particularly for the determination of halogens and sulfur. Another example is the analysis of samples required for the recycling of used woods. For this the determination of elements such as As, B, Cu, Pb, Cr and particularly Cl is required. Further new application areas will open up with the development of more low-priced and user friendly spectrometers.

In 1962 plasmas of several 10000 K were generated for the first time using a ruby laser. A pulsed, high energy laser beam is focused onto a small point of the sample thereby generating a free floating plasma sphere, which then *breaks down*. After a laser pulse of 5 to 20 ns duration the plasma cools down and the ions and atoms release their energy, similar to the plasma torch in ICP-OES, partly in the form of characteristic emission lines. In principle this novel procedure, termed **laser-induced breakdown** spectroscopy (LIBS) can be miniaturized but is inexpensive. Being the size of a pocket calculator, in the future portable LIB-atomic emission spectrometers may be useful for various applications in the field, particularly in the life sciences, which today are still probed by conventional atomic or ICP spectroscopy.

A still more advanced and sensitive method is inductively coupled plasma mass spectrometry (ICP-MS). ICP-MS is a very powerful tool for trace (ppb–ppm) and ultra-trace (ppq–ppb) elemental analysis. It is rapidly becoming the technique of choice in many analytical laboratories for the accurate and precise measurements needed for today's demanding applications. In ICP-MS, again, a plasma or gas consisting of ions, electrons and neutral particles is formed in argon gas. The resulting ions are then passed through a series of apertures (cones) into a high vacuum mass analyzer. The isotopes of the elements are identified by their mass-to-charge ratio (m/z) and the intensity of a specific peak in the mass spec-

Table 4.2 Detection limits ($\mu g\ l^{-1}$) of selected elements
(reproduced by kind permission of PerkinElmer Inc. All rights
reserved).

Element	AAS, flame Detection limits ($\mu g\ l^{-1}$)	AAS, graphite oven Detection limits ($\mu g\ l^{-1}$)	ICP-OES Detection limits ($\mu g\ l^{-1}$)	ICP-MS Detection limits ($\mu g\ l^{-1}$)
Ag	1.5	0.005	0.6	0.002
Al	45	0.1	1	0.005[a]
As	150	0.05	2	0.0006[b]
Au	9	0.15	1	0.0009
B	1000	20	1	0.03[c]
Ba	15	0.35	0.03	0.00002[d]
Be	1.5	0.008	0.09	0.003
Bi	30	0.05	1	0.0006
Br				0.2
C				0.8[e]
Ca	1.5	0.1	0.5	0.0002[d]
Cd	0.8	0.002	0.1	0.00009[d]
Ce			1.5	0.0002
Cl				12
Co	9	0.15	0.2	0.0009
Cr	3	0.004	0.2	0.0002[d]
Cs	15			0.0003
Cu	1.5	0.014	0.4	0.0002[c]
Dy	50		0.5	0.0001[f]
Ga	75		1.5	0.0002
Gd	1800		0.9	0.0008[g]
Ge	300	0.6	1	0.016[i]
Hf	300		0.5	0.0008
Hg	300	0.6	1	0.016[i]

Table 4.2 Continued

Element	AAS, flame Detection limits ($\mu g\ l^{-1}$)	AAS, graphite oven Detection limits ($\mu g\ l^{-1}$)	ICP-OES Detection limits ($\mu g\ l^{-1}$)	ICP-MS Detection limits ($\mu g\ l^{-1}$)
Ho	60		0.4	0.00006
I				0.002
In	30		1	0.0007
Ir	900	3.0	1	0.001
K	3	0.005	1	0.0002^d
La	3000		0.4	0.0009
Li	0.8	0.06	0.3	0.001^c
Lu	1000		0.1	0.00005
Mg	0.15	0.004	0.04	0.0003^c
Mn	1.5	0.005	0.1	0.00007^d
Mo	45	0.03	0.5	0.001
Na	0.3	0.005	0.5	0.0003^c
Nb	1500		1	0.0006
Nd	1500		2	0.0004
Ni	6	0.07	0.5	0.0004^c
Os			6	
P	75000	130	4	0.1^a
Pb	15	0.05	1	0.00004^d
Pd	30	0.09	2	0.0005
Pr	7500		2	0.00009
Pt	60	2.0	1	0.002
Rb	3	0.03	5	0.0004
Re	750		0.5	0.0003
Rh	6		5	0.0002
Ru	100	1.0	1	0.0002
S			10	28^j
Sb	45	0.05	2	0.0009
Sc	30		0.1	0.004
Se	100	0.05	4	0.0007^b
Si	90	1.0	10	0.03^a
Sm	3000		2	0.0002
Sn	150	0.1	2	0.0005^a
Sr	3	0.025	0.05	0.00002^d
Ta	1500		1	0.0005
Tb	900		2	0.00004

Table 4.2 Continued

Element	AAS, flame Detection limits (μg l^{-1})	AAS, graphite oven Detection limits (μg l^{-1})	ICP-OES Detection limits (μg l^{-1})	ICP-MS Detection limits (μg l^{-1})
Te	30	0.1	2	0.0008[k]
Th			2	0.0004
Ti	75	0.35	0.4	0.003[l]
Tl	15	0.1	2	0.0002
Tm	15		0.6	0.00006
U	15000		10	0.0001
V	60	0.1	0.5	0.0005
W	1500		1	0.005
Y	75		0.2	0.0002
Yb	8		0.1	0.0002[m]
Zn	1.5	0.02	0.2	0.0003[d]
Zr	450		0.5	0.0003

All detection limits are given in micrograms per liter and were determined using elemental standards in dilute aqueous solution. All detection limits are based on a 98 % confidence level (3 standard deviations).

All atomic absorption detection limits were determined using instrumental parameters optimized for the individual element, including the use of System 2 electrodeless discharge lamps where available. Data shown were determined on an AAnylist™ 800.

All ICP-OES (Optima 4300/5300) detection limits were obtained under simultaneous mutlielement conditions with the axial view of a dual-vies plasma using a cyclonicspray chamber and a concentric nebulizer.

Cold-vapor mercurydetection limits were determined with a FIAS-100 or FIAS-400 flow injection system with amalgamation accessory.

The detection limit without an amalgamation accessory is 0.2 μg/L with a hollow cathode lamp, 0.05 g/L with a System 2 electrodeless discharge lamp. (The Hg detection limit with the dedicated FIMS-100 or FIMS-400 mercury analyzers in < 0.005 μg/L without an amalgamation accessory and < 0.0002 μg/L with anamalgamation accessory.) Hydride detection limits shown were determiend using an MHS-15 Mercury/Hydride system.

GFAA detection limits were determined on an AAnalyst 800 using 50 μL sample volumes, an integrated platform and full STPF conditions. Graphit furnace detection limits can be further enhanced by th use of replicate injections.

Unless otherwise noted, ICP-MS detection limits were determined using an ELAN 6100/9000 equipped with Ryton™ spray chamber, Type II Cross-Flow nebulizer and nickel cones.

All detection limits were determiend using 3-second integration times and a minimum of 8 measurements.

Letters following an ICP-MS detection limit value refer to the use of specialized conditions or a different model instrument as follows:

[a] Run on ELAN DRC in standard mode usint Pt cones and quartz sample introduction system. [b] Run on ELAN DRC in DRC mode using Pt cones and quartz sample introduction system. [c] Run on ELAN DRC in standard mode in Class-100 Clean Room using Pt cones and quartz sample introduction system. [d] Run on ELAN DRC in DRC mode in Class-100 Clean Room using Pt cones and quartz sample introduction system. [e] Usin C-13. [f] Using Dy-163. [g] Using Gd-157. [h] Using Ge-74. [i] Using Hg-202. [j] Using S-34. [k] Using Te-125. [l] Using Ti-49. [m] Using Yb-173.

trum is proportional to the amount of that isotope (element) present in the original sample. Table 4.2 compares the sensitivity of the various methods in atomic spectroscopy.

4.3
Interferences

As pointed out in Section 4.1.3 on background correction, AAS is a relative method. Quantitative measurements require calibration curves such as shown in Figure 4.3. Nevertheless, various disturbances termed "interferences" cause deviations of the sample measurement from the measuring situation of the reference material. We can distinguish both chemical and physical interferences.

Any process that disturbs the quantitative atomization of the element of interest is termed a *chemical interference*. For example, if we want to determine calcium in solution in the presence of Na, Cl and SO_4, as a first step $CaSO_4$ is formed, which is eventually converted by the flame into CaO. On the other hand, using a $CaCl_2$ solution as a reference, no CaO is generated. CaO is only slightly atomized in a typical air–acetylene flame. This means that less Ca is determined in the presence of sulfate ions than without it. This is termed "chemical interference". A general method to avoid this interference is the addition of another cation in excess, which forms a more stable compound with the interfering anion. In the situation discussed here, we would add barium as $BaCl_2$ and the sulfate would be converted into insoluble $BaSO_4$, leaving Ca as CaCl.

All processes that influence the total number of gaseous atoms are termed *physical interferences*. The origin is almost exclusively the pneumatic nebulizer. The important parameters are the different density, viscosity and surface tension between the sample and reference. Only the most common method of avoiding physical interferences is mentioned here: a large dilution of the sample, if this possible.

Depending on its temperature, the absorption cell not only contains atoms in the ground state, but a fraction of the sample atoms, and other atoms, are thermally excited, or even ionized, and are capable of emitting light. Thus, the photo-detector not only receives (attenuated) light from the hollow cathode lamp as the primary light source but also light from excited atoms or ions not necessarily identical to the sample atoms. This causes significant false measurements termed *spectral interference*. If this additional emission light is not spectrally excluded by the monochromator, a smaller attenuation of light is detected, resulting in too low a concentration of the sample being measured. To avoid this disturbance, the exciting light from hollow cathode lamp is modulated, e.g., by 50 Hz. This allows the ac light signals from light source to be distinguished from the dc signals from thermally excited atoms, by processing the photomultiplier signal by a phase- and frequency-selective lock-in amplifier set to the 50 Hz signal.

4.4
Selected Further Reading

Becker-Roß, H., Florek, S. Echelle Spectrometers and Charge Coupled Devices, *Spectrochim. Acta, Part B*, **1991**, *52*.

Boumanns, P. W. J. M., ed. *Inductively Coupled Plasma Emission Spectroscopy – Parts 1 and 2*, in Elving, P. J., Winefordner, J. D., eds., *Chemical Analysis Vol. 90*, John Wiley & Sons, New York, **1987**.

Dean, J. R., Ando, D. J. *Atomic Absorption and Plasma Spectroscopy*, 2nd edn., Wiley-VCH, Weinheim, **1997**.

Ebdon, L., Evans, E. H., Fischer, A. S. eds. *An Introduction to Analytical Atomic Spectrometry*, John Wiley & Sons, New York, **1998**.

Haswell, S. J. *Atomic Absorption Spectrometry: Theory, Design and Applications*, Elsevier, Amsterdam, **1991**.

Heckmann, P. H., Trabert, E. *Introduction to the Spectroscopy of Atoms*, North Holland, Amsterdam, **1989**.

Hill, S. J., ed. *ICP Spectrometry and its Applications*, Sheffield Academic Press, Sheffield, **1998**.

Hollas, J. M. *High Resolution Spectroscopy*, John Wiley and Sons Ltd., **1998**.

Jenniss, S. W. *Application of Atomic Spectrometry to Regulatory Compliance Monitoring*, John Wiley & Sons, Chichester, **1997**.

Jackson, K. W. *Electrothermal Atomization for Analytical Atomic Spectroscopy*, John Wiley & Sons, Chichester, **1999**.

Kurfürst, U. *Solid Sample Analysis. Direct and Slurry Sampling Using GF-AAS and ETV-ICP*, Springer, Heidelberg, **1998**.

Laitinen, H. A., Ewing, G. W., eds. *A History of Analytical Chemistry*, American Chemical Society, Washington, D. C., **1977**.

Montaser, A., Golightly, D. W., eds. *Inductively Coupled Plasmas in Analytical Atomic Spectrometry*, 2nd edn., Verlag Chemie (VCH), Weinheim, **1992**.

Moore, G. L. *Introduction to Inductively Coupled Plasma Atomic Emission Spectrometry*, Elsevier, Amsterdam, **1988**.

Schlemmer, G., Radziuk, B. *Analytical Graphite Furnace Atomic Absorption Spectrometry*, Birkhäuser Verlag AG, Basel, **1998**.

Snedden, J., Lee, Y. I., eds. *Lasers in Analytical Atomic Spectroscopy*, John Wiley & Sons, New York, **1997**.

Svanberg, S. *Atomic and Molecular Spectroscopy. Basic Aspects and Practical Applications. Springer Series on Atomic, Optical, and Plasma Physics, Vol. 6*, Springer, Heidelberg, **1996**.

Varma, A. *Handbook of Furnace Atomic Absorption Spectroscopy*, CRC Press, Boca Raton, **1990**.

Welz, B., Becker-Ross, H., Florek, S., Heitmann, U. *High-Resolution Continuum Source AAS*, Wiley-VCH, Weinheim, **2005**.

5

Molecular Absorption Spectroscopy

Verba docent, exempla trahunt
(unknown ancient author)

Words teach, examples convince

In Chapter 3 we became acquainted with the general structure of a grating and a prism monochromator (Figure 3.22) and the various necessary optical components. On this basis, in Chapter 4 we discussed atomic absorption and emission spectroscopy. We will now focus our attention in this chapter on molecular absorption spectroscopy and in Chapter 6 on molecular emission spectroscopy.

5.1
The Bouguer–Lambert–Beer Law

5.1.1
Derivation

From the scheme in Figure 3.1 it is the task of absorption spectrophotometry to determine to what extent a sample transmits light of a specific wavelength λ. In this context "light" is defined as spectral radiant energy $\Phi_e (\lambda)$ (W nm^{-1}) or radiant flux density at a surface (E m^{-2} s^{-1}) (see Section 3.2). For simplicity and without defining the specific unit, we utilize the intensity of the incident light at a position $x = 0$ termed I_0, the intensity at position x is termed I (Figure 5.1). Bouguer (1729) and Lambert (1760) discovered that the attenuation of light that traverses a clear medium is proportional to the light intensity I and the traversed sample thickness dx (Bouguer–Lambert law):

$$\mathrm{d}I \ \propto \ I \times \mathrm{d}x$$

Introducing the absorption- or extinction coefficient $\alpha(\lambda)$ we obtain Eq. (5.1)

$$\mathrm{d}I = \alpha(\lambda) \ I \ \mathrm{d}x \tag{5.1}$$

Optical Spectroscopy in Chemistry and Life Sciences. W. Schmidt
Copyright © 2005 WILEY-VCH Verlag GmbH & Co. KGaA, Weinheim
ISBN 3-527-29911-4

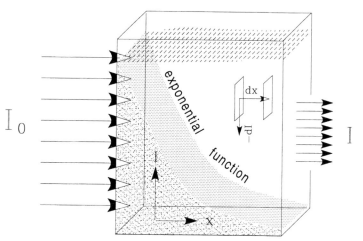

Figure 5.1 According to the Bouguer–Lambert–Beer law the light intensity of a parallel light beam decreases exponentially with the thickness of the sample.

Johann Heinrich Lambert (1728–1777)

The Bouguer–Lambert law only applies under specific conditions which are not always fulfilled, particularly not for biologically relevant samples such as proteins or various suspensions, i. e.:

• The incident light has to be monochromatic and collimated (parallel).
• The absorbing molecules have to be dispersed on a molecular basis, i. e., they have to be distributed homogeneously (cf. sieve effect, discussed below), they should not scatter or exhibit any mutual interactions.
• Scattering and reflection at the sample's surface cause – similar to absorption – light to be attenuated and have to be excluded.

In addition to this, in 1852 Beer showed that for most solutions of absorbing molecules the proportional factor α in Eq. (5.1) is itself is proportional to the concentration c of the molecule of interest. Incorporating this finding of Beer into Eq. (5.1), we finally obtain the Bouguer–Lambert–Beer law (usually abbreviated to the Lambert–Beer law):

$$dI = - \alpha(\lambda) \, c \, I \, dx \tag{5.2}$$

Integration of Eq. (5.) 2 over the total sample thickness x yields:

$$I = I_0 \, e^{-\alpha(\lambda) \, x \, c} \tag{5.3}$$

where the integration constant I_0 describes the incident light intensity and I the light intensity at any position x inside the sample, i.e., the intensity decreases exponentially with the sample thickness x (Figure 5.1). Conventionally Eq. (5.3) is written in the logarithmic form, with

$$\log \frac{I}{I_0} = - \alpha(\lambda) \, x \, c \; \log(e) \tag{5.4}$$

and the so-called decade or molar extinction coefficient (where $M^{-1} = 1 \times mol^{-1}$)

$$\varepsilon(\lambda) = \alpha(\lambda) \times 0.4343 \, \left[M^{-1} \, cm^{-1}\right]$$

It follows from Eq. (5.4) that

$$\log \frac{I_0}{I} = A(\lambda) = \varepsilon(\lambda) \, x \, c \tag{5.5}$$

where $\log I_0/I$ and $\varepsilon(\lambda) \times x \times c$ are defined as absorption or absorbance $A(\lambda)$. It is possible to describe the molecule as an opaque disk with the cross-section σ, which is seen by a photon of frequency ν. If the light frequency is a long way off from the resonance frequency then the effective diameter of the disk approaches zero and has its maximum at the resonance frequency. With $\varepsilon = \sigma$ $(6.023 \times 10^{23}/2.33) = 2.61 \times 10^{23}$ we obtain the typical cross-sections and extinction coefficients as given in Table 5.1.

Because a logarithm is a dimensionless entity, $A(\varepsilon)$ is of course is not a physical unit but merely a number (similar to the "molecular weight" when given in "dal-

Table 5.1 Examples of cross-sections and extinction coefficients.

Type of spectrum	σ (cm^2)	ε (M^{-1} cm^{-1})
Atomic absorption	10^{-12}	3×10^8
Molecular absorption	10^{-16}	3×10^4
Infrared absorption	10^{-19}	3×10
Raman scattering	10^{-29}	3×10^{-9}

Table 5.2 Extinction coefficients of some biologically relevant molecules at wavelengths of maximum absorption, including the attributed transition types.

Species	λ_{max} (nm)	ε (M^{-1} cm^{-1})	Transition
Flavin radical	600	5000	$\pi \rightarrow \pi^*$
Flavin (ox)	450	12000	$\pi \rightarrow \pi^*$
Carotene	450	120000	$\pi \rightarrow \pi^*$
Cytochrome c	420	126000	(Figure 5.31)
Chlorophyll a	420	85000	
Tryptophan	280	5600	$n \rightarrow \pi^*, \pi \rightarrow \pi^*$
Tryptophan	219	7000	$n \rightarrow \pi^*, \pi \rightarrow \pi^*$
Adenine	260	13400	$n \rightarrow \pi^*, \pi \rightarrow \pi^*$
Guanine	275	8100	$n \rightarrow \pi^*, \pi \rightarrow \pi^*$
Cytosine	267	6100	$n \rightarrow \pi^*, \pi \rightarrow \pi^*$
Thymine	264	7900	$n \rightarrow \pi^*, \pi \rightarrow \pi^*$
DNA	258	6600	$n \rightarrow \pi^*, \pi \rightarrow \pi^*$
RNA	258	7400	$n \rightarrow \pi^*, \pi \rightarrow \pi^*$
Benzene	ca. 256	215	$\pi \rightarrow \pi^*$
Benzene	ca. 180	60000	$\pi \rightarrow \pi^*$
Phenol	ca. 210	2200	$\pi \rightarrow \pi^*$
Methanol	177	200	$n \rightarrow \sigma^*$
Water	167	7000	$n \rightarrow \sigma^*$
Ethylene	163	15000	$\pi \rightarrow \pi^*$
Hexane	150	?	$\sigma \rightarrow \sigma^*$
Ethane	135	?	$\sigma \rightarrow \sigma^*$

tons"). Table 5.2 show the extinction coefficients and maxima of some representative molecules including the types of transition attributed to them (cf. Figure 2.20). With amino or nucleic acids, absorption bands often reflect various different transitions simultaneously.

The transmission T is defined as the portion I of the incident light intensity I_0 which is transmitted by the sample:

$$T(\lambda) = \frac{I}{I_0}$$

$$A(\lambda) = -\log T(\lambda)$$

(5.6)

Thus, the absorption A is – in contrast to the transmission T– proportional to the concentration of the sample to be measured. This allows a fast optical measure-

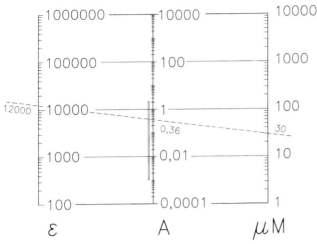

Figure 5.2 Nomogram for the determination of the absorption A of a sample with known extinction coefficient ε ($M^{-1}\ cm^{-1}$) and given concentration (mM). For example, 30 μm oxidized flavin with ε = 12 000 at a sample thickness of 1 cm gave an absorption value of A = 0.36.

ment of the concentration of the species to be measured. For this purpose the nomogram in Figure 5.2 is helpful.

5.1.2
Deviations

Apparent and real deviations from the Lambert–Beer law are often observed. Thus, in cases of doubt the law should be verified experimentally. Extinction E, optical density $[OD(\lambda)]$ and also $-\log T(\lambda)$ are then only identical with the absorption $A(\lambda)$ if the prerequisites of the Bouguer–Lambert law are fulfilled (Section 5.1.1). Light should only be attenuated by absorption, not by scattering or reflections.

Extinction is defined as the total light attenuation by the sample. It is caused both by absorption as well as by scattering and reflection (see below). However, in contrast the term extinction coefficient relates only to the pure absorption of a molecule.

Further apparent deviations are caused by a concentration-dependent aggregation and formation of complexes, as based on the interaction of transition dipole moments (hypo-/hyperchromism, see Section 5.4.2) or on changing dissociation constants, e. g., of an acid (HA $\leftrightarrow$ H^+ + A^-).

A very important, often overlooked apparent deviation depends on the so-called sieve effect, as described in Figure 5.3. It is caused by the non-homogeneous distribution of an absorbing substance. For homogeneous distribution the absorption is defined by $A = \log I_0/I$. For $I = 0.1 \times I_0$, A is also 1. If we assume that in a hypothetical experiment all particles are restricted just to the lower half of the cuvette (e. g., chlorophyll bound to chloroplast particles within photosynthesis cells), we obtain a maximum absorbance of $A = 0.3010$, if the lower half does not transmit any light – independent of the molecule investigated. As a result of an inhomogeneous distribution of the molecules to be measured the sample exhibits "holes", and the light quanta can pass through these holes and thus the sample without any chance of interacting with the absorbing species (a "sieve"). Thus, the presumptions of the Bouguer–Lambert [Eq. (5.1)] no longer hold (cf. Figure 5.18).

The extinction coefficient (ε or a), refraction index n, the Fresnel equations and also scattering spectroscopy (Chapter 8) and CD/ORD spectroscopy (Chapter 9) are only described theoretically with any satisfaction by the dispersion theories. This allows the scientifically elegant formulation of natural laws in a mathematically complex form. We are already acquainted with this procedure from the introduction of the complex wavefunction Ψ in Schrödinger's equation. Thus, for example, the complex refraction index $n' = n(1 - ik)$ contains the well known refraction index n as a real part as well as the "absorption index" k as an imaginary part – essentially the extinction coefficient ε – within a single complex number n'.

Finally, we have to take into account that the Lambert–Beer law itself is only limited to dilute solutions. Therefore even under ideal conditions its scope is limited, and beyond this scope leads to real deviations. Following the dispersion theory it is not actually ε but rather the so-called molar refraction

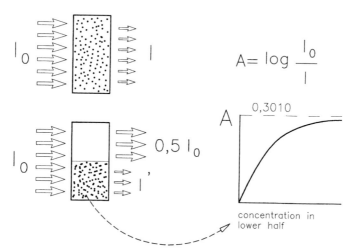

Figure 5.3 Illustration of the sieve effect, which leads to errors in concentrations measured for non-homogeneously distributed components. In a theoretical experiment if we force the absorbing species into the lower half of a cuvette, the optical absorbance will reach only a maximum density of $\log 2/1 = 0.3010$ even if the lower half of the cuvette is completely filled ($\varepsilon = \infty$).

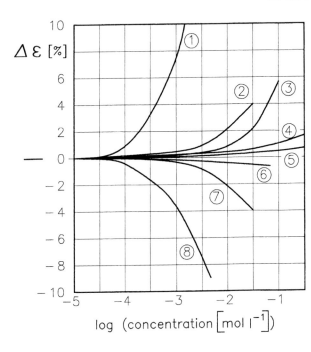

Figure 5.4 Deviations of Lambert–Beer law for various ions in aqueous solvents, compared with values at very low concentration (10^{-5} mol l^{-1}). (1) Eosin-anion at 366 nm, (2) Eosin-anion in KI solution at 546 nm, (3) 2,4-dinitrophenolate ion in aqueous phenolic solution at 436 nm, (4) 2,4-dinitrophenolate ion in KCl solution at 436 nm, (5) NO_2^-, (6) CrO_4^{2-} at 436 nm, (7) 2,4-dinitrophenolate ion in $La(NO_3)_3$ solution at 436 nm, (8) Eosin-anion in tetraalkyl ammonium bromide solution at 436 nm (modified from Kortüm, G. Z. Elektrochem., Angew. Physik. Chem. 1936, 42, 287).

$$R_{mol} = \varepsilon \, \frac{n}{(n^2 + 2)^2} \qquad (5.7)$$

which is concentration independent, where $n(\lambda)$ is the refractive index (Section 3.6.1) of the medium (buffer). As the refractive index n and thus also $n/(n^2 + 2)^2$ generally change with the concentration, we can expect the opposite change in the ε-value. Figure 5.4 demonstrates the deviations in the value of ε in percent for various substances for low concentrations (10^{-5} M). In some cases they are considerable and depend strongly on the individual molecule.

The extinction coefficient $\varepsilon(\lambda)$ which describes the absorption properties as a function of the electron distribution of the absorbing molecule, depends strongly on various factors such as temperature (freezing of certain degrees of freedom, cf. Figure 2.22), concentration (interaction) and solvent (micro-environment). Below we will discuss how these basic disturbing effects can be utilized for analytical purposes.

5.1.3
Terminology

We will summarize terms that are often used in the discussion of electronic spectra:

- *Chromophore*: A molecule or covalently bound, unsaturated chemical group that determines the electronic absorption spectrum and hence the color.

- *Auxochrome*: A saturated chemical group such as OH^-, NH_2 or Cl^-, which itself does not show any absorption in the observed spectral range. However, on binding to a chromophore it changes the wavelength absorption maximum of the chromophore as well as its extinction coefficient.
- *Photochromic molecule*: A molecule that exists in two different forms with different wavelengths for their absorption maxima. On irradiation with light of wavelengths of the respective maxima, both forms are converted into each other. Example: The phytochrome molecule which is found in all higher green plants controls numerous (photo-) physiological responses. It exists in the physiologically inactive red (P_r, 660 nm) and the active far red (P_{fr}, 730 nm) absorbing form.
- *Bathochromic shift*: Shift in the absorption wavelength towards longer wavelengths as a result of substitution or solvent effects.
- *Hypsochromic shift*: Shift in the absorption wavelength towards shorter wavelengths as a result of substitution or solvent effects.
- *Hyperchromic effect*: Increase in the extinction coefficient.
- *Hypochromic effect*: Reduction in the extinction coefficient.
- *Isosbestic point*: Wavelength λ_i, where two or more different molecular species of a mixture exhibit the same extinction coefficient ε_i.

Comment: Owing to the definition of absorption ($A = \log I_0/I$) we always measure the transmitted light (I) that has *not* interacted with the sample. If the presumptions of the Lambert–Beer law are valid the conclusion is correct that the "missing difference" ($I_0 - I$) has been completely absorbed by the sample. Thus, the measurement of absorption is basically an indirect measurement (measurement of the difference $\log I_0 - \log I$) – in contrast to measurements of fluorescence, phosphorescence (Chapter 6) and photoacoustic spectra (Chapter 7). In addition, absorption and emission of light is a quantum process, i.e., a singular event, which in itself does not lead to a "spectrum". A spectrum is simply the statistical distribution of wavelengths of a large number of individual molecular events. Indeed, today there are (mainly fluorescence) techniques available which allow the measurement of *single*, individual atoms and molecules. However, mainly for technical reasons, such as the high sensitivity required, these are restricted to fluorescence spectroscopy.

5.2
Monochromators

As discussed in Section 3.8, for dispersion in the visible wavelength range gratings are usually utilized, because prisms exhibit a distinctly wavelength-dependent resolution. A diode array spectrophotometer (OMA: optical **m**ultichannel analyzer) has significant advantages for various applications, however in the chemical–biological field its range is particularly restricted. Similar to the established use of the infrared spectral range today, in the long term Fourier spectroscopy will also become established (see Figure 5.8).

5.2.1
Monochromator Set-ups

Numerous principle light paths in monochromators have finally been realized which all verify the optical scheme shown in Figure 3.22. They are the result of an optimization which includes various factors such as expense of the optomechanical parts, type of dispersing element, price, mechanical size of the monochromator, robustness and specification such as stray light (cf. Section 5.2.7), compensation of various lens and grating errors, scan speed, wavelength range, photometrical precision, etc. We will only discuss the most common set-ups.

The so-called Rowland circle defines the required imaging properties of a concave grating. If the entrance slit is localized on the Rowland circle, then the dispersed spectrum is also imaged on the Rowland circle (Figure 5.5). If, however, according to Seya-Namioka, the grating revolves as indicated in the figure, the entrance and exit slits are held fixed, which allows the construction of a simple, low-cost monochromator of high light throughput, very few optical components and therefore a low stray light level. Only the angle between the entrance and exit slits as given by the manufacturer (defined by the holographic production process, e. g., in Figure 5.5: 61.6°) must be carefully met with a low tolerance to minimize defocusing.

In the most often utilized assembly, from Czerny–Turner (Figure 5.6), a flat dispersion grating is rotated, where the imaging of the entrance onto the exit slit is performed by two parabolic mirrors (Section 3.5.3). The advantage of this set-up is the elimination of coma errors.

If the sine condition (cf. Section 3.5.3) is not met, images are distorted particularly outside the optical axis which is called a coma.

Instead of surface mirrors, in spectrophotometers reflecting prisms are often preferred as beam deviating elements. Their action is based on total reflection and thus they are more efficient over a larger wavelength range than mirrors. Replacing the two focusing mirrors by a common, larger one is called an "Ebert set-up", which is preferred for smaller monochromators in particular.

An array spectrophotometer, also called a simultaneous spectrophotometer as all wavelengths of the spectrum are monitored at the same time, has no exit slit. The dispersed spectrum is spread across a diode array of 512 or 1024 microdiodes with a total width of typically 25 mm (*IC*-normalized case) and detected simultaneously (Figure 5.7). An array spectrophotometer does not have any movable parts and therefore is very robust, which represents its main advantage.

On this basis, complete spectra are monitored with high velocity (taking readings out of the diode array is sequential and hence this is the time limiting step). However, these advantages are a trade-off for the annoying disadvantages:

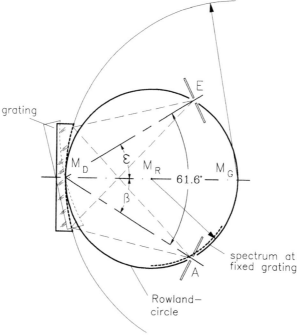

Figure 5.5 The Rowland circle defines the imaging properties of a spherical concave diffraction grating. Its diameter corresponds to the curved radius of the grating. If the entrance slit E is located somewhere on the Rowland circle, the spectrum is imaged just on this circle. In addition, holographic gratings in the rotary arrangement shown, suggested by Seya–Namioka, require a well defined, fixed angle between the entrance and exit slits (here 61.6°).

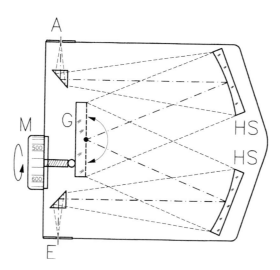

Figure 5.6 The most frequently used set-up, from Czerny–Turner, uses a flat grating with mechanical sine-correction. Using a suitable prism arrangement entrance and exit beams are just on a straight line; concave mirrors (HS) allow largely coma-free imaging of the entrance (E) onto the exit slits (A).

(A)

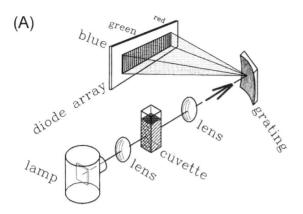

(B)

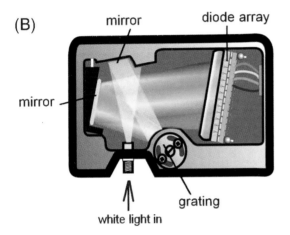

Figure 5.7
(a) Schematic diagram of a diode array spectrophotometer (optical multichannel analyzer, *OMA*). Its predominant feature is the absence of any mobile parts. The spectral resolution is accomplished by a diode array (from Hewlett Packard, modified). (b) A miniaturized version of the optical multichannel analyzer (USB2000, Ocean Optics), size approximately $80 \times 63 \times 34$ mm, 190 g.

- Typically, the sample is irradiated with intense white light, i.e., with additional wavelengths outside the monitored wavelength range. This leads to an enhanced bleaching particularly of photosensitive samples (photolysis).
- The sample is part of the imaging light path. This requires clear and non-light scattering samples, otherwise wavelength resolution and light throughput are significantly decreased, stray light increases and thus the dynamics and sensitivity of the system are reduced.
- The wavelength resolution is not defined by a slit but by the grating and the diode array, and is therefore fixed. This requires various, removable and expensive dispersion gratings (cf. Section 5.2.2),
- The active area of the photodiode array per wavelength is extremely small (e.g., $25~\mu m \times 2.5$ mm), and thus the sensitivity is

small as compared with conventional spectrometers. The advantages of the photomultiplier, such as sensitivity, speed, large detection area are missing.

- Sensitivity of below 250 nm is typically low.
- The stray light level of approximately 3 % in the *survey mode* (spectrum with low spectral resolution) is high and can only be reduced to 0.05 % if we reduce the high monitoring speed of, e. g., 5 ms per spectrum down to 1 s per spectrum or lower (averaging, *high precision mode*), losing the great advantage of the array spectrometer.
- The monitoring speed for a complete spectrum in the *survey mode* is typically 0.1 s, in the precision mode 2–14 s (e. g., PerkinElmer, Lambda Array 3840). Monitoring below 10 ms per spectrum is restricted to specific problems (cf. fast scanning spectrometers, Section 5.6.2).
- The typical dynamic working range of diode array spectrometers (5×10^{-2} to 1 A) is small compared with a scanning spectrophotometer with a photomultiplier (10^{-3} to 3 A).
- When measuring absorbance spectra the logarithm of the signal is determined by calculation *after* digital storage. This is indispensable and leads to a logarithmic scale for photometric resolution. For example, if the total photometric range is $A = 2$ and the linear signal is stored with a typical resolution of 16 bit, then the lower quarter $A = 0$ to $A = 0.5$ is represented by only 16 points (4 bits), the upper quarter by 61 440 points (16 bit–12 bit).

The diode array spectrometer was originally derived from the "multielement detecting atomic spectrometer" with several detectors on the Rowland circle arranged "in-line". Today the linear diode array is being replaced more and more by a two-dimensional CCD detector (charge-coupled device), which was originally developed for imaging detectors in the field of television and video devices. The most common chip format in optical spectroscopy is 1024×256 pixels. Columns of 256 pixels each corresponding to the image of the linear entrance slit are collected on a common shift register and read out as a unit. This lowers the memory space and speeds up the read-out process tremendously. CCD detectors are approximately two orders of magnitude more sensitive and possess higher dynamics than diode arrays. However, the latter offer, compared with CCDs, a higher level of linearity and a good dynamic range. The pixels are usually significantly larger than those of CCD chips. Typical sizes are 20–50 μm by 500–2500 μm. They exhibit a much better signal to noise ratio (SNR) and allow higher light intensities. The selection of the detector type depends on the individual application, if indeed a simultaneous spectrometer is required at all (Table 5.3).

In summary, fast survey spectra are easily obtained by array spectrometers if a reduced precision and dynamics are acceptable. However, precision measurements are best performed on the basis of a conventional spectrophotometer

Table 5.3 Comparison of diode array and CCD detectors.

Parameter	Diode array	CCD
Dynamic range	15 bit	16 bit
Signal to noise ratio (SNR)	$10\,000:1$	$900:1$
Detection limit	<3.3 pJ cm^{-2}	< 3.8 fJ cm^{-2}
Saturation	>107 nJ cm^{-2}	>250 pJ cm^{-2}

with a photomultiplier detector. Thus, prior to buying a new spectrophotometer you have to carefully balance the advantages of simultaneous and conventional scanning instruments. Test measurements are recommended.

So far Fourier analysis, for technical reasons, has been limited to infrared spectroscopy. However, the number of spectrophotometers available that are capable of also working in the UV–VIS spectral range is increasing steadily. In contrast to the so-called simultaneous spectrometers, i.e., diode array and CCD-based devices where the read-out is sequential, Fourier spectrometers are true simultaneous spectrometers. The wavelength dispersion is based on an interference procedure originally introduced by Michelson, thereby being the first American to win a Nobel Prize in 1907. There are no longer any prisms or gratings (Figure 5.8). A white light beam is split by a semitransparent mirror into two beams which follow different paths. After reflection by mirrors 1 and 2 the two beams are reunited and focused onto the sample. As the two light beams are produced by the same light source they are capable of interfering with each other. Through a minute displacement Δx of the two beams relative to each other an interferogram is generated. This represents the Fourier-transformed absorption spectrum of the sample (see Section 5.4.4.5). Calculating the inverse Fourier transformation (using an on-line PC) yields the required original absorption spectrum. Compared with a dispersive spectrometer where the incident light passes a very small slit, the light throughput is substantially enhanced and the whole spectrum monitored in the subsecond time range (and with tricks even down into the nanosecond

Albert Abraham Michelson (1852–1931)

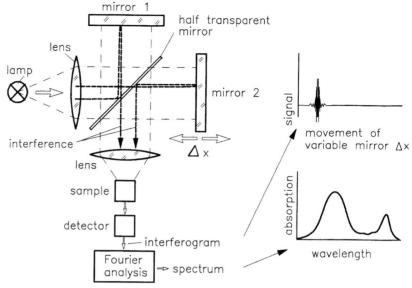

Figure 5.8 Schematic of the Michelson interferometer. A collimated (parallel) light beam is split by a semi-transparent mirror and after reflection by a fixed mirror (1) and a movable mirror (2), respectively, is reunited and focused onto the sample. The two reunited light beams are capable of interfering and generate, upon shifting of mirror 2 by Δx, a characteristic interferogram. By means of an inverse Fourier transformation this is concentrated in the required spectrum. Logarithmic conversion and correction of the dispersion of the particular apparatus are performed in a conventional manner.

time range). Parallel to the measuring light a weak laser of well defined wavelength is co-detected (e. g., an HeNe laser), defining individual *sampling points*. The absolute wavelength calibration is performed with specific calibration samples of known wavelength characteristic – or, for example, in the NIR range (Chapter 10) simply with the extreme sharp rotational lines of the "contaminating" water vapor as these are practically always present (Bruker).

5.2.2
Types of Dispersion Gratings

Diffraction gratings have been known for over 180 years (see Table 1.1). Because of their importance in spectroscopic instruments, they have been the subject of considerable and sustained research. Originally, dispersion gratings were transmission gratings. The grating can be a square of 50 × 50 mm, typically with 1200 grooves per mm; the total groove length is then 3 km. The grooves are formed mechanically by the burnishing action of a diamond tool on a glass matrix without the removal of any material. Nowadays gratings are mostly manufactured as reflecting, often holographic gratings ("interference gratings"). These are generated using two widened, crossing, e. g., argon, laser beams of 488 nm that form an interference pattern in the form of the required grating. In a subsequent photochemical process the grooves are etched and a protecting layer of SiO_2 deposited by vapor. Holographic gratings can be produced with up to 6600 grooves mm^{-1} (frequency doubled argon line at 257 nm) and on any curved matrix (concave gratings).

Once a *master grating* has been manufactured, it can be replicated to produce many exact copies of the original. The invention of various replication techniques may be seen as one of the most important factors in the development of spectroscopy. A replica blank of high optical quality is coated with a thin epoxy layer (0.1–1 μm) and sandwiched together with the master. When the epoxy is cured the master and replica are separated with the epoxy layer staying attached to the replica. The epoxy layer is now an exact (inverse) copy of the grooves of the master. Finally, depending on the envisaged wavelength range, a special reflection coating of MgF_2, Au or Pt is deposited, and, again, a protective layer of SiO_2.

Interference gratings do not have a real "blaze angle" as is possible with ruled gratings (cf. Figure 3.39), and they do not reach the efficiencies of ruled gratings of up to 95 %. However, they show a lower, more even efficiency throughout the whole spectral range (Figure 5.9). Even at a very high density of grooves they show practically no errors. Thus, there are no "ghosts" and the stray light level is at least 100 times below that of high quality ruled gratings. This increases their dynamics, i. e., the photometric range dramatically.

In 1902 R. W. Wood discovered peculiar irregularities in the efficiency curves of gratings (see Figure 5.9), which to date are not completely understood. These are particularly strong if the incident light is polarized perpendicular to the grooves. In spectroscopy it is a good idea to keep this phenomenon in mind as it often leads to misinterpretations of spectra, particularly in non-compensated spectra such as single beam spectrometers and fluorimeters: e. g., often there is a strong Wood's anomaly near 550 nm, just at the α-peak of cytochrome c (see Figure 5.30).

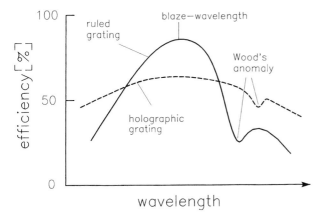

Figure 5.9 Wavelength dependent efficiency of holographic and ruled diffraction gratings.

In 1883 H. A. Rowland invented the concave reflecting grating with the double function of dispersing and focusing a light beam, similar to a concave mirror (Section 5.2.1). This is the basis today of various, particularly low priced spectrophotometers using holographic concave mirrors. Thus, the number of optical components is reduced and the wavelength range covered compared with ruled gratings is considerably widened, particularly towards the short wavelength, 185 nm. It should be noted that for concave gratings the efficiency exhibits a strong dependency on the direction of polarization of the incident light. Small deviations may lead to peak shifts (cf. polarization spectroscopy and fluorimetry, Chapter 6) and ellipsometry (Section 8.7).

Particularly for diode array spectrometers (Section 5.2.1) or imaging CCD spectrometers special dispersion gratings have been developed which image the entrance slit sharply after dispersion onto the whole area of the diode array (instead of onto the Roland circle: so-called *flat gratings*).

5.2.3
Linearization of Wavelength

From the theoretical standpoint the presentation of a spectrum on the basis of energy units has many advantages. This is accomplished by plotting absorption as a function of wave number. However, for technical/historical reasons in the ultraviolet, visible and near-infrared spectral range the presentation of wavelength is common (A versus λ). The reason is that the rotation angle q of a dispersion grating is approximately proportional to wavelength; but only "approximately". If we "scan" a spectrum by rotating the grating according to the grating formula [Eq. (3.33)] the entrance angle ε as well as the dispersion angle β changes. From Figures 3.34 and 5.5, $\varepsilon-\beta$ = const. (e. g., 61.6°, this angle depends on the individual grating and the set-up). Utilizing the addition theorems of trigonometry we convert the grating formula and obtain, e. g., for the special grating in Figure 5.5:

$$\lambda = 2 \, \cos \, (30.8°) \times a \times m \times \sin \, (30.8° - \varepsilon) \tag{5.8}$$

With the constant $c = 2 \times \cos(30.8°) \times a \times m$ and the rotational angle of the grating $q = (30.8° -\varepsilon)$ we obtain the optical function

$$\lambda(\Theta) = c \, \sin \Theta \tag{5.9}$$

Accordingly, the wavelength changes proportionally with the sine of the rotation angle q. Figure 5.10(b) shows the principle of the sine drive which allows the sine-dependency of a grating to be compensated mechanically. For the rotational angle q the mechanical function holds:

$$\sin \Theta = \frac{x}{l} \tag{5.10}$$

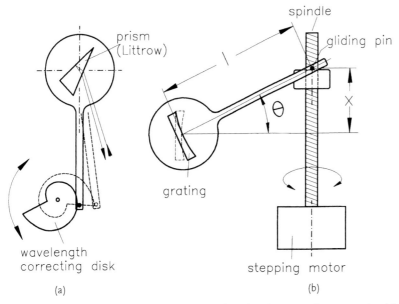

Figure 5.10 Linearization of wavelengths of (a) a prism and (b) a grating monochromator. The curved disk in prism monochromators is adapted to that particular prism only, while the sine correction is suitable for any gratings (change of grating is possible).

where the x-movement is typically realized by a motor driven spindle. If we insert Eq. (5.10) into Eq. (5.9), we obtain a linear relationship for λ and x. In modern spectrophotometers the spindle is replaced by a computer-controlled stepper motor. It appears safe to assume that in the near future the mechanically complex and expensive spindle drive will be completely replaced by a mathematical sine correction by the on-line computer.

For the Littrow set-up (Figure 5.11) we have: $\varepsilon = \beta = q$, and according to the gratings equation it follows that

$$\lambda = \frac{2a}{m} \sin \Theta$$

$$2a \cos \Theta \, \delta\Theta = m \, \delta\lambda \tag{5.11}$$

$$\frac{\delta\Theta}{\delta\lambda} = \frac{m}{2a \cos \Theta} = \tan\frac{\Theta}{\lambda}$$

where $\delta\theta/\delta\lambda = \delta\beta/\delta\lambda$ just defines the dispersion angle [Eq. (3.3.5)]. The dispersion angle increases proportionally to the scattering order. However, the Littrow set-up is rarely utilized in grating spectrometers in molecular spectroscopy. It rather serves as theoretical reference for diverse specifications, it is however more often found in prism spectrometers – as, e.g., in the classical MIV-Q3 monochromator from Zeiss.

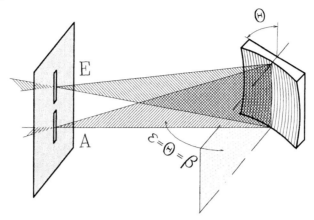

Figure 5.11   Littrow arrangement of a diffraction grating.

Considering Eq. (3.28), prisms show a strongly non-linear dependency on the entrance angle as well as on the refractive index of the prism material (Figure 3.34). A reasonable linearization of prism monochromators is obtained by deflection through appropriately curved disks [see Figure 5.10(a)]. However, resolution with a constant slitwidth varies strongly with the wavelength.

Owing to the inherent stray light of single monochromators their dynamic range is limited to 4 or 5 orders (see Section 5.2.7). Combining two monochromators in series ("double monochromator") the stray light level is reduced to $<10^{-9}$ (approximately multiplication of the stray light levels), with the consequence of an equally enhanced photometric dynamic range. In this case both the grating drives have to be carefully synchronized mechanically.

5.2.4
Types of Scanning Absorption Spectrophotometers

When measuring the absorption of a substance we determine its absorbency at a specific wavelength λ_1. Setting the monochromator to this wavelength (Figure 5.6) we determine the difference between the values obtained with and without the sample inserted:

$$A(\lambda_1) = \log I_0(\lambda_1) - \log I(\lambda_1)$$

Similarly, we can scan a whole wavelength range $\Delta\lambda$ between λ_1 and λ_2 with and without a sample inserted (double beam spectrophotometers utilize two beams in parallel, a sample and a reference cuvette) and provide – by means of an on-line computer – a corrected absorption spectrum [Figure 5.12(a)]:

$$A(\lambda) = \log I_0(\lambda) - \log I(\lambda) \tag{5.12}$$

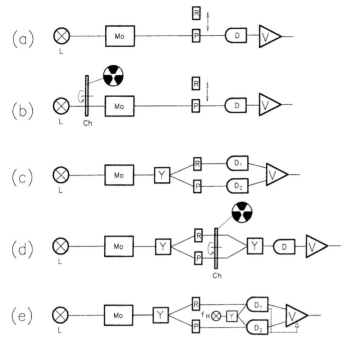

Figure 5.12 Types of scanning absorption spectrophotometers. Details in text (from Naumann and Schröder, 1987, modified).

So far, the logarithm has usually been achieved using a logarithmic amplifier, i. e., a hardware component. In order to reduce costs, the process of determining the logarithm is increasingly being transferred to the software part of the spectrophotometer where the original signal is stored in linear terms. However, this leads to some difficulties. (1) Prior to calculating the logarithm the *baseline* (i. e., the "real zero signal") has to be determined with high precision; particularly when measuring small absorption values, minute deviations cause striking changes in the absorption values and spectral shapes. (2) Subsequent "off-line calculation" of the logarithm of the linearly stored signals results in logarithmic photometric resolution (see Section 5.2.1). (3) Measurements of absorption spectra with *rapid scan* spectrophotometers and on-line linear to log conversion (Section 5.6.2) requires very high conversion rates, which are only attainable with special hardware components.

For technical reasons ac signals (alternating current) are easier to amplify than dc signals (direct current) (Section 5.2.5). Thus, prior to amplification the original dc light signal is converted into an ac signal by means of a (mechanical) *chopper* [Figure 5.12(b)].

In contrast to sequential measurements, the sample and reference can be measured simultaneously after an appropriate beam split (beam splitter Y) and with separate detectors D_1 and D_2. Finally, both independent signals are converted into the absorption spectrum [Eq. (5.5)] [Figure 5.12(c)]. This set-up allows for compensation for fluctuations in the light source, however does not compensate for differences in detector sensitivity.

The arrangement according to Figure 5.12(d) requires only a single detector. The measuring light is split into two beams (beam splitter Y). After the respective beams pass through the sample, reference and chopper they are reunited (reverse beam splitter Y). By means of a phase sensitive amplifier (lock-in amplifier), which receives the reference signal from the chopper motor, the corrected absorption spectrum is calculated [Eq. (5.5)]. This set-up is the most common, but, has several disadvantages. Owing to the limited chopper frequency (e. g., 60 Hz) and according to the sampling theorem, the scan velocity of the wavelengths is scarcely more than 30 s for a spectrum between 400 and 800 nm. If this were not the case, photometrical and wavelength errors would amount to unacceptably high values. Typically, the (mechanical) distance between the sample/reference and detector D is greater than about 20 cm. Thus, the accepted solid angle of the light emanating from the sample is as small as 0.001 sr. This, however, excludes the measurement of turbid, scattering samples such as biological *in vivo* samples or low temperature samples (which are not converted into a glass form).

In order to collect as many light quanta as possible, we have to choose a solid angle as great as possible – close to 2π sr (see Figure 5.19). Even the cathode area of a single photomultiplier is very inhomogeneous in terms of efficiency; if the two rays of a double beam spectrophotometer hit only slightly different areas of the same photocathode – which is the case even with optimum adjustment – the baseline correction is insufficient and it deviates significantly from the ideal horizontal line. In conventional spectrophotometers these deviations are corrected by serial "multipots", a time-consuming and deficient procedure. Following the Gaussian rules of error propagation, the errors of sample and reference rays are additive in the final measured result. Figure 5.13 shows an exploded drawing of a conventional double beam spectrophotometer (Kontron Instruments GmbH).

If we reintroduce a second detector according to Figure 5.12(e) and compensate for the amplification differences of both detectors by a second ac light source (*frequency* f_H), we get rid of the limitations of scan velocity. The sample/reference and detector can be mounted very close to each other, which allows measurement of highly scattering, turbid samples (cf. Figure 5.19).

If the requirements of optical performance and wavelength resolution (e. g., $\Delta\lambda = \pm0.5$ nm) are not too strict – which generally is the case in chemical and biological optical molecular spectroscopy – the set-up of Seya–Namioka as discussed in Section 5.2.1 on the basis of a holographic concave grating is close to ideal (Figure 5.14).

Fast spectral scanning, turbid and strongly scattering (*in vivo*) samples are acceptable, and a compact design, a small stray light level, i. e., a large dynamic measuring range and – most important – a properly programmed on-line microcomputer (PC) allow all types of spectral manipulations. This low-cost arrangement is small and robust. Sequential scanning of

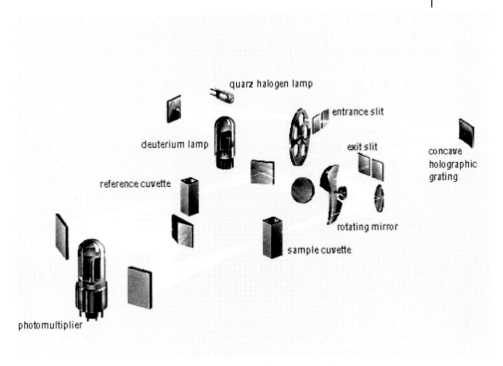

Figure 5.13 Exploded drawing of a conventional double beam spectrophotometer (Kontron Instruments GmbH).

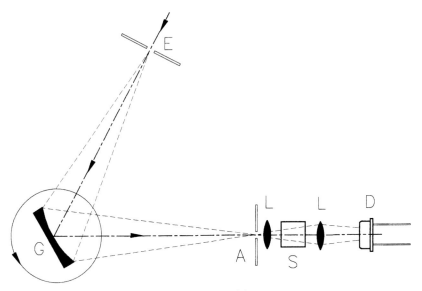

Figure 5.14 Simple arrangement of a small, powerful mono-chromator, after Seya–Namioka, based on a holographic con-cave grating.

the reference spectrum is not required for each individual sample. The reference spectrum as a "correction curve" is stored "once and for all" in the on-line computer and spectra are automatically corrected while scanning a spectrum – hidden from the user.

5.2.5
Noise

The signal to noise ratio (SNR) is the most important specification of a spectrophotometer and determines the photometric resolution. On the other hand, the wavelength resolution is defined by the slitwidths as well as by the quality of optical imaging of the entrance slit onto the exit slit (Figure 3.22). We can distinguish several origins of noise. These depend both on the incident light intensity I_0 and on various physical effects of the components of the signal transduction chain (Figure 5.15).

- *Photon noise*: Each light source (including unwanted background radiation) emits photons with a certain statistical probability. A black body radiator (Section 3.4.1), e.g., at 300 K reveals an average quantum flux of $n = 4 \times 10^{18}$ s^{-1} cm^{-1}, which, however, can undergo strong temporal variations. The ratio of quanta n hitting the detector divided by the number of quanta m detected is termed quantum efficiency $\eta = m/n$. The theoretical signal to noise ratio $SNR = \sqrt{(\eta n)}$, as calculated for completely noise free detectors, is practically never reached. The sample itself often causes immense temperature noise, which is observed particularly in strongly optically scattering samples. This effect can be reduced largely by cooling the sample, down to liquid nitrogen or even liquid helium temperature.
- *Detector noise*: We have to distinguish between various causes. The Johnson noise (often called Nyquist noise) depends on the thermal motion of charged particles in electrical resistors, the

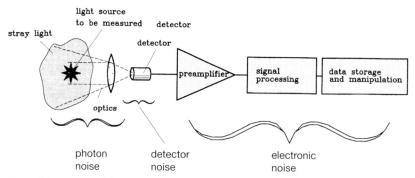

Figure 5.15 Survey of the origin of various noise components of a single beam spectrophotometer.

noise energy being $1/2\varkappa \times T$ ($\varkappa$ = Boltzman-constant, T = absolute temperature) and is independent of current (Johnson, J. B.,
Phys. Rev. 1928, *32*, 97). Cooling the photodetector diminishes
this noise component. Schott noise (Schottky, W., *Ann. Physik.*
1918, *57*, 54) in general is caused by the motion of discrete charge
carriers across interfaces and boundary layers. An important
example is the charge flow of photoelectrons between the cathode
and anode of photomultipliers. The current i caused by the Schott
noise is proportional to the square root of the averaged photon
current i_M, i.e., $I \sim \sqrt{i_M}$. Recombination noise occurs in photoconductors (Section 3.7.2) by recombination processes after a
preceding charge separation, and is inversely proportional to the
square root of the total number of free charge carriers. $1/f$ noise is
defined by an energy which decreases reciprocally with the signal
frequency. The theoretical causes are not particularly well
understood. However, when very small dc signals have to be
measured, this component is efficiently reduced if the signal is
correspondingly modulated [cf. Figure 5.12(b) and (d)]. Temperature noise has to be distinguished from Johnson noise and
takes place in detectors with a thermoeffect (Section 3.7.3); it
depends on the statistical temperature changes. Microphonic
noise is caused by mechanical vibrations which are converted into
electrical signals. A clear-cut theoretical description is not really
possible. This type of noise is avoided by stable construction of
the analytical device. Interference noise is caused by interference
processes with the frequency of the commercial power line and
its harmonics (in Europe 50, 100, 150, in the USA 60, 120, 180...
Hz): Practically all electrical devices are connected to a power line
because there are motors, transformers and in particular fluorescent and gas discharge tubes, and are sources of extraneous
radiation. Therefore, these frequencies are avoided in phase- and
frequency-sensitive lock-in amplifiers commonly found in dou-
ble-beam (Section 5.2.4) or dual-wavelength spectrophotometers
(Section 5.5). In addition, all (Thyristor-controlled) pulse genera-
tors which are hidden in electronic power supplies and even in
cordless telephones are a source of unacceptable extraneous
radiation.
- *Electronic noise*: This occurs in all parts of the amplifier that follow
  on from the photodetector. Photon and detector noise as such
  feed into the amplifier. In addition, these components generate
  the various noise components discussed previously. Finally, if we
  use a computer to process the amplified spectroscopic signals,
  digital noise is observed which is caused by an analog to digital
  signal conversion that is always limiting. Particularly in difference and absorption spectroscopy, after subtraction of two large

signals from each other, the photometric resolution of the small difference may not be acceptable. For example, for a (smaller) resolution of 10 bit (1024) an absorption signal of $A = 1$ is sub-divided into approximately 1000 steps; thus a reasonable differ-ence signal of 0.01 A corresponds to 10 counts: too small a number for an acceptable resolution of the structure of a spec-trum. Finally, white noise is defined as frequency-independent noise – e. g., in contrast to $1/f$-noise. It contains noise signals equally distributed over all frequencies. However, it is only of theoretical interest, and does not occur in real experiments.

Finally, in optical spectrometers all sources of noise contribute to the overall noise signal, and we define, under specified conditions, a total signal to noise ratio (SNR), e. g.: "$SNR = 0.0005$ at a response time of 50 ms, 2 nm wavelength resolu-tion at the wavelength of 500 nm".

Noise can be expressed as standard deviation N_σ of many discretely measured data points y_i, which are scattered around an average value $\bar{y}$:

$$N_\sigma = \sqrt{\frac{\sum\limits_{i=1}^{n}(y_i - \bar{y})^2}{n-1}} \quad \text{with} \quad \bar{y} = \frac{\sum\limits_{i=1}^{n} y_i}{n}$$

The standard deviation describes the scattering of data points and is required for the rating of any analytical procedure. However, it is too cumbersome to deter-mine the standard deviation of continuously scanned spectra. It is much easier to measure the peak to peak noise N_{P-P} which is simply the difference between the largest and smallest value within a certain wavelength range. As based on cal-culations by G. Liek (University of Münster, Germany) the smallest detectable sig-nal corresponds to $3/5 N_{P-P}$. Referring to his theory, it is not acceptable to define the noise itself or twice the noise signal as the detection limit.

5.2.6
Photometric Error

As each individual signal carries an error, when determining the absorption ac-cording to the formula $A = \log (I_0/I) = \log I_0 - \log I$ the total error is calculated by the law of error propagation of Gauss. At high optical densities the transmitted light value I becomes very small, i. e., δI is relatively large. At small optical den-sities, however, we subtract two comparably large signals from each other. Thus the resulting difference signal is strongly influenced by the error δI. Therefore it is reasonable that "somewhere around the medium absorbance values" the error will be the smallest. This minimum is determined by the dependency of the dif-ferential δI on I. For example, with PbS detectors (as used in the near-infrared, Chapter 10) the error of I, i. e., δI, does not depend on I at all (Johnson noise, see Section 5.2.5) and is constant. Therefore

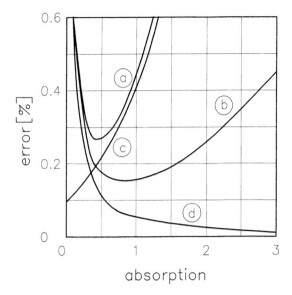

Figure 5.16 Photometric error of an absorption photometer (a) with a solid state photodetector and (b) with a photomultiplier. Dark current and stray light become more relevant with increasing sample thickness [curve (c)]. Drift and noise of the light source are more important at smaller sample densities [curve (d)] (modified from Pardue et al., *Clin. Chem.* 1975, 20, 1028).

$$dA = -0.43 \, \frac{dI}{I}, \quad \text{and}$$

$$\frac{dA}{A} = -\frac{0.43}{A} \frac{dI}{I}$$

and we obtain a relative minimum error of [Figure (5.16a)]

$$\log \frac{I_0}{I} = A = 0.43 \tag{5.13}$$

With photomultipliers, however, Schott noise determines the signal to noise ratio (Section 5.2.5), and this increases proportionally with the square root of I, i.e., $dI = p \times \sqrt{I}$. Thus we obtain

$$\frac{dA}{A} = -\frac{0.43}{A} \frac{p}{\sqrt{I}} \tag{5.14}$$

with a minimum at [Figure 5.16(b)]

$$\log \frac{I_0}{I} = A = 2 \times 0.43 = 0.86$$

In conclusion, photomultipliers allow higher absorption values than solid state detectors for optimal measurements.

Errors such as dark current (Section 5.2.5) and stray light (Section 5.2.7) become dominant with increasing optical density and extinction (i. e., including scattering) [Figure 5.16(c)]. If the error increases proportionally with I, which is the case for drift and noise of the light source itself, we obtain curve (d) in Figure 5.16 (after Pardue et al., *Clin. Chem.* 1974, 20, 1028).

5.2.7
Stray Light

A very important criterion for a monochromator is the so-called stray-light. This is light which is scattered around the monochromator case, possibly traversing the sample. Hence the formula defining absorption (log I_0/I) has to be modified:

$$A = \log \frac{I_0 + I_S}{I + I_S} \qquad (5.15)$$

Assuming that the sample and reference cuvettes are equally involved this leads to additional photocurrents I_S in both light paths. According to Eq. (5.15) for $I \ll I_S$, A is constant. Or, for a stray light level of 10 % ($I_S = 0.1 \times I_0$) the measurable absorption is at the maximum:

$$A = \log \frac{I_0 + 0.1I_0}{0.1I_0} = 1.04 \qquad (5.16)$$

For a stray light level of 0.01 % – a typical value for a single monochromator – a maximum absorption range of $\Delta A = 4.00$ is calculated. With increasing extinction and constant stray light levels I_S, the relative measuring errors increase tremendously (Figure 5.17). For a stray light level of 1 % and an absorption of $A = 0.5$ the error amounts to 2 %.

The essential result of stray light is the compression of absorption peaks: the spectrum becomes distorted and, for example, displays the wrong absorption values, i.e., too low concentration values.

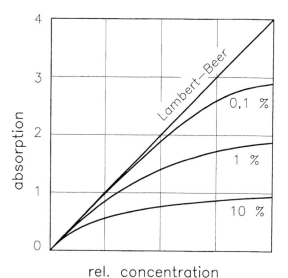

Figure 5.17 Deviations of absorption values from of the Lambert–Beer law with increasing stray light (%).

According to Eq. (5.15) the stray light level can be quantified by elimination of the respective I_0 signals I by means of absorption. Cut-off or cut-on filters (Section 3.6.1) which completely absorb light of specific wavelengths are suitable components. The remaining signal is caused by stray light. Unfortunately, cut-off filters typically exhibit a relatively strong fluorescence only in the wavelength range where they are supposed to be non-transmitting. Therefore, in practice, solutions of special molecules exhibit excellent cut-off properties in various wavelength ranges – without fluorescence:

- Potassium chloride (12 g L^{-1}): $\leq$ 200 nm
- Sodium iodide (10 g L^{-1}): 220 nm
- Sodium nitrate (50 g L^{-1}): 340 nm
- Methylene blue (0.01 %): 600–700 nm

5.2.8
Measurement of Turbid Samples

So far we have discussed dissolved molecular sample solutions which are optically clear and thus do not disturb straight light propagation in spectrophotometers. They simply attenuate the light intensity by pure absorption. However, samples in various fields of science are often turbid and highly scattering (see Chapter 8), which strongly disturbs the light path (Figure 5.18) and thus the assumptions of the Lambert–Beer law (Section 5.1.2). The theoretical situation becomes rather complex and is almost unsolvable. However, in practice, useful solutions are available.

The incident light flux I_0 is either transmitted in the normal straight way (rays 1), scattered once (rays 2, 6) or numerous times (rays 3, 4), regularly (ray 5, mirror-like surface reflection: entrance angle = exit angle) or diffusely (ray 6, Lambert radiator) reflected. The exiting light flux I is more or less non-directional, depending on the scattering properties of the sample, i.e., particle size and form and the interaction and arrangement of the particles. In addition, at strong absorption bands the phenomenon of anomalous dispersion takes place: the refrac-

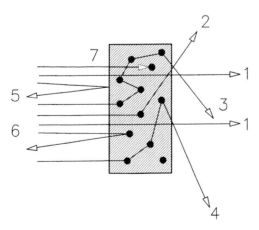

Figure 5.18 Schematic representation of the light path of a strongly scattering sample. (1) Straight transmission, without interaction, (2) single scattering process, (3 and 4) multiple scattering, (5) surface reflection (*specular reflection*), (6) diffuse reflection and (7) absorption.

tive index n changes abruptly close to the absorption maximum – and at the same time the reflection capacity is diminished (R_∞, Section 8.5.1). This essentially complicates the situation.

Two parameters play a vital role in the theory of transmission, scattering and (diffuse) reflection: the refractive index n and the absorption A. Both strongly depend on the wavelength, the collimation and polarization direction of the incident light flux and are concomitantly treated in theory as the "complex refractive index".

If the size of the particles involved is comparable or smaller than the wavelength of light (λ), the intensity contribution of reflection, refraction and scattering can no longer be taken separately and the phenomenon is simply referred to as "scattering" (Chapter 8). However, if the particles are significantly larger than the wavelength of light, interference phenomena occur leading to deviations with diffuse emission. Various theories such as that of Kubelka–Munk (Z. Tech. Physik 1931, 12, 593,) or Mie (Ann. Physik 1908, 25, 377,) are capable of describing with great success the scattering properties of different types of sample. Essential contributions and rigorous theoretical advances in this field were delivered by G. Kortüm (1962). In Chapter 8 the problems of scatter and reflection measurements are discussed in more detail.

If the sample thickness and the number of scattering particles are sufficiently large, the light distribution inside the sample and also the emanating light is essentially diffuse and isotropic, i. e., independent of the light direction. This is suggested by the approximate validity of the Lambert cosine law – the sample appears equally bright from all directions of view – and reflection and absorption spectra become nearly identical in shape (Section 8.5.2).

In double beam spectrophotometers with conventional light paths [Figure 5.12(d)] the solid angle as detected by the photodetector is only as small as 0.001 sr, so diffusely emitted light is barely detectable. A solution to this problem was suggested by Norris and Butler, mainly for biological/biochemical applications: an *end-on* photomultiplier with a large photocathode (e. g., 1 inch diameter) is attached tightly to the sample, collecting light from a solid angle of approximately 2π sr (Figure 5.19). Thus, highly scattering samples with absorbances up to $A = 8$ are measurable, if the stray light level can be kept sufficiently small – e. g., by using a double monochromator (see Section 5.2.7). As an example of a measurement, Figure 5.20 shows the absorption spectrum of a full-grown leaf of a beech tree with an extinction of approximately $E = 4.5$ and a true absorption of up to $A = 1.8$. The scattering properties of the leaf are readily corrected by a white, non-fluorescing and non-absorbing piece of filter paper with an identical extinction at 760 nm. Note, at this wavelength there is essentially no absorption by the leaf, however there is comparable scattering of the paper and leaf. The spectral difference of the uncorrected extinction of the beech leaf minus the extinction of the paper (reference spectrum) yields the true absorption spectrum of the beech leaf. For the spectroscopy of highly scattering samples, derivative spectroscopy is virtually indispensable. Thus, single biochemical components are detectable in living tissue in a semi-quantitative way (Section 5.4.3).

According to Figure 5.18 the average medium light path x through strongly scattering samples is larger than in non-scattering samples with the consequence of an increased absorption ($A = \varepsilon \times c \times x$). The measured absorption value in

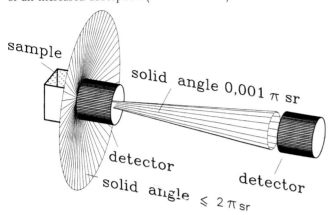

Figure 5.19 Schematic representation of the solid angle covered by the detector as a function of distance of the light emitted and transmitted by the sample.

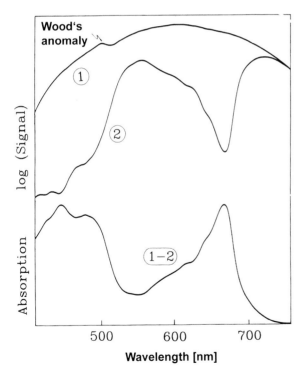

Figure 5.20 Acquisition of the absorption spectrum of a beech leaf with a single beam spectrophotometer. Curve 1, correction curve, obtained using a white, non-fluorescent filter paper. Curve 2, uncorrected absorption spectrum of a beech leaf. Lower curve (1–2), "true" (corrected) absorption spectrum of a beech leaf ($E = 4.5$; $A = 1.8$).

strongly scattering samples can be easily increased by an "intensity factor" of as much as $\beta = 100$ ($A = \beta \times \varepsilon \times c \times x$), compared with a measurement of the same concentration of the absorber in a solution. The plant photoreceptor phytochrome, mentioned in Section 5.1.3, is typically present *in vivo* in extremely small, hormone-like concentrations. These are measurable – without any enrichment – *in vivo*, because the intensity factor β of natural plant tissues such as coleoptile or hypocotyl amplifies the signal dramatically. In order to measure extracted, purified phytochrome, this is embedded in highly scattering media such as $CaCO_3$ (*slurry*: 1 mL of the medium to 1 g $CaCO_3$, see Section 8.4.2). However, absolute measurements of small concentrations such as this are always difficult, but are usually not required. For example, in measurements of physiological interest we are concerned with the kinetics or relative concentration or the respective concentration changes.

There are two basic prerequisites for the validity of the Lambert—Beer law: (1) the molecular dispersive distribution of the sample molecules and (2) the *linearity* (collimation) of the incident measuring light beam. However, with completely diffuse incident measuring light the theory predicts the doubling of the effective light path within the sample – and thus of the measured signal. The reason that this effect is not typically adopted in absorption spectroscopy is due to the additional diffuse emanating light from the sample.

With conventional double beam spectrophotometers, the spectral measurements at low temperatures (e. g., at –196 C) are very time-consuming and not applicable routinely. Samples and references are measured within special cuvettes and dewars in the form of glasses because these do not scatter and thus do not disturb the light paths. However, on the basis of a vertical light path and a single-beam set-up where the photomultiplier is tightly attached to the sample (Section 3.8, cuvettes), very turbid and/or highly scattering, non-glassy samples are measured easily and fast (cf. Figure 3.45, Schmidt, W., *Anal. Biochem.* 1982, *125*, 162).

5.2.9
Specifications

If a new spectrophotometer has to be purchased, first of all we have to ascertain the exact specifications required, which determine its price, size and sensitivity. Moreover, it is not a good idea – as formerly practiced – to buy a "universal" spectrophotometer with many options suitable for various applications with superb specifications for each. For particular applications, it makes more sense and it is the better choice, to use specific but moderately priced apparatus, for example for "fast scanning of a large spectral range", "measuring calcium ions", "measuring proteins at low temperature", "measuring in the dual wavelength mode" or "measuring in the far-ultraviolet C". However, experience has shown that in many cases there are no such spectrophotometric systems available commercially, because the manufacturers do not anticipate sufficiently large numbers being sold. Thus, nowadays in many research laboratories we find "homemade" spectrophotometric set-ups, often produced for specific requirements necessary in diploma or Ph. D. work. The following specifications might help in the selection of the most appropriate spectrophotometer:

- Scanning spectrophotometer (spectra) or fixed wavelengths (kinetics)?
- Dual-beam, dual-wavelength or single-beam spectrophotometer? Simultaneous spectrophotometer (optical multichannel or array analyzer)?
- Double or single monochromator? Light throughput (*F*-number), stray light?
- Required wavelength range, scan velocity? Is the UV range (<350 nm) required (<200 nm, vacuum-UV: flushing with nitrogen gas)? Is the NIR range (near-infrared) required (red-sensitive photomultipliers are expensive)? NIR between 1000 and 2500 nm is used for specific applications (see Chapter 10).
- Resolution of wavelength and photometric value (dynamics)? Blaze wavelength (main wavelength range utilized)? Order of separation filters? Photometric range? Absolute wavelength precision and precision of repetition? Signal to noise ratio? Linearity of baseline?
- Long- and short-term stability of the measuring signal? Warm up time?
- Data acquisition and processing? Single photon counting or analog signal processing (sensitivity)? On-line or integrated computer, standard software? Does additional software have to be purchased? Can the user program the system themselves? Data storage and presentation? Standard files? External high quality laser printer supported? Quality of printouts (for publication!)?
- Clear and/or (e. g., biological/biochemical) scattering samples measurable? Low-temperature measurements (liquid nitrogen) of glasses and particularly turbid samples? Scan speed? Temperature controlled, multiple sample holder?
- Price category? Size, weight, ozone removal required? Thermostated location required? Mobile, battery powered field spectrometer?
- Service? Introduction course offered (often expensive)? Warranty? Test measurements offered?

Currently it would appear that it is not realistic to plan more than five years in advance, because of the fast technological advances in electronics and optics. The "latest developments" today are possibly obsolete and impracticable tomorrow.

5.3
Absorption Properties of Molecules

5.3.1
Fundamental Electronic Transitions

Owing to their inherent electronic structure, without exception all molecules exhibit absorption. For smaller, non-polymeric molecules in most cases this is restricted to the near (300–400 nm), the far- (200–300 nm) or even the so-called vacuum-ultraviolet wavelength range (<200 nm). The term "vacuum" is used because below 200 nm the absorption of the air in the surrounding environment disturbs the measurement, and therefore the complete optical part of the spectrophotometer has to be operated in a vacuum or at least flushed with nitrogen (which shows no absorbency in the far-UV). There is a large body of literature that treats the practice, theory and interpretation of far-ultraviolet spectra.

Table 5.2 shows several examples of either colored or typical UV-absorbing molecules together with the transitions responsible. The energetic properties of various transition types have been discussed previously for Figures 2.18 and 2.20. $n \rightarrow \pi^*$ transitions of individual groups are symmetry forbidden thus yielding extinction coefficients of less than $\varepsilon = 100$ M^{-1} cm^{-1}. With increasing solvent polarity the spectra show a hypsochromic shift. $\sigma \rightarrow \sigma^*$ transitions as in ethane are located in the vacuum-UV and thus are of minor interest to the optical spectroscopist. $\pi \rightarrow \pi^*$ transitions are the most important ones, particularly with conjugated molecular structures, e. g., in the spectra of aromatic molecules. According to Hückel, aromatic molecules are those with $4n + 2$ conjugated π-electrons ($n = 1$, 2,...). They are extraordinarily stable and have even been discovered in extra-galactic objects by means of optical spectroscopy. They are energetically low lying and exhibit pronounced absorption bands; examples are aromatics such as benzene ($n = 1$), naphthalene ($n = 2$), anthracene ($n = 3$), naphthacene ($n = 4$, Figure 5.21) or, e. g., the oxidized flavin molecule ($n = 4$, see Figure 5.23). $\pi \rightarrow \pi^*$ transitions of conjugated polyenes exhibit a bathochromic shift with an increasing number of double bonds (Figure 5.22). The spectra are complex, however they are understood in principle. Thus, the flavin band in the near-UV exhibits a bathochromic shift with increasing polarity of the solvent, while the "blue band" at 450 nm is slightly shifted hypsochromically with a concomitant blurring of the vibration bands (Figure 5.23). Experience has shown that the shift of the band in the near-UV is proportional to the square root of the dielectric constant ε of the solvent. This behavior offers a simple probe as to the polarity of the microenvironment of unknown flavin systems (Figure 5.22).

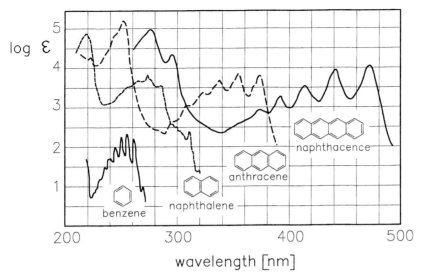

Figure 5.21 Absorption spectra of selected aromatics exhibiting an increasing red shift with increasing number of conjugated double bonds (modified from Silverstein, R. M., G. C. Bassler, T. C. Morrill, 1974, *Spectrometric Identification of organic compounds*, John Wiley & Sons, Inc., New York, London, Sidney, Toronto).

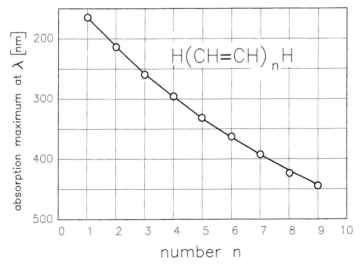

Figure 5.22 Bathochromic shift of $\pi \rightarrow \pi^*$ transitions of conjugated polyenes with increasing number of double bonds.

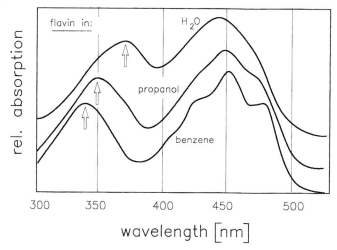

Figure 5.23 Absorption spectra of flavins in solvents of different polarities.

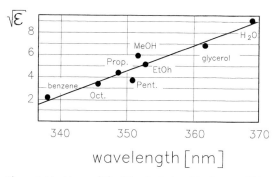

Figure 5.24 Linear shift of the flavin band in the near-UV range with the square root of the dielectric constant of the solvent.

5.3.2
Charge-transfer Complexes

Mixtures of electron donors (D) and acceptors (A) in solution often lead to broad, structureless absorption bands devoid of vibration structure, which are not shown by the original molecules themselves: in contrast to the single components D and A the mixture is strongly colored, even if no new, additional chemical species is created. These bands are attributed to certain donor–acceptor entities, called charge-transfer-complexes (DACs) (Figure 5.25). Acceptors are most often Lewis acids such as poly- and nitro-aromatics, quinine and iodide, donors are Lewis

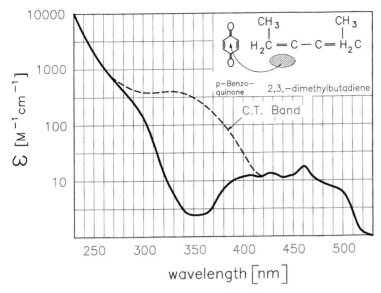

Figure 5.25 Dotted line, absorption spectrum of the charge-transfer-complex (CTC) of *p*-benzoquinone (Lewis acid) with 2,3-dimethylbutadiene (Lewis base). Solid line as control, *p*-benzoquinone solubilized in cyclohexane.

bases such as aromatic hydrocarbons, dienes and amines. The "stronger" the donor and acceptor the more the charge-transfer band is shifted to longer wavelengths. The bonding properties and thus the spectral characteristics of charge-transfer complexes were first described and explained by Mulliken.

The wavefunction ψ_{AD} of the two molecules D and A, which form a *sandwich*-conformation with a distance of approximately 0.34 nm, describes the mutual dipole–dipole or hydrogen-bond interaction. On the other hand, $\psi_{A^-D^+}$ describes the common wavefunction of A^- and D^+ after the electron exchange is complete. The common linear combination of both wavefunctions ψ_{AD} and $\psi_{A^-D^+}$ leads to two so-called charge-transfer orbitals, where the ground state is stabilized, however, the (first) excited state is destabilized:

$$\psi_{S0} = a\psi_{AD} + b\psi_{A^-D^+}$$

and

$$\psi_{S1} = a\psi_{A^-D^+} - b\psi_{AD}$$

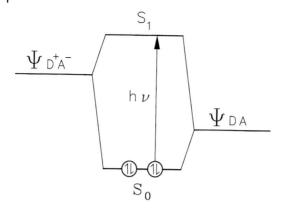

In most cases is $b \ll a$ and we obtain $\psi'_{S0} \approx a \times \psi_{AD}$ and $\psi'_{S1} \approx a \times \psi_{A \cdot D^+}$ (Figure 5.26). Thus, the spectroscopic transition is energetically virtually identical with the light-induced electron transfer from donor to acceptor, which is the basis for the terminology *charge-transfer transition*. Therefore, in these cases it is not acceptable to use the term *charge transfer complex* in the ground state (i. e., without light excitation), because at this point *no* charge transfer takes place.

In this context the so-called contact-charge transfer needs to be mentioned; e. g., between O_2 and aromatic hydrocarbons. In contrast to the usual charge-transfer complex, this complex possibly has an intermediate character as the bonding energy is much smaller than the kinetic energy $\varkappa T$ (Boltzmann constant $\times$ absolute temperature).

5.3.3
Transition Metal Complexes

With the transition metals in the fourth period of the periodic system, d-orbitals are involved in chemical bonding. There are many so-called transition metal complexes of interest, particularly in the life sciences. These are molecules or ions in which "ligands" (atoms, ions or chemical groups) share their lone electron pairs with the central transition metal. The classic example is the cobalt hexachloride ion ($CoCl_6^{3-}$), where the cobalt ion Co^{3+} is surrounded by six Cl^- ions, which occupy the top and bottom corners of an octahedron (Figure 5.27).

The six ligands influence the five energetically equivalent, but geometrically distinct d-orbitals of the central atom differently (see Figure 2.8). Two of the five d-orbitals have their lobes on the x-, y- and z-axes ($d_{x^2-y^2}$, d_{z^2}), while the other three orbitals – orthogonal to each other – occupy the corresponding bisecting axes (d_{xy}, d_{xz}, d_{yz}). In the following hypothetical experiment all orbitals can by occupied by electrons. If now – coming from infinity – the six negatively charged ligands approach the central atom along the three directions in space $\pm x$, $\pm y$ and $\pm z$ (any other direction can be decomposed into these three space directions), the electrons in the orbitals mentioned first experience a stronger repulsive force than

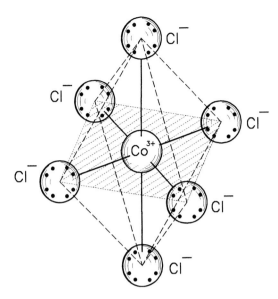

Figure 5.27 In cobalt hexachloride ($CoCl_6^{3-}$) the central cobalt ion (Co^{3+}) is surrounded by six Cl^- ions which occupy the corners of an octahedron.

the electrons in the other three orbitals, which occupy the bisecting axes. Thus we obtain a typical 3:2 energy splitting of the size of the so-called crystal split energy Δ (Figure 5.28). The upper two states are termed "e" (*excited orbitals*), the lower three states "t" (*three orbitals*). In which order these five orbitals are filled depends on the number of available d-electrons, on Δ and on the mutual electrostatic repulsion of the paired electrons (pairing energy, cf. Hund's rule, Section 2.3.2.2). For example, the Cr^{3+} ion has three outer electrons, which, of course, occupy the three low lying t-orbitals. Co^{3+} on the other hand has six electrons. With a smaller Δ energy, e. g., in cobalt hexafluoride, the electrons will be distributed evenly over all five orbitals (Figure 5.28). On the other hand, in cobalt hexacyanide the crystal splitting energy Δ is so large that the complex absorbs in the UV range, i. e., it appears colorless.

With low crystal split energy Δ the total spin S of the system is, according to the vector addition, large (*high spin*), in the opposite case small (*low spin*). Thus, these properties can be – independent of optospectroscopic analysis – checked by electron spin resonance spectroscopy (ESR).

As a rule of thumb the crystal split energy Δ increases with more concentrated ligand charges in space. Largely independent of the type of central atom, the spectrochemical series holds true where the splitting for the following order of ligands increases and – concomitantly – the spectra exhibit a hypsochromic shift:

$$I^- < Br^- < Cl^- < F^- < C_2H_5OH < H_2 < NH_3 < NO_2^- < CN^-$$

The example of Co^{3+} demonstrates the color properties of transition metal complexes (Table 5.4).

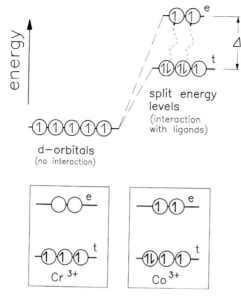

Figure 5.28 Split of the five d-orbitals of the central atom within an octahedral ligand field. Depending on the ratio of crystal splitting energy (Δ) and electron pairing energy the available electrons are distributed either within the lower three t-orbitals or over all five orbitals (2e + 3t).

Table 5.4 With "stronger" ligands the crystal splitting energy Δ increases, thus shifting the absorption maximum hypsochromically to where the solution of molecules appears in the corresponding complementary color.

Co^{3+} complex	Absorption (nm)	Complementary color (observed)
$[CoF_6]^{3-}$	700	Green
$[Co(CO_3)_3]^{3-}$	640	Green–blue
$[Co(H_2O)_6]^{3+}$	600	Blue
$[Co(NH_3)_5Cl]^{2+}$	535	Purple
$[Co(NH_3)_5OH]^{2+}$	500	Red
$[Co(NH_3)_6]^{3+}$	475	Yellow–orange
$[Co(CN)_5Br]^{3-}$	415	Citrus yellow
$[Co(CN_6)]^{3-}$	310	Pale yellow

Transition metal complexes exhibit an extraordinary richness in color. They often absorb in the long wave visible spectral range. They are prerequisites as photoreceptors in photophysics, photobiology and photochemistry. Recently they have even been utilized in the so-called photodynamic therapy for cancer treatment.

Within a transition metal complex we can distinguish three types of electronic transitions (Figure 5.29):

- Transitions between molecular orbitals (MOs) within the central metal atom which, in spite of the bonding of the ligands, retain their metallic character. These transitions are very weak (ε = 10-100 M^{-1} cm^{-1}), as they are symmetry forbidden.
- Transitions between MOs, which are mainly localized at the ligands and barely influenced by the complex bonding.
- Transitions between MOs, which are localized both at the central atom and the ligands: charge-transfer transitions (see Section 5.3.2). These transitions are very intense (ε = 10^3 – 10^4 M^{-1} cm^{-1}) and therefore are often found in the UV or in the visible spectral range. They represent the opto-spectroscopic properties of many biologically/biochemically relevant molecules.

A biologically very important group of metal complex bonds are the porphyrin pigments, such as hemoglobin (pigment of the blood, central ion Fe^{2+}), the various redox-active cytochromes of the respiratory chain or the chlorophyll (green molecules in leaves, central atom Mg) of all photosynthetically active organisms. In these molecules the octahedron structure with a central atom is incorporated into particular proteins. The four ligand positions of the base of the pyramid are

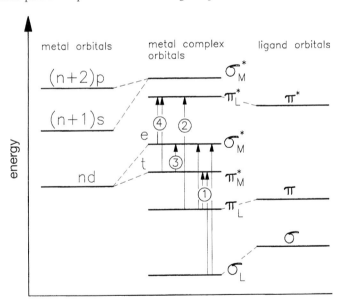

Figure 5.29 The three types of electronic transitions for a transition metal complex. (1, 4) Transitions between ligands and the central atom; this type is responsible for the characteristic color and the essential properties of many transition metal complexes and porphyrins. (2) Transitions within the ligand's sphere. (3) Transitions that are located at the central atom (modified from Neubacher and Lohmann. In: W. Hoppe, W. Lohmann, H. Marckl, H. Ziegler, *Biophysik*. Springer Verlag Berlin, Heidelberg, New York, 1982).

occupied by the lone electron pairs of the nitrogen atoms of the plane porphyrin ring system. The two corners of the pyramid are occupied by specific amino acids (histidine) and/or by an oxygen molecule (hemoglobin, Figure 5.30). If the central atom is iron, we call this porphyrin complex heme or a heme-group. For example, in cytochrome c the pyramid corners of the heme are occupied by the N-atom of a histidine residue rather than the S-atom of a methionine residue of the embed-

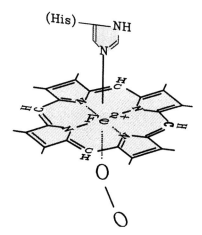

Figure 5.30 Schematic representation of heme with a bound O_2 molecule, which occupies the top corner of the octahedron pyramid; the bottom corner is occupied by a histidine of the apoprotein.

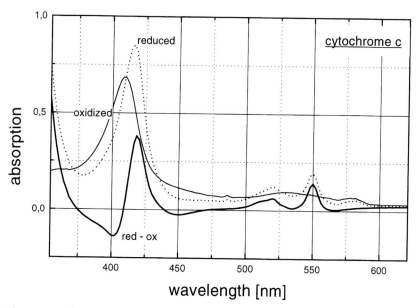

Figure 5.31 Absorption spectra of cytochrome c in its oxidized and reduced states, and the calculated difference spectrum. Isosbestic points are marked by dotted lines. A characteristic is the so-called α-peak at 550 nm, which, e.g., is utilized for the enzymatic assay of mitochondria (respiratory chain).

ding protein. The redox change of cytochromes predominantly occurs at the central iron atom $[(Fe^{2+}) \leftrightarrow (Fe^{3+})]$; the corresponding absorption spectra of cytochrome c are depicted in Figure 5.31. The absorption changes of cytochromes in the so-called α-peaks (Cyt c at 550, Cyt b at 560, Cyt a at 610 nm) are very specific, and are therefore utilized for their identification, while the respective β- and γ-bands (Soret-bands) are less pronounced.

In hemoglobin the fifth ligand is also a nitrogen atom from histidine, the sixth position remains free for optional bonding of an O_2 molecule. In heme the iron is always found in the divalent form. Arterial, oxygen-loaded blood is known to be light red, while blood in veins is deep red, as the spectra also illustrate (Figure 5.32). This absorption change caused by oxygen bonding in the red (660 nm) and near-infrared (940 nm) spectral range which is utilized by so-called *pulse oxymeters* in medicine to measure and monitor oxygen saturation of blood (at the earlobe or at the tip of a finger). Even if the absorption change of hemoglobin between 560 and 575 mm is an order of magnitude larger, the red spectral range is preferred as light scattering in the tissue is less critical (Chapter 8).

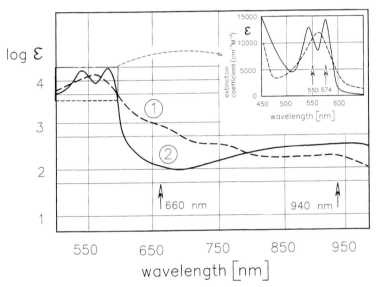

Figure 5.32 Absorption spectra of hemoglobin in (1) an oxygen free and (2) an oxygen loaded state. For increased clarity the total spectrum is depicted on a logarithmic scale, and the inset is on a normal linear scale. The oxygen loaded, arterial blood is known to be red, as caused by a 10 times lower absorption in the red region (higher reflectivity).

5.4
Modification of Absorption Spectra

The term modification describes (i) the alteration of absorption spectra due to experimental effects and (ii) the mathematical treatment of already existing spectra in order to extract hidden information.

5.4.1
Difference Spectrophotometry

As discussed previously, all molecules have their own specific absorption spectra. Various parameters such as temperature, polarity of the solvent (microenvironment), motility (i.e., the viscosity of the microenvironment and the embedding in it), pH values ($pK!$), hydrogen bonds and many others lead to a variety, even if often only minor spectral differences. Therefore it is important, when measuring absorption spectra, to keep track of all relevant boundary conditions such as solvent, ionic strength, slit width, temperature, etc. However, redox changes and other, seemingly minor, intramolecular variations will result in more pronounced spectral changes. For example, this holds true for the spectra of the "colorless" leucoflavin or the flavin radical compared with the yellow, fully oxidized flavin (Figure 5.33), or the spectra of the reduced–oxidized cytochromes c with several

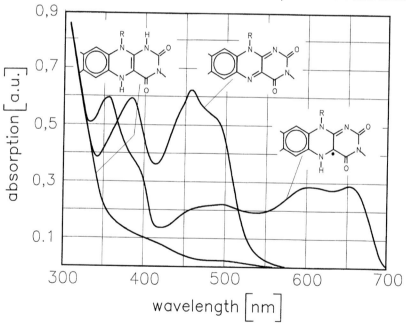

Figure 5.33 The absorption spectra of the three oxidation states of the flavin chromophore (from the left): reduced, oxidized and radical form (blue radical). The oxidized chro- mophore is an aromatic compound with a correspondingly strong absorption ($n = 4$, cf. Figure 5.11).

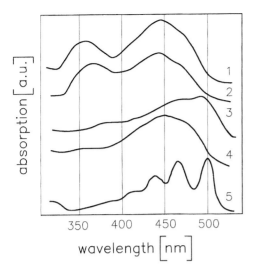

Figure 5.34 Absorption spectra of oxidized flavin in water: (1) at room temperature and (2) at −196 °C (liquid nitrogen); (4) absorption spectrum of β-carotene in ethanol at room temperature and (5) at −196 °C. If flavin in its solid state is measured as a powder (corrected for scattering, cf. Figure 5.20), we obtain spectrum (3) (the peak in the near-UV is missing!).

isosbestic points (Figure 5.31). By definition, difference spectra intersect the 0-line at a wavelength where the extinction coefficients of the two redox forms are identical. Therefore difference spectra serve as a basis for the wavelength selection in dual wavelength spectroscopy (Chapter 5.5).

An extreme example for the high temperature sensitivity of spectra is β-carotene. At room temperature its absorption spectrum – except for the missing near-UV peak – is very similar to the absorption spectrum of flavins. However, at −196 °C its spectrum is altered completely, which is essentially caused by the inhibition of rotatory freedom. On the other hand, flavin spectra are scarcely changed upon refrigeration down to −196 °C (Figure 5.34).

5.4.2
Hyper- and Hypochromism

If we decompose proteins into their amino acids (denaturation) or nucleic acids into their individual bases (melting), the absorption properties of the samples are changed significantly. This can only be caused by a changed interaction of the individual components: extinction maxima are shifted, extinction coefficients ε are changed. If ε becomes larger we call it hyper-, if ε becomes smaller hypochromism (Section 5.1.3). This observation suggests that the interaction of the individual components of a molecule with undisturbed secondary and tertiary structures is different compared with the situation when the individual amino acids of nitrogen bases have immediate contact with the solvent. A crude qualitative explanation could be that a transition dipole moment m polarizes the electrons of the concomitant (or even different) neighboring chromophores and thereby induces (transient) dipole moments m'. A parallel assembly of dipole moments m' [Figure 5.35(a)] diminishes m, the probability of absorption decreases: thus a case of

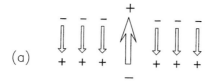

(a)

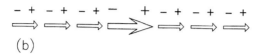

(b)

Figure 5.35 Simplified representation of hypo- and hyperchromism by induced (transition) dipole moments, which interact with the inducing dipole moments. (a) A parallel arrangement causes hypochromism, and (b) a serial arrangement hyperchromism.

hypochromism. However, if the induced dipoles m' are assembled "in series" [Figure 5.35(b)], m increases: thus a case of hyperchromism.

Hypochromic effects are rarely used to analyze the structure of macromolecules, however they serve as a simple and fast assay to monitor the crude state of order of macromolecules both quantitatively and kinetically. Figure 5.36 shows the pronounced hypochromism (concomitant with a slight hypsochromic shift) of single- and double-stranded polyriboadenylic acid (poly A) compared with the monomer (equal concentration of the monomers). Figure 5.37 shows the pronounced absorption increase at 260 nm upon melting of deoxyribonucleic acid (DNA, Table 5.2). The conspicuous absorption increase at 80 °C is typical for a cooperative process, which can therefore be analyzed spectrophotometrically. However, upon cooling the process is not completely reversible.

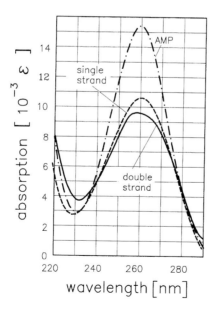

Figure 5.36 Hypochromism of polyriboadenylic acid (poly A). The monomer (AMP) shows the strongest absorption, while single or even double stranded poly A show significantly lower absorption (identical concentrations of the monomers, modified from v. Holde).

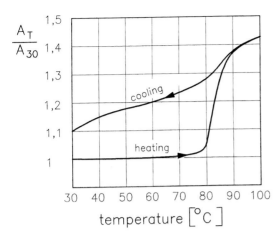

$\dfrac{A_T}{A_{30}}$

cooling

heating

temperature $\left[^\circ C\right]$

Figure 5.37 Absorption increase at 260 nm on melting of deoxyribonucleic acid (DNA). The abrupt absorption increase at 80 °C reflects a typical cooperative process, which is not fully reversible on cooling.

5.4.3
Derivative Spectrophotometry

Derivative spectrophotometry was introduced in 1953 by Hammond and Price (*J. Opt. Soc. Am.* 1953, *43*, 924) and Morrison (*J. Chem. Phys.* 1953, *21*, 1767) and is utilized increasingly in all areas of optical spectroscopy. In spite of spectacular analytical accomplishments its results have to be interpreted with extreme care as we will see below.

The first, second and higher mathematical derivatives (differentiations) of an absorption spectrum $A(\lambda)$ are defined according to the differential quotient $dA/d\lambda$, $d^2A/d\lambda^2$,.... $d^nA/d\lambda^n$ (Figure 5.38). For example, the first derivative is equivalent to the evaluation of the slope of a spectral curve at each individual point, as a function of wavelength; the maxima of a spectrum correspond to intersections of the first derivative with the zero-line. Optima of the first derivative correspond to reversion of the slope in the original spectrum. If the second derivative is large than zero then the spectrum at these wavelengths is curved with the concave side upwards ($+y$ axis), and vice versa. Without discussing the radii of the curves, reversion points, etc., in detail, the (artificial) example teaches us that both individual components become more pronounced with the increasing order of the derivatives, while they are completely hidden in the original spectrum. Particularly in the fourth derivative, both original components are even better pronounced than in the second derivative – and the maxima point in the same (positive) direction as the original components rendering the interpretation easier. Starting from the Lambert–Beer law [$A = c \times x \times \varepsilon$, Eq. (5.5)], we obtain as the first derivative:

$$\frac{dA}{d\lambda} = c\,x\,\frac{d\varepsilon}{d\lambda}$$

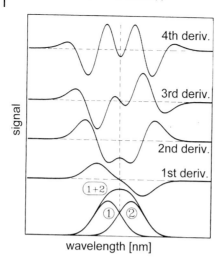

Figure 5.38 First and fourth derivative of the additive spectrum (1 + 2) composed of two similar Gaussian curves slightly shifted relative to each other (1, 2).

and as the second derivative:

$$\frac{d^2 A}{d\lambda^2} = c \, x \, \frac{d^2 \varepsilon}{d\lambda^2}$$

and finally the most valuable fourth derivative:

$$\frac{d^4 A}{d\lambda^4} = c \, x \, \frac{d^4 \varepsilon}{d\lambda^4} \quad \text{etc.} \tag{5.17}$$

Consequently, the height of a derivative spectra is directly proportional to the concentration, which increases their value analytically. Of course, in practice the differential quotients have to be substituted by difference quotients of wavelengths, e.g., $d\lambda \rightarrow \Delta\lambda = 1$ nm.

On increasing the order of the derivative, minute components hidden in the original spectrum are increasingly resolved. However, in the same way the probability of introducing artifacts increases. For example, in the fourth derivative of the example (Figure 5.38) "submaxima" (satellites) appear near the "border lines" to the left and right of the peaks, which do not correspond to real components of the original spectrum.

Figure 5.39 shows the absorption spectrum at −196 °C of the mycelium of a white mutant of the fungus *Neurospora crassa*, and its fourth derivative. A surprisingly large number of individual components show up (not noise!), which are predominantly attributed to various components of the respiratory chain.

Derivative spectroscopy allows the precise determination of absorption maxima of single components as required for mathematical multiple components analysis (see Section 5.4.5). Figure 5.40 again shows an artificial additive spectrum composed of two individual, largely different Gaussian curves. The first striking result

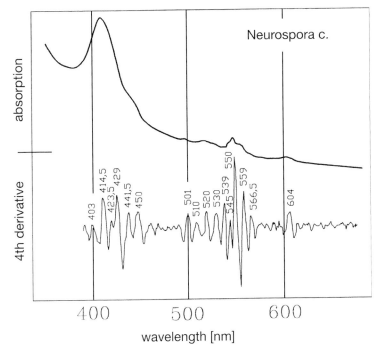

Figure 5.39 Absorption spectrum of the mycelium of a white mutant of the fungus *Neurospora c.*, measured at –196 °C. The spectrum is an average of 32 individual spectra as scanned with a single beam spectrophotometer. The derived fourth derivative exhibits a series of well defined peaks which are attributable to various individual components.

we observe is a shift of the peak in the additive spectrum, which corresponds to component 2, in the direction of the maximum of component 1. This is a very general phenomenon observed in composite spectra. The maximum of a relatively narrow peak (1), on top of a broader base spectrum (2) is always shifted "uphill". It is interesting to note that the peak maxima are again correct in the fourth derivative.

We will summarize the advantages of derivative spectrophotometry:

- Peak bands will be narrower thereby increasing the wavelength resolution.
- "Flat" signals are suppressed (depending on the choice of $\Delta\lambda$).
- The number of peaks increases with each subsequent derivative. The first derivative of a one-peak spectrum has two peaks, of a two-peak spectrum three, of an *n*-peak spectrum *n* + 1. The satellite-artifacts mentioned above can be utilized for the identification of a substance (*fingerprints*).
- Derivative spectroscopy allows simple and sensitive (routine) trace analysis (e. g., 200 ng Zn^{2+} or Cd^{2+} as dithizonate), in

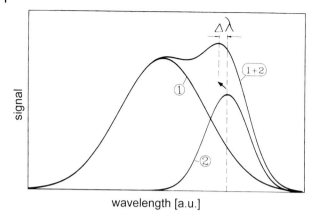

Figure 5.40 Summed spectrum (1 + 2) of two single Gaussian curves (1, 2) of different widths. In the sum spectrum the peak of the narrow component is shifted by $\Delta\lambda$ in the direction of the maximum of the broader component.

environmental analysis (e. g., <1 ppm aniline or phenol in drinking water), in pharmaceutical analysis (cortisone, vitamin, barbiturates) and in food analysis (milk, juices, coffee, tea, cocoa, meat, cheese).

- Derivative spectroscopy, of course, is not restricted to absorption spectroscopy but can be used in all types of spectral analyses, e. g., fluorescence spectra.

5.4.4
Correlations

5.4.4.1 Theory

Even if no analytical method allows more information to be extracted from a signal than is inherent, the reverse argument holds true that typically there is more information hidden within a signal (spectrum) than is typically recovered. This is the essence of valuable applications of various "correlation methods". Correlations represent a very general concept and are indispensably connected to various analytical procedures and processes in science and technique even if it is not always obvious. In the broadest sense the term *correlation* is defined as the interaction or concomitant influence of two functions.

Mathematically a correlation is described as a "multiplication–integration procedure". The terms "folding" or "convolution" are used as synonyms.

Mathematically this identification is not quite correct. *Folding* and *correlation* differ in the loci of g(x) [Eq. (5.18)]: if g(x) is reversed = "folded", i. e., with x and x' at the positions as indicated in Eq. (5.18), we talk of "convolution", however, if g(x) is only shifted, i. e., g(x'−x), we refer to "correlation".

Procedure: A function $f(x)$ is multiplied by another function $g(x)$ which is – with respect to the x-coordinate (time or space) – slightly shifted by a certain value. Finally, this product is averaged across the whole x-range (integrated), a process described by the so-called "folding integral". The result of this folding of the two functions $f(x)$ and $g(x)$ is another function $h(x)$:

$$h(x) = \int_{x=-\infty}^{+\infty} f(x') \, g(x-x') \, d'x \tag{5.18}$$

where this relationship has to be calculated for each individual position of x (here wavelength λ) separately. Each individual point (wavelength) of the folding integral requires the integration over the whole available space (in this example only one dimension), requiring considerable time on calculations.

The bulky form of the folding integral [Eq. (5.18)] is conventionally abbreviated by:

$$h(x) = f(x) * g(x) \tag{5.19}$$

where "*" indicates the folding operation.

For a real application with computer support the theoretical integral given above [Eq. (5.18)] is of course replaced by the (finite) sum:

$$h(n) = \sum_{m=0}^{n} f(n) \, g(n-m) \tag{5.20}$$

5.4.4.2 Smoothing

The averaging property of a correlation function is often used to smooth noisy signals. Figure 5.41 (1) shows the noisy absorption spectrum of a mature beech leaf generated by an extremely small slitwidth of only 0.1 nm and a decrease in voltage of the light source. By folding with the simple rectangular function as indicated we obtain an excellent smoothing [Figure 5.41 (2)].

In specific cases like this the calculation can often be dramatically shortened. Here $f(x)$ describes the spectrum from 360 to 760 nm, $g(x)$ a rectangular function (often termed the "kernel") of 22 nm width for smoothing. The required calculation needs only fractions of a second on the microcomputer (PC), as each individual wavelength only has to be integrated across the (small) width of the kernel, and not the whole wavelength range.

Of course, this process only takes place during the actual measurement in the spectrophotometer, and, if we set the slitwidth to 22 nm – in order to reduce noise – prior to the measurement. However, the effective slit function of a spectrophotometer with typical identical rectangular entrance and exit slits represents *not* a rectangle but rather a triangle; see Figure 5.43.

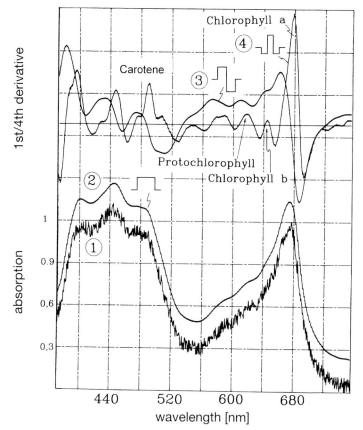

Figure 5.41 Correlations of the (noisy) absorption spectrum of a beech leaf (1) with a rectangular kernel results in smoothing (2), with the kernel given in spectrum (3) in the first derivative, and with the "W"-shaped kernel in spectrum (4) to the fourth derivative.

The correlation procedure for a spectrum with a certain kernel function can be visualized as an investigation of the structural similarities of the kernel and the spectrum to be analyzed. "Noise" (fast, small and irregular changes of a signal) is not contained in a smooth, rectangular function as is used for smoothing: it is not investigated, therefore suppressed.

5.4.4.3 Derivatives

Let us take another simple kernel, consisting of a positive and a negative rectangular function as shown in Figure 5.41 (3). Its shape reminds us of the first derivative of a single band of a molecular spectrum. Indeed, the correlation of the absorption spectrum of the beech leaf with this kernel yields the first derivative exactly, as can be verified by comparison with the conventional calculation according to the formula $g(x) \sim f(x) - f(x - dx)$. The width of the kernel corresponds to "dx" in conventional derivative spectroscopy. The mathematical derivation of this finding could be left to the theoreticians. Of course, if we replace the kernel by its mirror image, the first derivative will be reversed, which is equivalent to multiplication by "−1".

If we choose a W-shaped kernel [Figure 5.41 (4)] we obtain the 4th derivative of the absorption spectrum. On this basis the identification of two pigments is given to demonstrate the analytical possibilities of derivative spectroscopy. Summarizing: folding integrals with kernels covering only a small spectral range (corresponding to "slitwidth") offer an elegant assessment of derivatives as well as smoothing functions. The kernel width is easily adjusted to the particular requirements (depending on the spectrum to be modified). In addition, even specific components in complex spectra can be extracted with proper kernel shapes: Is there a specific molecule, e. g., a cytochrome b, hidden in the spectrum?

5.4.4.4 Deconvolution

As discussed earlier, the process of folding takes place during measurement with the spectrophotometer where the effective kernel (obtained with identical entrance and exit slits) represents a triangular function (see Section 5.4.4.5). If we look closer at the formula of the folding integral in Eq. (5.20), a simple reversion of this formula appears to be feasible. On this basis we could recover spectral resolution which was lost by the practically unavoidable folding of the spectrum during the measurement with the slit function:

$$f(n) = h(n) - \sum_{m = 1}^{n} g(n) f(n - m) \tag{5.21}$$

Through this algorithm the "unfolded spectrum" $f(n)$ is derived from the measured spectrum $f(n - m)$ using the intrinsic, known kernel $g(n)$ of the spectrometer. The initial value $f(0)$ is simply a factor of normalization. For simplicity and the purpose of demonstration we choose for $g(n)$ a rectangular function of defined width instead of the correct but more complex triangular function. Thus, "defolding" the absorption spectrum of Didymium (filter BG 32, Schott, Germany) which was intentionally measured with large entrance and exit slits of 10 nm, yields the spectrum in Figure 5.42 (1). For comparison, the Didymium spectrum was scanned a second time with smaller widths of only 1 nm [Figure 5.42 (2)]. The low resolution spectrum [Figure 5.42 (1)] unfolded with a 10 nm kernel, i. e., back-calculated to a theoretical slitwidth of 0 [Figure 5.42 (3)], shows – expectedly – significantly more structure than the spectrum measured at 1 nm resolution [Figure 5.42 (2)]. Finally, Figure 5.42 (4) shows the 4th derivative of spectrum 1, which however does not offer more structure than spectrum 3 calculated back to slitwidth "0".

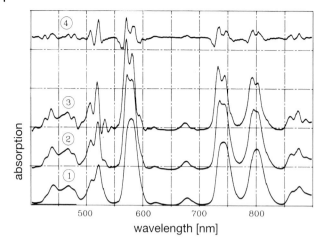

Figure 5.42 Absorption spectrum of Didymium (filter BG 32, Schott, Germany), measured with 10 (1) and 1 nm (2) slitwidths. "Unfolding" of spectrum 1 with a rectangular kernel of 10 nm yields spectrum 3. Curve 4 is the fourth derivative of spectrum 1 (dx = 10 nm), the fine structure of which is comparable to that of the unfolded spectrum (3); i.e., the methods are equivalent.

5.4.4.5 Fourier versus Correlation

In mathematical terms, Fourier analysis is also a type of correlation method. According to Fourier, curves of any shape $f(x)$ (which of course have to meet some postulations such as continuity) can be decomposed into an (infinite) series of sine and cosine functions. This is described by the Fourier theorem which is written in its most general, complex form as:

$$h(x) = \int_{x' = -\infty}^{+\infty} f(x')\, e^{-2\pi i x x'}\, dx' \tag{5.22}$$

with the Euler relationship $e^{ix} = \cos x + i \sin x$, with $I = \sqrt{(-1)}$. The analogy to Eq. (5.18) is easily recognized, the kernel $g(x)$ is "simply" replaced by the exponential function. If we omit the imaginary part of the formula that contains only information of the phase, Eq. (5.22) is largely simplified to the basic equation of Fourier transform spectroscopy that is easily applicable in practice:

$$h(x) = \int_{x' = -\infty}^{+\infty} f(x')\, \cos(-2\pi x x')\, dx' \tag{5.23}$$

An extremely important theorem of spectroscopy connecting Fourier transformation and the folding integral is documented in Figure 5.43. As mentioned pre-

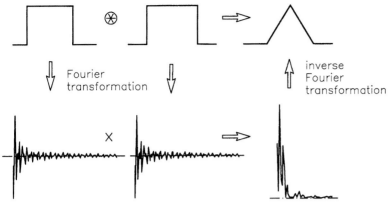

Figure 5.43 Representation of an important theorem of spectroscopy that connects Fourier transformation and a folding integral. The correlation of a rectangle with itself (autocorrelation) yields a triangular function, the Fourier transformation of a rectangular function sin(x)/x. Folding (*) of two functions corresponds to the product (x) of their Fourier transforms {in the example [(sin x)/x]²}. The inverse Fourier transformation of the latter again leads back to the triangular function.

viously, the correlation of a rectangular function with itself (autocorrelation) yields a triangular function, and the Fourier transformation of a rectangle according to Eq. (5.23) gives the function (sin x)/x. Now, the basic statement of the Fourier theorem is that the folding of two functions is equivalent to the product of their Fourier transforms. In the case of the rectangular function the product of their Fourier transforms yields [(sin x)/x]². The inverse Fourier transformation of the latter again leads back to the triangle which closes the circle of the arguments. Summarizing, theoretically it is does not matter in which space spectra are manipulated. However, in practice it could be significant.

Figure 5.44 compares smoothing of the absorption spectrum of a Didymium filter performed both (i) by the folding method and (ii) by Fourier analysis. Curve 1 shows the largely noise free spectrum and curve 2 the same spectrum with additional white noise added. Curve 5 is the Fourier spectrum of curve 2 but obviously with strong high frequency contributions. Removing the high frequencies corresponding to curve 6, the back transformation (inverse Fourier transformation) of curve 6 yields spectrum 3. Convolution of spectrum 2 with the given rectangular function generates spectrum 4. Thus, in practical work for the purpose of smoothing, the method of convolution appears preferable to the Fourier method, particularly for the reason that significantly less calculation time is required.

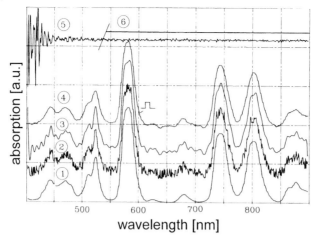

Figure 5.44 Comparison of smoothing of the absorption spectrum of a Didymium filter by folding and by Fourier analysis. Curve 1, Didymium absorption spectrum; curve 2, a similar spectrum measured with a strong noise component; curve 5, Fourier spectrum of curve 2 (with high frequency components). If we cut out the higher frequency components as indicated in curve 6, the inverse Fourier transformation yields curve 3 (the only contribution left of the cut-off point in curve 5). Folding of spectrum 2 with the given rectangular kernel yields spectrum 4.

5.4.5
Spectral Multicomponent Analysis

Often we have the task of determining the presence and/or concentration of individual components in an unknown sample mixture. If these components do not show any interaction this task is simply a mathematical problem. Including, for simplicity, the sample thickness into the absorbance value A, according to the Lambert–Beer law [Eq. (5.5)] for j components in the mixture we obtain for wavelength λ_i:

$$A(\lambda_i) = \sum_{j=1}^{n} \varepsilon_j(\lambda_j) \times c_j$$

or in short

$$A(\lambda_i) = \sum_{j=1}^{n} \varepsilon_{ij} \times c_j \tag{5.24}$$

or even shorter written in the matrix/vector form

$$A_i = \varepsilon_{ij} \ c_j \tag{5.25}$$

Solving the system of equations in Eq. (5.25) is simplified by Cramer's rule of matrix algebra: assuming three components "j" and three wavelengths "i" (checkpoints), for the concentration of the first component we obtain:

$$
c_1 = \frac{\begin{vmatrix} A_1 \varepsilon_{12} \varepsilon_{13} \\ A_2 \varepsilon_{22} \varepsilon_{23} \\ A_3 \varepsilon_{32} \varepsilon_{33} \end{vmatrix}}{\begin{vmatrix} \varepsilon_{11} \varepsilon_{12} \varepsilon_{13} \\ \varepsilon_{21} \varepsilon_{22} \varepsilon_{23} \\ \varepsilon_{31} \varepsilon_{32} \varepsilon_{33} \end{vmatrix}}
\tag{5.26}
$$

The given numerator-determinant, which in the first column corresponds to the three absorbance values, is divided by the determinant of the original extinction coefficient matrix. Corresponding equations hold for concentration c_2 and c_3. For reasons of space, we will not delve further into the mathematical methods here, and also will not discuss numerous alternative methods of multicomponent analysis, such as the iteration procedure. Extensive experiments and complex calculations have yielded impressive results, however in most cases as types of "model analysis", with "suitable" components and well defined concentration ratios.

In practical work we have to be more modest: of course, the simplest and most precise determination is that of only two components. Typically, a three component determination is not expected to yield errors of less than 5%. Only in very special cases is the determination of more than five components feasible, at all. For all practical possibilities, use of the "method of determinants" as described above, a favorable spectral distribution of the individual components and appropriate selection of wavelengths and checkpoints where the total absorption is measured are all important.

5.5
Dual Wavelength Spectrophotometry

5.5.1
Introduction

In the early 1950s Britton Chance (*Sci. Instrum.* 1951, *22*, 634) suggested a novel method for measuring very small absorption changes in highly scattering and turbid samples, which is an essential analytical procedure for particular applications in biochemistry and physiology; important results and discoveries are based on the technique termed *dual wavelength spectroscopy*. The basic idea is simple. While in double beam spectroscopy two cuvettes – sample and reference – are irradiated with one single but variable wavelength {$A = \log [I_0(\lambda)/I(\lambda)]$, double beam, Figure 5.12}, in dual wavelength absorption spectrophotometry only one cuvette – the sample – is irradiated with two different wavelengths, and the difference in absorbance between λ_1 and λ_2, i.e., $\Delta A = A(\lambda_1) - A(\lambda_2)$ is determined. The method is typically restricted to two predetermined but fixed wavelengths, where one of these corresponds to an isosbestic point of the pigment system to be measured, the other to the absorbance maximum of a system component. The measuring signal is – so to speak – anchored to a reference signal of the same sample thus representing the required absorbance change. Thus, all other uncertainties such as long term and short term shifts or noise components common to both wavelength positions are largely compensated for. The closer the measuring and reference wavelengths λ_1 and λ_2 are chosen to each other the better the compensation process. In contrast to double beam spectrophotometry, in dual wavelength spectrophotometry we are not so much interested in the absolute values of absorption as in their relative change with time, i.e., the kinetics. Depending on the sample, the method is more sensitive than conventional double beam spectrophotometry by a factor of 100 to 1000.

5.5.2
Methods

The standard set-up is shown in Figure 5.45 (Schmidt, W. J. *Biochem. Biophys. Methods* 1980, *2*, 171). In contrast to light throughput, the wavelength resolution is of secondary significance. Thus narrow banded interference filters are excellent "monochromators" in dual wavelength spectrophotometry. They allow higher light throughput than grating monochromators. Two light beams with λ_1 and λ_2 illuminate the sample alternatively via a vibrating mirror (vibration frequency between 30 and 100 Hz). Finally, the signals $I(\lambda_1)$ and $I(\lambda_2)$ corresponding to wavelength positions λ_1 and λ_2 are converted by means of a phase sensitive amplifier (*lock-in*, see Section 5.2.4), and further processed by an on-line computer.

The lock-in amplifier allows a valuable signal of up to 20 000 times smaller than the noise contribution to be filtered out of the measured total signal, when the signal to be measured can be sequentially repeated and synchronized with a reference frequency, as in the dual wavelength spectrophotometer. In principle, the lock-in amplifier is an ac rectifier, which filters the "real signal" out of the bulk signal by phase selective detection. The pulsating dc signal is finally converted into a stable dc signal and amplified (Figure 5.46). Lock-in amplifiers work between 0.2 and 200 kHz and are used either in form of a large, expensive, high quality and versatile research device or, for special applications, such as miniaturized and low cost modules with a fixed phase and can be as small as a cigarette box.

The absorption changes to be detected are typically very small. Then absorption and transmission changes are practically identical and we could simply omit the logarithmic amplifier obligatory in conventional double beam spectrophotometry.

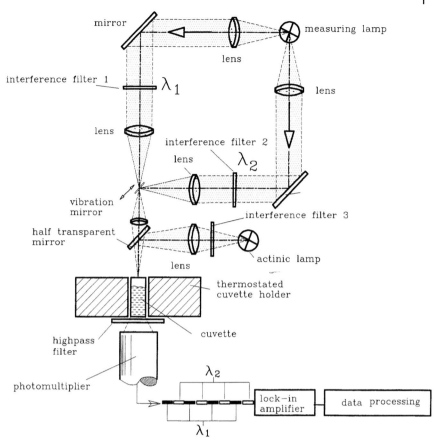

Figure 5.45 Set-up of a conventional double beam spectrophotometer. Two orthogonal beams emanating from the measuring lamp are separated, collimated and dispersed by interference filters λ_1 and λ_2, and finally focused onto a small vibrating mirror (typical vibration frequency 120 Hz). The generated wavelength sequence $\lambda_1, \lambda_2, \lambda_1, \lambda_2...$ of light pulses is largely absorbed particularly by optically dense samples, and the low intensity, transmitted light detected by a photomultiplier. Sub- sequently the signal is further processed by a *lock-in* amplifier and the final computer. Using, e.g., a half mirror and appropriate blocking filters between the sample and detector, extremely small light-induced (by an actinic lamp, interference filter 3) absorbance changes (ΔA < 0.0001) are detectable on a strong optical background ($E \gg 4$). A temperature controlled cuvette holder guarantees a constant temperature.

Differentiation of Eq. (5.6) yields

$$dA = - d \log T = -0.43 \frac{dT}{T}$$

Finally replacing differentials by differences we obtain

$$- \Delta A \propto 0.43 \frac{dT}{T}$$

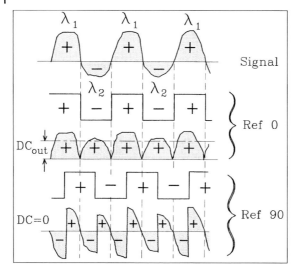

Figure 5.46 Representation of the operation of phase- and frequency-selective amplifiers (*lock-in*), as used in double beam and dual wavelength spectrophotometers. Signals at wavelengths λ_1 and λ_2 are attenuated differently by the sample and thus generate the signal shown (top). With corresponding phase selection of the amplifier (here 0°) we obtain the dc difference signal required ($I_{\lambda 1} - I_{\lambda 2} = dc_{out}$). With a detection phase of the amplifier that is shifted in relation to the measuring signal by 90° the exit signal disappears (dc = 0). Lock-in-amplifiers most often work as phase- and frequency-selective.

As predicted $\Delta T \ll T$, i.e., $\Delta A \ll 1$. Therefore, the transmission change ΔT is proportional to the absorption change (ΔA).

One important application of dual wavelength spectroscopy in photochemistry and photobiology needs to be mentioned. Because the measuring light signal is modulated precisely with one defined frequency, the phase and frequency selective lock-in amplifier can be adjusted exactly to just this one frequency while all other frequencies and particularly dc signals are not amplified at all. Thus, in parallel to the detected weak measuring light beam, a much stronger actinic light beam capable of inducing photochemical processes could impinge onto the (*in vivo*) sample if the detector is protected from the actinic light by suitable blocking filters. In this way, e.g., very small *light-induced absorption changes* (*LIACs*) have been continuously and kinetically detected in living cells in order to study the primary steps of sensory transduction in photobiology.

The **m**icro **d**ual **w**avelength spectrophotometer (MDWS, Figure 5.47, Schmidt, W. J. *Biochem. Biophys. Methods*, **58**, 2004, 15–24) is a miniaturized but highly versatile modification of the set-up described above. The two light passes including interference filters defining the two wavelengths of the observing and vibrating mirror are replaced by electronically chopped light emitting diodes (LEDs) of appropriate wavelengths. Utilizing a chopper frequency in the kHz range and a function generator chip allow the response time to be further decreased consider-

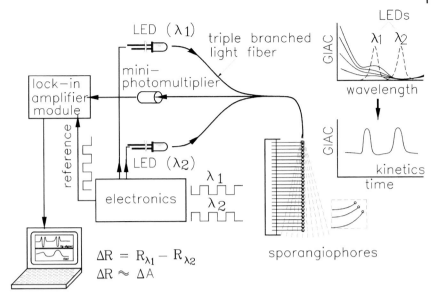

Figure 5.47 Micro-dual wavelength spectrophotometer (MDWS), demonstrating the kinetic measurement of minute reflection changes. Alternating light of two color-selected LEDs with emission wavelengths of λ_1 and λ_2 is fed into two arms of a triple branched light fiber and illuminates a "hedge" of sporangiophores of the fungus *Phycomycetes*. The reflected light is collected by the third arm of the light fiber bundle and guided back to the photocathode of a miniature photomultiplier. After pre-amplification the signal is fed to a lock-in amplifier (LIA), which receives its reference voltage from the electronics generating the alternating voltage operating the LEDs. After conversion into a dc signal, the signal is fed to a computer.

ably (<ms) and increase sensitivity (by decreasing the $1/f$ noise: see Section 5.2.5). The device is small, robust and without any mechanically moving parts and thus allows measurements under the rugged conditions, such as in parabolic flights.

5.5.3
Measurement of Gravity-induced Absorbance Changes *In Vivo*

Minute gravity-induced absorption changes (GIACs) were measured for the first time in sporangiophores of the fungus *Phycomycetes*. Measurements were made on an Airbus 300-ZERO G in the millisecond time range by reflection spectroscopy, utilizing LEDs of 460 (blue) 665 nm (red), respectively. This special airplane operated by Novespace in Bordeaux and mainly financed by DLR (Deutsche Luft- und Raumfahrt) and ESA (European Space Agency) performs parabolic flights over the Mediterranean and the Atlantic Ocean, facilitating gravity changes

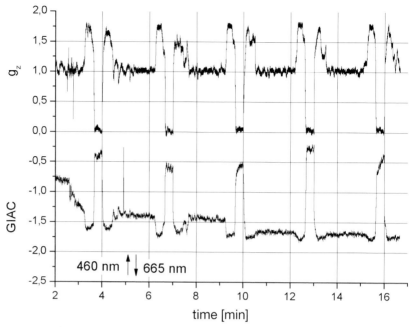

Figure 5.48 Measurement of gravity-induced absorbance changes (GIACs) of sporangiophores of the fungus *Phycomyces* by reflection (Figure 5.47) monitored for about 15 min and 5 parabolas on the "Airbus 300 ZERO G" (ESA campaign 32, March 2002, in Merignac, Bordeaux, France). The top part of the figure shows the measured values of gravity expressed in g (acceleration on the earths surface; 1 g equals 9.81 m s^{-2}). The lower part of this figure shows the accompanying GIACs which either reflect absorption increase in the blue or decrease in the red spectral range.

between 0 and 1.8 g (1 g = 9.81 m s^{-2}). As shown in Figure 5.48 the GIACs were correlated with simultaneously measured gravity changes during parabolic flights, yielding new insight into the primary reactions of graviperception in fungi (Schmidt, W., Galland, P. *J. Plant Physiol.* 2004, *135*, 183–192). For conversion of reflection into absorption spectra see Section 8.5.1.

5.6
Spectrophotometers for Specific Applications

Since the end of the last world war various optical spectroscopic analytical devices have been developed for many applications. The technical and patent literature in this field is overwhelming. Many splendid ideas –such as Fourier or correlation analysis – were developed early on, however their broad verification had to await today's technical capabilities.

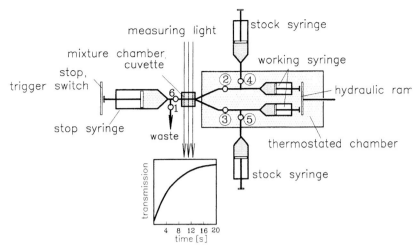

Figure 5.49 *Stopped-flow* spectrophotometer for monitoring
fast chemical reactions. For details see text.

Particular recent advances in optoelectronics and microcomputer techniques
have led to a continuously spreading palette of spectrophotometric applications,
last not least stimulated by custom made set-ups that are not available commercially. Below we briefly discuss some specific devices in absorption spectroscopy
for fast spectral and kinetic measurements.

5.6.1
Stopped-flow Spectrophotometry

The most often used method for monitoring "fast kinetics" in chemistry and biochemistry is the *stopped-flow* technique (Figure 5.49). Two components that are to
react with each other are placed in separate "stock syringes" (ca. 25 mL). By means
of a hydraulic system the liquids are first transferred into the smaller "working
syringes". Using a hydraulic ram the liquids are then shot into the measuring
chamber, which, in turn, is connected to a "stop syringe". While the "old" liquid
is poured into the "stop syringe" the measuring chamber/cuvette is filled with
fresh components (mixing), as the ram of the stop syringe is abruptly "stopped"
by a stop and thereby the data acquisition is started (computer, oscilloscope).

Air bubbles are the most difficult problem with this technique and have to be
rigorously avoided. In order to control the temperature the whole set-up is in a
temperature controlled water bath. Well designed stopped-flow spectrophotometers allow time scales between the mixing and stopping (start of measurement)
of less than 1 ms. In the example in Figure 5.49 the (relatively slow) transmission
kinetics of the reaction of deuteroferrihem with hydrogen peroxide (H_2O_2) is
shown (modified from Jones et al., *Biochem. J.* 1975, *135*, 361).

5.6.2
Fast Spectral Monitoring

For investigations of fast, e. g., enzyme-catalyzed, chemical reactions monitoring at only one specific wavelength is no longer sufficient as the composition of the reaction mixture changes in unknown way: fast *spectral* monitoring of the kinetics is required. In the patent literature, manifolds, often with sophisticated set-ups, working down to the millisecond range are presented. We can distinguish three types of "fast" spectrophotometers:

- Array, also termed "simultaneous" spectrophotometers such as the **o**ptical **m**ultichannel **a**nalyzer (OMA) discussed previously which do not need any mobile parts and which cover a whole spectral range simultaneously. The optical multichannel analyzer in its modern form is highly mimiaturized (Figure 5.7).
- The *Michelson interferometer* (Section 5.4, Figure 5.8), which monitors spectra simultaneously, however it requires refined optomechanical scanning components.
- *Rapid scanning spectrometers*, in which the conventionally dispersed spectrum is sequentially and mechanically scanned. This is accomplished either by rapid rotation of a dispersive grating or prism, or of a secondary deflecting element in the light path such as a polygon or scanner mirror, or of a rotating slit. In contrast to simultaneous spectrometers these offer the greatest possible sensitivity as high quality detectors with a large sensitive area and gratings or prisms with high light throughput can be used (cf. Section 5.2.1), and the photodetector can be adjusted with respect to price and performance to the individual requirements.

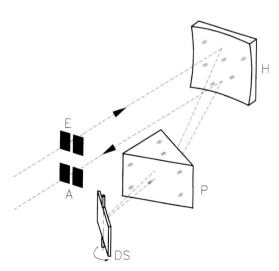

Figure 5.50 Rapid-scanning spectrophotometer based on a prism-monochromator in a Littrow arrangement. The rotational or galvanometer mirror (DS) scans the spectrum as generated by the prism (P) (double transmission → doubling of dispersion). The concave mirror (H) allows the correct imaging of the entrance (E) onto the exit slit (A).

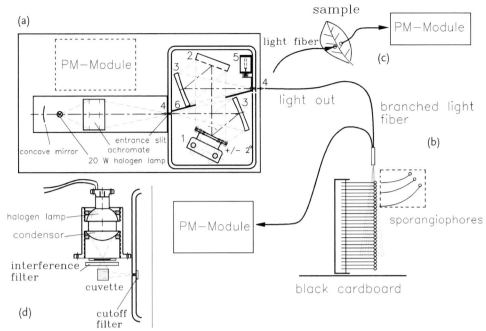

Figure 5.51 A modified Czerny–Turner rapid-scanning spectrometer (RSS) with vibrating mirror. Light is focused onto the entrance slit (4). Emanating from the entrance slit (4) it is collimated by the concave mirror (3) and by the scanning mirror (1) and deflected onto the plane grating (2). Finally, the dispersed light is focused onto the exit slit (4) by means of a second concave mirror. The position of the grating in the idle position of the scanning mirror defines the midpoint of the spectrum, the amplitude of the scanning mirror the spectral range. The monochromator (a) is operated in various spectroscopic modes. Absorbance of a typical glass cuvette of 1×1 cm is measured by a simple light fiber arrangement (not shown). Reflection measurements are easily performed using a bifurcated light fiber system (b). A light fiber clip is used for measuring absorption spectra of solid samples such as leaves, flowers or earlobes (c). For measuring fluorescence (emission) spectra the sample is placed in front of the entrance slit and the excitation wavelength is determined by an appropriate interference filter. The excitation light is blocked from entering the monochromator by a suitable high path filter (d).

Figure 5.50 shows a rapid scanning spectrometer based on a prism monochromator in a Littrow set-up, where a rotating mirror (DS) is scanning the spectrum. This set-up has the disadvantage that on a complete rotation of the mirror only a minute angle range (e. g., 5°) is utilized for spectral acquisition; the dead time of the spectrometer, i. e., a long time gap between spectra, of approximately 95 % is significant. The same holds true in rapid-scanning spectrophotmeters, where the grating rotates.

This disadvantage is overcome by the grating set-up with a vibrating mirror (laser scanner) in Figure 5.51, particularly when the spectrum is monitored in both directions of motion of the scanning mirror (Schmidt, *Biochem. Biophys. Methods* **58**, 125–137, 2004). This is a modified Czerny–Turner set-up (Figure 5.6), where the scanning mirror (1) is located in the collimated (parallel) light path and the fast variation of the entrance angle ε [Eq. (3.33)] on the grating (2) causes the sequential dispersion at the exit slit (4). The vibrational amplitude defines the spectral range, the position of the grating the spectral mid-point. Scan times for a whole spectrum in the sub-millisecond range are feasible. The monochromator (a) is operated in various spectroscopic modes. Absorbance of a typical glass cuvette of 1×1 cm is measured by a simple light fiber arrangement (not shown). Reflection measurements are easily performed using a bifurcated light fiber system (b). A light fiber clip is used for measuring absorption spectra of solid samples such as leaves or earlobes (c). For measuring fluorescence (emission) spectra the sample is placed in front of the entrance slit and the excitation wavelength is determined by an appropriate interference filter. The excitation light is blocked from entering the monochromator by a suitable high path filter (d).

For demonstration purposes, Figure 5.52 shows the spectral time course of the onset light pulse of a small neon bulb operated with a power frequency of 50 Hz. The scanning frequency of the freely swinging scanning mirror (1, Figure 5.52) is slightly higher at 50.5 Hz with the result of a beat frequency of approximately 0.5 Hz covering a whole onset pulse of 10 ms duration.

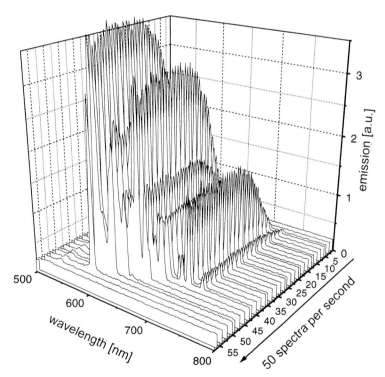

Figure 5.52 Series of 50 subsequent emission spectra of a neon-gas lamp operated with a 50 Hz line frequency. The scanning frequency of about 50.5 Hz causes an interference pattern, which allows a single "on-pulse" lasting approximately 10 ms to be resolved spectrally.

Another *rapid-scan* method is used with great success by the manufacturer OLIS in the USA. Two conventional Czerny–Turner monochromators are set up in series, in a similar way to the double monochromators, with the difference being that both dispersion gratings are arranged as mirror images of each other. The exit slit of the first monochromator is simultaneously the entrance slit of the second monochromator and is represented by a rapidly rotating "slit disk" (62.5 Hz, 16 slits). Thus the spectrum appears – according to the linear dispersion – as a colored band located in the intermediate image area. It is scanned spectrally within 1 ms and, independently of the wavelength, imaged onto the fixed exit slit of the second monochromator.

In contrast to slow scanning double beam monochromators, rapid scanning single beam monochromators allow monitoring of individual spectra with a high repetition rate of 50 to 1000 spectra per second. Again, lock-in amplifiers significantly increase the signal to noise ratio. In addition, owing to the high signal frequency, condenser coupling significantly reduces $1/f$ noise and the large temperature drift of PbS detectors being utilized in near-infrared spectroscopy (see Chapter 10).

5.7
Selected Further Reading

Blass, W. E., Halsey, G. W. *Deconvolution of Absorption Spectra*, Academic Press, London, **1981**.

Bonsquet, P. *Spectroscopy and its Instrumentation*, Adam Hilger, Bristol, **1971**.

Britton, G., ed., *Carotenoids: Spectroscopy*, Birkhäuser Verlag, Basel, **1995**.

Egan, W. E. *Phase-Lock Basics*, John Wiley & Sons, New York, **1998**.

Görög, S. *Ultaviolet-Visible Spectrophotometry in Pharmaceutical Analysis*, CRC Press Boca Raton Fl 33487, **1995**.

Gottwald, W., Heinrich, K. H. *Spektroskopie für Anwender*, Wiley-VCH, **1998**.

Howell, J. A. Ultraviolet and Absorption Light Spectrometry, *Anal. Chem.* **1998**, *70*, 107R-118R (review article).

Jansson, P. A., ed., *Deconvolution of Images and Spectra*, 2nd edn., Academic Press, London, **1996**.

Karlberg, B., Pacey, G. E. *Flow Injection Analysis, A Practical Guide*, Elsevier, Amsterdam, **1989** (general introduction to flow techniques).

Kortüm, G. *Kolorimetrie, Photometrie und Spektrometrie*, Springer-Verlag, Berlin, **1962** (an excellent textbook on the theory of optical spectral analysis, in German).

Lang, L., ed. *Absorption Spectra in the Ultraviolet and Visible Region*, Akadémiai Kioadó, Budapest, **1966**.

Malmstadt, H. V., Enke, C. G., Crouch, S. R., Horlick, G. *Optimization of Electronic Measurements*, The Benjamin Publishing Company, Menlo Park, CA, **1974**.

Norris, K. H., Butler, W. L. Technique for Obtaining Absorption Spectra in Intact Biological Samples, *Inst. Radio Engs. Trans. Bio-Med. Electron*, **1961**, *8*, 153-157 (one of the mostly cited publications on the *in vivo* optical spectroscopy of biological specimens).

Orchin, M., Jaffé, H. H. *Symmetry, Orbitals, and Spectra (S.O.S.)*, John Wiley & Sons, New York, **1971**.

Pavia, D. L. *Introduction to Spectroscopy: A Guide for Students of Organic Chemistry*, Saunders College Publishing, **1997**.

Peiponen, K.-E., Vartiainen, E. M., Asakura, T. *Dispersion, Complex Analysis and Optical Spectroscopy: Classical Theory*, Springer Tracts in Modern Physics, 147, Springer-Verlag, Berlin, **1998**.

Perkampus, H.-H. *UV-VIS-Spektroscopy and its Applications*, Springer-Verlag, Berlin, **1992**.

Pretsch, C., Seible, S. *Tables of Spectral Data for Structure Determination of Organic Compounds (NMR, IR, MS, UV/VIS)*, 2nd edn., Springer-Verlag, Berlin, **1989**.

Smith, B. C. *Fundamentals of Fourier Transform Infrared Spectroscopy*, Lewis Publishers, **1995**.

Smith, K. C., ed., *Photochemical and Photobiological Reviews*, Vol. 1–7, Plenum Press, New York, **1983** (offers a range of optospectroscopic methods).

Steward, E. G. *Fourier Optics, An Introduction*, Ellis Horwood, Chichester, **1987**.

Struve, W. S. *Fundamentals of Molecular Spectroscopy*, John Wiley & Sons, New York, **1989**.

Sommer, L. *Analytical Absorption Spectrophotometry in the Visible and Ultraviolet, The Principles*, Elsevier, Amsterdam, **1989**.

Svanberg, S. *Atomic and Molecular Spectroscopy*, Springer-Verlag, Berlin, **1991**.

Sweedler, J. V., Ratzlaff, K. L. *Charge-Transfer Devices in Spectroscopy*, John Wiley & Sons, New York, **1994**.

van Holde, K. E. *Physical Biochemistry*, foundation of modern biochemistry series, Prenctice-Hall, Inc., Englewood Cliffs, New Jersey.

Wilson, R. G. *Fourier Series and Optical Transform Techniques on Contemporary Optics: An Introduction*, John Wiley & Sons, New York, **1995**.

6

Luminescence Spectrophotometry

Nihil tam difficile est, quin quaerando investigari posset
(Terence)

Nothing is so difficult that it cannot be investigated

6.1
Introduction

The term luminescence describes the emission of light by electronically excited atoms and molecules. Electronic excitation requires the input of energy, which can occur by different modes. A prefix is usually added that defines the excitation mechanism in the system: e. g., the light emission of glow-worms is due to biochemical processes, and is therefore termed bioluminescence. Chemical reactions that are accompanied by light emission give rise to chemiluminescence. There are also certain pigments that emit light upon exposure to ionizing radiation, most often induced by tritium (^{3}H): they exhibit radioluminescence. A series of further phenomena such as mechano-, thermo- or electroluminescence will simply be mentioned here (cf. Figure 1.1). Many molecules and molecular systems emit light upon the absorption of light and this is termed photoluminescence. Fluorescence and phosphorescence are particular types of photoluminescence and will be the focus of the current chapter. The term fluorescence is derived from the mineral fluorspar, the material in which the phenomenon was first observed in 1565. Measurements of phosphorescence were first performed in 1858 by E. Becquerel (1820–1891). Advancing the first observations of P. Heinrich (1758–1825) he demonstrated that after excitation by light, all alkali and alkali earth metals exhibited phosphorescence. There are also other forms of light emission, such as that of an incandescent light bulb (black body radiation), or synchrotron or Cerenkov radiation that are not the result of electronic excitation.

Luminescence can be analyzed in different ways. Molecules and materials can be identified, and in some situations, quantified by the characteristics of their luminescent emissions. Luminescence can also be used to probe the characteristics of the environment of the emitting lumophore. The luminescent probe can be extremely sensitive to even the smallest changes in a complex system such as a

Optical Spectroscopy in Chemistry and Life Sciences. W. Schmidt
Copyright © 2005 WILEY-VCH Verlag GmbH & Co. KGaA, Weinheim
ISBN 3-527-29911-4

Henri Becquerel (1852–1908)

living cell. The analysis of these signals is the task of the spectroscopist. The excellent, but somewhat old, book by Rendell (1987) with its additional computer-guided supplement (1991), is highly recommended for practical aspects of fluorescence and phosphorescence. For a theoretically more profound introduction to the basics of fluorescence the textbooks by Parker (1968), Lakowicz (1983), Demas (1983) and Guilbault (1990 ed.) are highly recommended. In addition, fluorescence microscopy allows the spatial localization of fluorescent components in the cells of plants and animals to be imaged, as discussed in great detail in the textbooks of Göke (1988) and Piller (1977).

The basic difference between absorption and luminescence spectroscopy is reflected in the measuring method. The absorption of a species is defined in Eq. (5.5) by $A(\lambda) = \log I_0/I = \varepsilon(\lambda) \times c \times l$, where I_0 is the intensity of the analyzing light incident on the sample, I the intensity of light transmitted through the sample, c the concentration of the absorbing species and $\varepsilon(l)$ the molar absorption coefficient. What is actually *measured* is the portion I of the incident light I_0 that is not absorbed. Accordingly, the determination of A is an indirect analytical method. On the one hand this allows the elegant elimination of the intrinsic dispersion of the absorption spectrometer through comparison with a reference sample and the *absolute* determination of absorption values (A or OD in the old nomenclature of optical density). However, absorption measurements, due to the differential nature of their determination, ($A = \log I_0 - \log I$) are about 2–3 orders less sensitive than luminescence measurements. With weakly absorbing samples one is faced with the problem of detecting a small difference between two large signals: with strongly absorbing samples the intensity I of the transmitted light is very small and has to be subtracted from a large reference signal (cf. Section 5.2.6).

In contrast, luminescence measurements are *relative* but direct measurements: only light emitted from the sample is detected, yielding an immediate spectroscopic fingerprint. However, as the technique does not utilize a reference in the apparatus, the correction of a luminescence spectrum for the dispersion (non-linear wavelength sensitivity) of the apparatus is somewhat problematic because an objective procedure – analogous to absorption spectroscopy – does not exist. Therefore, fluorescence data inherently exhibit large errors with respect to wavelength (± 5 nm) as well as to signal size ($\pm 10\%$). Nevertheless, compared with absorption spectroscopy, fluorimetry is more informative with respect to the photophysics of a molecule. While in absorption spectroscopy the only measurement parameter is the absorption $A(\lambda)$, fluorescence spectroscopy allows the measurement of a series of independent parameters, such as quantum yield (also termed quantum efficiency) Φ, distance R_c for radiationless energy transfer, lifetime τ, polarization p or anisotropy r, Stokes shift $\Delta\lambda$, and many others, each of which provides some information about the system.

6.2
Mechanisms Involved in Fluorescence

6.2.1
Origins of Fluorescence and Phosphorescence

The origin of fluorescence and phosphorescence has already been discussed in Chapter 2. They differ merely in the electronic nature of the excited states from which the light is emitted, i. e., they originate in the first excited singlet (S_1) and triplet state (T_1), respectively (cf. Figure 2.31). As a more general definition, fluorescence is the radiative transition between two states of identical multiplicity and phosphorescence is between two states of different multiplicity; thus, one can observe triplet–triplet fluorescence, doublet–doublet fluorescence and other exotic emissions. If a molecule is excited to a higher electronic and vibrational state it typically undergoes a radiationless relaxation to the S_1 state within approximately 10^{-14} s. The electronic ground state S_0 can be then reached from this state via various decay channels, of which the fluorescence channel is of interest to us in the present context (Figure 6.1), i. e., the energy difference ($S_1 - S_0$) is emitted in the form of a quantum or photon of light.

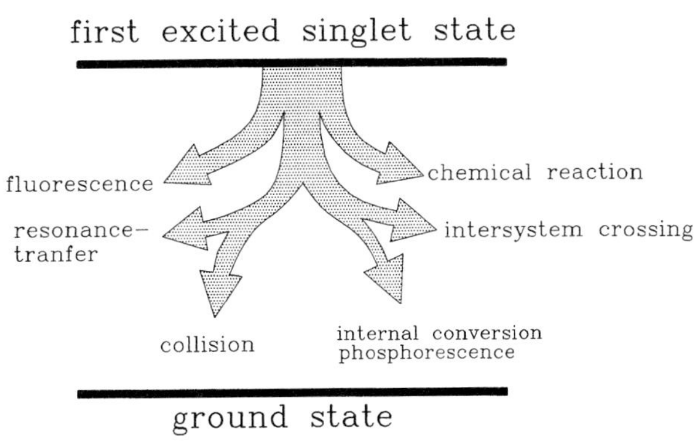

first excited singlet state

fluorescence

resonance–tranfer

chemical reaction

intersystem crossing

collision

internal conversion
phosphorescence

ground state

Figure 6.1 Possible relaxation channels of the first excited singlet state.

Azulene is the best known exception where fluorescence emission emanates from S_2 rather from S_1.

If a molecules does not fluoresce upon transition to the electronic ground state the excitation energy must be dissipated in some other way. Possibilities include: (1) radiationless energy transfer to other molecules (fluorescence quenching), (2) photochemical reactions, but most importantly (3) by internal energy transfer. The last involves internal energy redistribution within the same molecule. Here we can distinguish the so-called *internal conversion* (*IC*), where the molecule maintains its spin state but merely falls to the lowest vibrational level of the same electronically excited state dissipating vibrational energy (heat), and the so-called *intersystem crossing* (ISC) from the first excited singlet to the first excited triplet state, accompanied by a spin flip ($S_1 \rightarrow T_1$). Finally, from here it relaxes again either by light emission, i.e., phosphorescence ($T_1 \rightarrow S_0$), or radiationless by collision, possibly connected with chemical reactions, back to the ground state S_0 (see Figure 2.31).

These processes can be, greatly simplified, visualized as follows. In a rigid molecule the energy, as introduced into the molecule by a photon, redistributes immediately across the whole system and is then largely emitted again as light. However, within a flexible molecule the energy is predominantly dissipated through radiationless transitions, as shown by the following crude macroscopic analogy. If a steel ball falls onto a solid steel plate its kinetic energy immediately distributes across the whole plate and is re-emitted by a load bang (sound energy). However, if the ball hits a lead plate its energy is completely and largely soundless as it is absorbed without distribution of energy across the plate (cf. fluorophores in Figure 6.10).

Fluorescence is therefore merely the inverse of absorption. This statement is based on the famous law of Gustav Robert Kirchhoff (1859). According to this law, each atom that is a strong absorber at a specific wavelength is in principle an equally strong emitter at the same wavelength (Figure 1.4). This law was later proven theoretically by Albert Einstein. In principle, absorption and emission spectroscopy are capable of delivering the same information. However, in the field of molecular spectroscopy the practical differences are tremendous. In order to understand these and to utilize the advantages of fluorescence spectroscopy we must discuss the theory in more detail.

6.2.2
Potential Diagrams

Potential energy diagrams (see Figure 2.26) can be used to illustrate the photochemical and photophysical properties of individual molecules and more complex systems with respect to light emission. Thus, Figure 6.2 shows the potential energy of the ground state and the first singlet and triplet states as a function of nuclear distance for a hypothetical diatomic molecule. Within the resulting potential

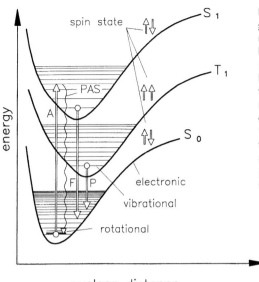

spin state

S_1

T_1

PAS

A

S_0

F | P — electronic

vibrational

rotational

energy

nuclear distance

Figure 6.2 Typical potential diagram of a hypothetical two-atomic molecule. Excited states as marked by their electron configurations (singlet S_0, S_1; triplet T_1), exhibit a large distance for the nuclei. The total energy is the sum of electronic, vibrational and rotational contributions, leading to typical "bandspectra" of molecules. The transitions corresponding to absorption (A), fluorescence (F), phosphorescence (P) and photoacoustic (PAS) are indicated.

well the possible energy states are in the form of discrete vibrational levels. Associated with each individual vibrational level – with even smaller energetic separations – are a series of rotational levels. Electronic transitions take place so rapidly that the nuclear separation distance does not change and the transitions are "vertical" on the potential energy diagrams. This is the Franck–Condon principle. These correspond to absorption (A), fluorescence (F) and phosphorescence (P). Shape, size and position of molecular luminescence spectra are largely modified by the fact that electronic transitions take place predominantly to higher vibrational excited target states

In a more general sense, photoacoustic spectroscopy (PAS) can by visualized as a type of emission spectroscopy, where the energy emitted on electronic transition is not detected in the form of light but heat (indicated by the wavy curve). Owing to its growing importance, particularly in the life sciences, it will be discussed in a separate chapter (Chapter 7).

6.2.3
From Potential Diagrams to Spectra

Utilizing the potential diagrams in Figure 6.2 we can construct the corresponding Jablonski diagrams, to show the basis of absorption and fluorescence emission spectra (scheme in Figure 6.3; phosphorescence follows an analogous behavior). The energy spacings of the vibrational levels on the lower electronic levels S_0 and S_1 (and T_1) are very similar. In this hypothetical example the Franck–Condon principle dictates that the $0 \rightarrow 2$ transition is the strongest, i. e., maximum absorption. Following excitation, molecules undergo rapid radiationless relaxation

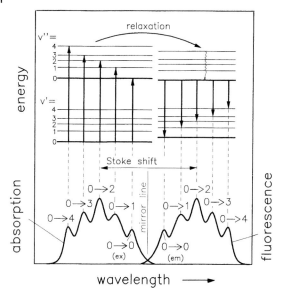

Figure 6.3 Deduction of absorption and fluorescence spectra of a molecule from the corresponding Jablonski energy diagram (schematically). The vibrational levels of various electronic states are comparable and explain the mirror image character of both spectra. The electronic states S_0 and S_1 are, as a function of the transition direction ($S_0 \times S_1$), "high" to different extents (cf. Figure 6.4). Thus their mirror image is only an approximation. The *Stoke's shift* is defined as the wavelength difference of maximum signal height of absorption and fluorescence spectra (here $v = 0 \rightarrow v = 2$).

via internal conversion and heat dissipation to the vibrational ground state of S_1 from where the electronic ground state S_0 is finally reached. However, fluorescence is only observed when the first electronic state (S_1) is not depopulated in a time frame shorter than that corresponding to the intrinsic radiative lifetime τ (see below). This can result from certain external influences or intrinsic molecular processes (see Figure 6.1). Figure 6.3 demonstrates why absorption (here S0 $\rightarrow$ S1) and emission spectra ($S_1 \rightarrow S_0$) are often mirror images: on electronic transitions the corresponding vibrational levels are reached in "reverse" order. Compared with absorption spectra, fluorescence and phosphorescence spectra are always bathochromically shifted (i.e., towards longer wavelengths). This is known as a Stokes shift. The Stokes shift denotes the observed difference in wavelength maxima of excitation and emission (Figure 6.3), and can differ from molecule to molecule. It is named after its discoverer Stoke, who formulated this law in 1858: the wavelength maximum of emission is always shifted to longer wavelengths compared with the absorption maximum.

6.2.4
Solvent Effects

Wavelengths and intensity of fluorescence emission strongly depend on extrinsic factors such as temperature, chemical bonds, viscosity and polarity of the solvent. With increasing solvent polarity the emission tends to get weaker and the emission spectra exhibit a red shift, as observed with flavin, for example (see Figure 6.6). Two reasons are responsible for these characteristics (Figure 6.4). On the one hand the electron cloud expands on electronic excitation, which typically increases the polarizability (change of polarity by extrinsic fields) and lowers the excitation energy. On the other hand, the process of electronic excitation takes place within approximately 10^{-15} s, while the reorientation of the nuclei and the solvent molecules takes much longer (approximately 10^{-10} s, Franck–Condon principle). For water with its very high dipole moment, this phenomenon is particularly pronounced. Fluorescence emission then occurs from this re-organized state on a picosecond to nanosecond time scale. This explains why the so-called $0 \rightarrow 0$ transitions of absorption and emission are energetically *not* identical (cf. different electronic levels in Figures 6.3 and 6.5). For flavin in water this energy difference equates to a wavelength shift of approximately 25 nm (Figure 6.6), but for anthracene in propanol the shift is only approximately 5 nm (Figure 6.5). The larger the polarity of the solvent and the polarizability of the fluorophore, the larger in general is the red shift in the spectra, as demonstrated by the near-UV band of the absorption spectrum of flavin as well as by its emission spectra in various solvents (Figure 6.6). Therefore, the mirror image character of emission and absorption spectra ($S_0 \rightarrow S_1$) is typically only approached, but is not perfect.

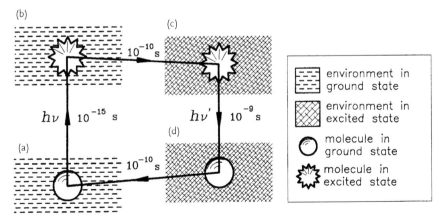

Figure 6.4 Schematic diagram of the interaction of a fluorophore with the solvent, both in the ground state and the first excited state; typical reaction times are included. During the "fast" transition the electrical dipole moment is changed causing, in turn, a "slow" reorientation of the solvent molecules. Correspondingly, the energy levels are changed, and as a consequence the spectra.

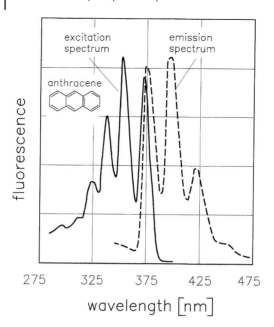

Figure 6.5 Absorption and fluorescence emission spectra of anthracene showing an approximate mirror character. The Stoke's shift is approximately 45 nm, both 0 × 0 transitions are significantly different (cf. Figure 6.4).

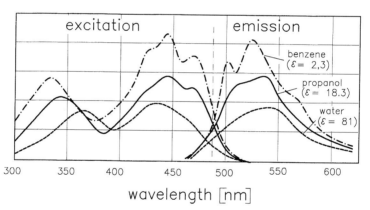

Figure 6.6 Absorption and corrected emissions spectra of flavin (excitation wavelength 380 nm, spectral resolution 5 nm) in various polar solvents (indicated as dielectric constant ε). Note the dependency of fluorescence quantum yield, vibrational bands and wavelength maxima on the polarity of the solvents.

However, spectral shifts are not always necessarily interpreted in terms of polarity changes. A hindering of the reorientation of solvent molecules around the excited molecule will result in a lower shift. Therefore, fluorescence spectra, as well as absorption spectra, at the lower temperature of –196 °C are hypsochromically (blue) shifted compared with those taken at room temperature. In addi-

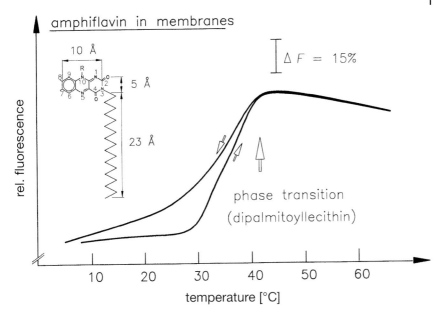

Figure 6.7 Fluorescence of amphiflavin being anchored via a long aliphatic hydrocarbon chain in position 7 in single shelled dipalmitoyl-lecithin membrane vesicles (cf. Figure 6.22), as a function of temperature (excitation wavelength $\lambda = 370$ nm, emission wavelength $\lambda = 540$ nm). Note the marked hysteresis, which reproduces well with cyclic temperature scans.

tion, a (partial) inhibition of molecular rotation can result in a sharpening of vibrational bands (cf. β-carotene, Figure 5.34, curves 4 and 5).

The effects of the microenvironment on fluorophores are utilized particularly in biological analysis. Thus, localization and the type of membrane binding of fluorophores can be investigated in more detail. For example, Figure 6.7 shows the temperature dependence of fluorescence of flavin (vitamin B2), which has been anchored by means of a long aliphatic C18-chain ("amphiflavin": "heads" hydrophilic, "tails" hydrophobic) in a model membrane of dipalmitoyl-lecithin. Upon heating to approximately 41 °C the membrane undergoes a reversible phase transition from crystalline to liquid crystalline, accompanied by a strong increase in fluorescence, in contrast to the expected fluorescence decrease in solution (quenching by collision, see below). This finding suggests that the flavin chromophore is shifted from the polar membrane surface into the more hydrophobic inner region of the membrane (cf. Figure 6.20).

6.2.5
Fundamentals of the Quantum Mechanics

In Chapter 2 we became acquainted with two types of electronic transitions in molecules, which are particularly important in luminescence spectroscopy, namely $n \rightarrow \pi^*$ and $\pi \rightarrow \pi^*$ transitions (Figure 2.20). Generally, for smaller molecules $n \rightarrow \pi^*$ bands are less energy rich than $\pi \rightarrow \pi^*$ bands, which does not necessarily hold true for more complex systems such as aromatic ketones. Thus, the lowest excited state in acetone corresponds to an $n \rightarrow \pi^*$ transition (270 nm) of a lone pair of electrons on the oxygen atom, which has a fairly low intensity ($\varepsilon = 20$ M^{-1} cm^{-1}), while the $\pi \rightarrow \pi^*$ transition (180 nm) corresponds to a π-electron of the C=O bond and gives rise to a stronger absorption band ($\varepsilon = 1000$ M^{-1} cm^{-1}). The quantum mechanical reason for this behavior is based on the differences in spatial overlap between the interacting orbitals; there is a strong spatial overlap of π- and π^*-orbitals whereas the overlap of n- and π-orbitals is much less significant, as presented for the carbonyl group of acetone (Figure 6.8) or the nitrogen atom in pyridine (Figure 6.9). For the same reason $n \rightarrow \pi^*$ transitions have a much longer lifetime than $\pi \rightarrow \pi^*$ transitions. Therefore, the probability of *intersystem crossing* into the triplet ladder with subsequent relaxation into the singlet ground state is significantly increased. For the same reason fluorescence generally takes place from $\pi \rightarrow \pi^*$ states, while ketones or nitrogen heterocycles, due to lower lying $n \rightarrow \pi^*$ transitions, do not show fluorescence, but more often phosphorescence. Phosphorescence and luminescence exhibit complementary behavior in analysis: with increasing probability of *intersystem crossing* fluorescence decreases, but phosphorescence increases.

Of more than 2000 pigments in the US *Color Index* only 200 are fluorescent! Neither hue nor "strength" of a color are coupled in any way to fluorescence capability. However, strongly fluorescent pigments are restricted to only a few classes including the xanthene-group (fluorescein, rhodamine), the acridine (euchrysine) group, cyanine dyes, tetrapyrroles or condensed aromatics There are no known cases of any azo-pigment being fluorescent.

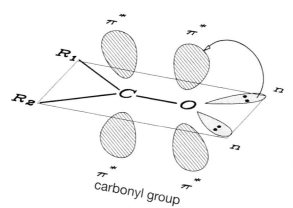

Figure 6.8 Three-dimensional representation of an energetically low lying $n \rightarrow \pi^*$ transition of the lone electrons of a carbonyl group. R_1 and R_2 are organic residues.

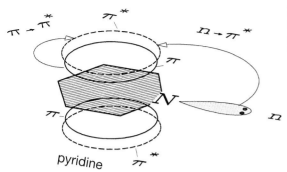

Figure 6.9 Equivalent to the carbonyl group, an energetically low lying n → π* transition in pyridine inhibits its fluorescence.

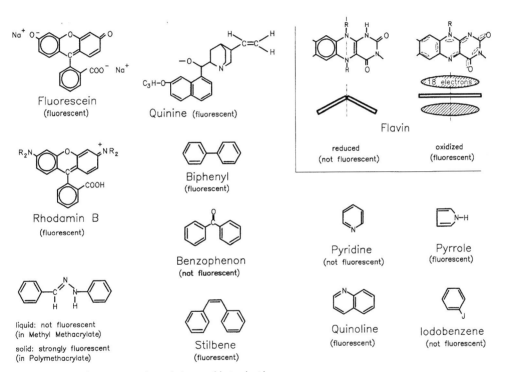

Figure 6.10 With some experience it is possible to decide on the basis of the structure of a molecule whether it is fluorescent or not (for details see text).

Figure 6.10 demonstrates with some examples the structural properties that are a prerequisite for fluorescence, even in the case where it is not always clear-cut, whether the lowest excited state (S_1) corresponds to a favorable π → π* or a n → π* transition.

- *Flavin* in its oxidized form is a rigid, flat and aromatic molecule with an extended π-electron system, while in its reduced form it

is non-aromatic and bent in the N5–N10-direction (into a but-
terfly shape). The low lying lone electron pairs at the nitrogen
atoms are responsible for a low energy $n \rightarrow \pi^*$ transition. In this
form, the molecule does not fluoresce.

- As shown in Figure 6.10, pyridine does not fluoresce while, surprisingly, quinoline does. The reason is probably its more
extended π-electron system.

- In pyrrole the nitrogen atom is bound by three single bonds to
the ring and to a hydrogen atom by hybrid sp^2-orbitals, while the
two "lone pair electrons" occupy the remaining p-orbital and thus
form a conjugated system with the ring electrons. The S_1-state is
a $\pi \rightarrow \pi^*$ transition and the molecule fluoresces.

- Iodobenzene does not fluoresce, because the heavy iodine atom
promotes intersystem crossing by spin-orbit coupling thus rapidly
depopulating the potentially fluorescent first excited singlet state.

- Benophenone obviously does not fluoresce due to its low lying
$n \rightarrow \pi^*$ transition.

- Fluoresceine, rhodamine B and quinine are classic fluorophores.
Obviously "lone pair" electrons of the oxygen bridge in rhodamine lead to an extended π-electron system, i. e., an S_1-state with
$\pi \rightarrow \pi^*$ character.

- Stilbene in isomeric *cis-* and *trans*-forms is highly conjugated and
therefore fluoresces.

- A clear-cut dependence of fluorescence quantum yield on the
steric rigidity of a molecule is generally observed, independent of
the molecular structure (Figure 6.10, bottom left). In liquid
environments this molecule is virtually non-fluorescent, but
becomes strongly fluorescent in rigid environments.

6.2.6
Fluorescence Quantum Yield

While the magnitude of absorption is defined by $\log (I_0/I)$, the analogous magnitude of fluorescence is defined by the *quantum yield* or *fluorescence efficiency* Φ_F:

$$\Phi_F = \frac{number\ of\ emitted\ photons}{number\ of\ absorbed\ photons} = \frac{I_F}{I_A} \tag{6.1}$$

By definition, the values for Φ_F are between 0 and 1 (or between 0 and 100 %) and
are largely defined by the individual structures of the molecules. The measurement of the quantum yield of fluorescence Φ_F will be discussed in more detail
in Section 6.5.3.

6.3
Fluorescence Measurement

6.3.1
Fluorimeter

Figure 6.11 shows the typical set-up of a fluorimeter (also known as a *fluorometer*) as used for the measurement of fluorescence spectra. First of all we distinguish between the excitation and emission light path (fluorimeter set-up modified from Jobin-Yvon, Model JY3; cf. light path Figure 3.22). A broad band excitation source is passed through the excitation grating to generate monochromatic exciting light, of which a small portion is sent via a partially transparent mirror to a quantum counter (see below). This signal is utilized for electronic compensation of variations in excitation light intensity during measurement (particularly important for UV lamps). The fluorescence emission spectrum of a compound is obtained by exciting the molecule at its absorption band and scanning the emitted wavelengths onto the detection system. Alternatively, in many devices one can set the detection wavelength to correspond to fluorescence emission and scan the excitation monochromator to generate the fluorescence excitation spectrum. The latter is useful in determining whether fluorescence emission arises from a single compound in a sample as it will correspond in profile to the absorption spectrum of that compound.

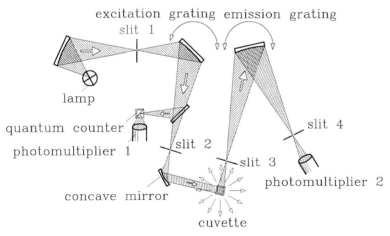

Figure 6.11   Principle set-up of a fluorimeter including excitation and emission light paths. Part of the excitation light is used to correct for lamp noise. Exchangeable fixed slits of differ- ent widths instead of expensive slits whose widths can be continuously changed are suffi- cient in fluorescence spectroscopy.

In contrast to the excitation light the emitted light is in principle isotropic and diffuse and therefore more difficult to control in its path through the optical components to the photodetector. Therefore, it is advantageous to position the cuvette as close as possible to the entrance slit (slit 3 in Figure 6.11) in the detection pathway.

The detection system utilizes slits of varying size, which have an impact on the quality of the data gathered. The wider the slits the more light is collected, but at the expense of spectral resolution. Slitwidths allowing a wavelength resolution $\Delta\lambda$ of between 5 and 20 nm are generally sufficient, as molecular emission spectra are comparably broad with only little fine structure. The slitwidth of the excitation monochromator can be chosen to be significantly smaller as the excitation light is well collimated ($\Delta\lambda \cong 235$ nm). Nevertheless, the pronounced structure on the emission spectra of UV gas discharge lamps (Figure 3.7) results in photometric uncertainties in the range of approximately $\pm 10\%$ in spite of correction of the excitation spectra.

6.3.2
Correction of Fluorescence Spectra

As mentioned previously, uncorrected fluorescence spectra are still heavily distorted by the instrumental dispersion factors, including the spectral efficiencies of the lamp, the monochromator and the photodetector. There are various ways of correcting for these influences. For theoretical reasons fluorescence excitation spectra and absorption spectra should be identical (where a single pure compound is present in the sample), so a correction curve $A(\lambda)$ for the excitation light (lamp + excitation monochromator) is obtained by normalizing the measured uncorrected excitation spectrum of a molecule with its absorption spectrum. In another method the correction curve $A(\lambda)$ for the excitation monochromator can be obtained by using a so-called quantum counter (Taylor, D. G., Demas, J. N. *Anal. Chem.* 1979, *51*, 717–722). A most effective quantum counter is a highly concentrated solution of rhodamine B in ethylene glycol (3 g L^{-1}). It absorbs virtually all impinging photons independent of their wavelength λ (cf. Figure 6.14), and a triangular cuvette is used in which excitation and emission take place at the same side of the surface in the so-called "front face" geometry. As the emission is observed at a fixed wavelength λ, the dispersion of the emission light path does not play any role and the measured signal is proportional to the impinging number of quanta. Unfortunately, the quantum counter rhodamine B is only useful in the wavelength range $\Delta\lambda = 200–600$ nm. Beyond 600 nm there is no suitable quantum counter, which makes correction of larger wavelengths more difficult.

For the full determination of a corrected emission spectrum a correction curve $E(\lambda)$ with respect to the emission part of the fluorimeter (emission grating + photomultiplier) is also required. The absolute determination of $E(\lambda)$ is difficult, but comparative correction is again easily accomplished. A measured emission spectrum of a standard substance is compared with its real (corrected) emission

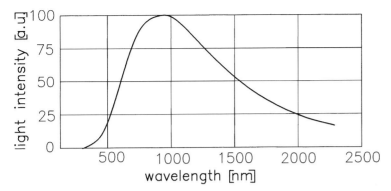

Figure 6.12 The emission spectrum of a halogen lamp that is operated at its correct voltage corresponds, with a good approximation, to the emission spectrum of a black body at a temperature of 3200 °C.

spectrum. A series of such model spectra such as quinine sulfate, ß-naphthol or 3-amino phthalimide has been published. Similarly, calibrated standard incandescent lamps can be adopted (black body radiator, see Figure 3.6). The measured signal divided by the expected (documented) "true" signal yields the correction curve $E(\lambda)$ for the emission part of the fluorimeter. A simple but useful approach is to use a halogen lamp with a filament temperature of 3200 °C (i. e., at its correct operating voltage), which approximately possesses the emission spectrum of a black body at that temperature (Figure 6.12, Section 3.4.1). Alternatively $E(\lambda)$ can be determined without external resources, using only a quantum counter and a white, non-fluorescent, diffusely reflecting body (e. g., coated with magnesium oxide), which is inserted in place of the triangular cuvette into the sample holder (reflection set-up). Firstly, the excitation light path $A(\lambda)$ is corrected via a quantum counter. Subsequently the total sensitivity $G(\lambda)$ of the fluorimeter is monitored as a function of wavelength λ. This is accomplished by a synchronous scan of excitation and emission wavelengths (an important option in modern fluorimeters!). Finally, the correction spectrum for the emission part of the monochromator $E(\lambda)$ is obtained by division of the total sensitivity curve $G(\lambda)$ by the previously determined correction curve of the excitation part $A(\lambda)$.

Figure 6.13 shows the uncorrected and corrected fluorescence excitation and emission spectra of flavin in aqueous solution. The great importance of the correction of the excitation spectra is obvious. Emission spectra are modified less significantly by correction, but typically undergo a bathochromic shift (to longer wavelengths) of 10–20 nm, depending on the particular spectrometer. In conclusion, detailed comparisons of uncorrected fluorescence spectra obtained with different fluorimeters are questionable.

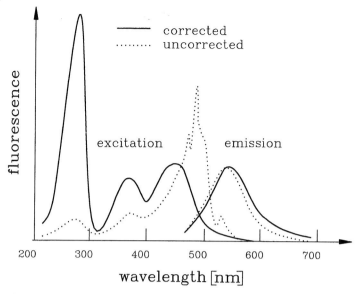

Figure 6.13 Fluorescence excitation and emission spectra of flavin in an aqueous solvent, as measured and after appropriate correction. Excitation spectra are significantly distorted by the optical system and require correction in any case, while emission spectra are modified only slightly, and include a small hypsochromic shift.

6.3.3
Linearity of the Fluorescence Signal

A practical and highly relevant but often overseen fact needs to be discussed here. All fluorescent substances, without exception, exhibit a strong non-linearity between absorbance and fluorescence emission at absorbances above approximately $A = 0.05$ (see below) (under the preconditions for the Lambert–Beer law to be valid, absorption is equivalent to concentration) (Figure 6.14). At very high absorbances the fluorescence versus absorbance curve actually bends asymptotically downwards. This is the concentration range in which counting of quanta works (inset of Figure 6.14). The reasons are as follows.

If Eq. (6.1) for the definition of fluorescence quantum yield is rewritten we obtain,

$$I_F = \Phi_F \, I_A \tag{6.2}$$

I_A is the number of absorbed photons, as opposed to the transmitted photons I_T. If we again call the number of incident photons I_0, with Eq. (6.3) we obtain,

$$I_F = \Phi_F(I_0 - I_T) \tag{6.3}$$

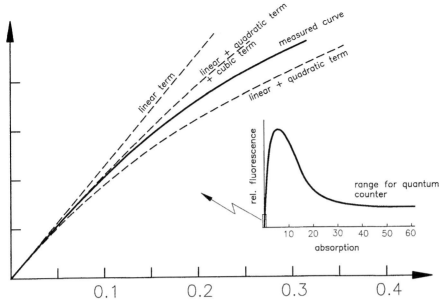

Figure 6.14 Dependence of fluorescence on the absorption (concentration) of a fluorophore. Up to an absorption of approximately A = 0.05 the measured curve is linear, largely independent of the specific molecule measured. The dashed curves show the corresponding mathematical approaches from Eq. (6.8). Inset: Extremely high concentrations of specially selected molecules such as rhodamine B result in a wavelength-independent, but quantum-flux dependent absorption/quantum emission. This effect is utilized for photon counting and correction of fluorescence.

and

$$I_F = \Phi_F \left(1 - I_T/I_0\right) \tag{6.4}$$

Now I_T/I_0 is only the transmission T, and from Eqs. (4.4) and (4.5) the absorption is given by

$$A = -1/T = e \cdot \times x \times \cdot c$$

If the absorption is written as a natural logarithm

$$\ln I_0/I_T = \alpha(\lambda)x \; c \tag{6.5}$$

we obtain in exponential form

$$I_T/I_0 = e^{-\alpha(\lambda)xc} \tag{6.6}$$

and substitution into Eq. (6.5) gives,

$$I_F = \Phi_F I_0 \left[1 - e^{-\alpha(\lambda)xc} \right] \tag{6.7}$$

This equation shows that the emission signal amplitude is proportional to the quantum yield Φ_F and the incident quanta I_0, but *not* to the concentration c! Now, the series expansion of an exponential function yields,

$$e^x = 1 + \frac{x}{1} + \frac{x^2}{2!} + \frac{x^3}{3!} + \dots \tag{6.8}$$

Note, for negative x-values subsequent members of the series are alternately added and subtracted.

For "small" values of x the higher exponentials can be omitted and we can write

$$e^{-x} = 1 + (-x)/1 = 1 - x \tag{6.9}$$

Substitution into Eq. (6.7) gives,

$$I_F = \Phi_F I_0 \, xc\alpha(\lambda) \tag{6.10}$$

Finally, if from Eq. (5.3) we replace (λ) by the molar extinction coefficient $\varepsilon(\lambda)$, we obtain the formula known as the *Parker law*:

$$I_F = 2.303 \, \Phi_F I_0 \, x \, c \, \varepsilon(\lambda) \tag{6.11}$$

In summary, the linearity of absorption and fluorescence holds true for adequately diluted samples, and it is only these that are acceptable for quantitative fluorescence measurement! This precondition is often ignored in practice. However, what is "adequately dilute"? In practice, within measurement precision, for fluorophores with highly different molar absorption coefficients (ε), linearity is guaranteed up to an absorption of approximately $A = 0.05$. Of course, a precise value cannot be given (Figure 6.14). Corresponding maximum concentrations of fluorophores can be determined using the Lambert–Beer law:

$$c_{max} = \frac{0.05}{2.303 \, x \, \varepsilon(\lambda)} \tag{6.12}$$

For example, the fluorescence amplitude of chlorophyll, with ε (670 nm) = 85 000 M^{-1} cm^{-1} and a path length of $x = 1$ cm increases linearly up to a concentration of $c_{max} = 0.05/(2.303 \times 1 \times 85\,000) = 0.255$ mM.

If we take higher terms of the series expansion in Eq. (6.9) into account, which is not a big problem with today's computing power, the construction of non-linear calibration curves for the fluorimetric measurement of higher concentrations of fluorophores is feasible. Nevertheless, as in fluorescence analysis, the upper limit will be defined by the *inner filter effect* discussed earlier. This is caused by reabsorption of the emitted light.

6.3.4
Turbid Samples and Measurement at Low Temperatures

Particularly problematic are fluorescence measurements with turbid, highly absorbing samples (suspensions) as, e. g., in chloroplasts, in which the photosynthesis of green plants takes place. Even if the *bulk* suspensions as such are significantly diluted, the *local* concentration of chlorophyll amounts to several moles per
liter, i. e., linearity of concentration and fluorescence are not guaranteed at all. In
addition, not even the assumptions of the Lambert–Beer law are fulfilled (Section
5.1). Therefore, in these cases, only relative, qualitative and kinetic measurements
are justified, which are performed with a sample of the same absorbance and
scattering properties. Quantitative conclusions are possible only on the basis of
calibration curves that have been constructed using this specific sample type.

In some cases low temperature-fluorimetry [e. g., −196 °C (77 K) liquid nitro-
gen] offers advantages. As the fluorescence intensity is strongly temperature dependent (radiationless transitions $S_1 \rightarrow S_0$ decrease in probability with decreasing
temperature), fluorescence from some substances can only be detected at low
temperatures. The set-up with a branched light fiber in the reflection mode, as

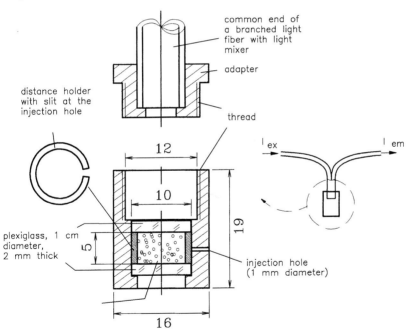

Figure 6.15 Cuvette for fluorescence measurements at −196 °C (liquid nitrogen) in the
reflection mode. The sample is applied (with
no air bubbles!) via a syringe through the 1
mm hole and measured by an attached
branched light fiber. An adapter receives the
common part of the light fiber. One arm of the
light fiber points to the exit slit, the other to the
entrance slit of a normal fluorimeter. The whole
set-up including the light fiber adapter is
placed in a dewar outside the fluorimeter and
covered with a black cloth (from Senger, Marburg, Germany).

shown in Figure 6.15, has proven to be very useful, as most samples become turbid and highly scattering in the deep-frozen state.

Monochromatic light is directed through one branch of the bifurcated optical fiber onto a sample of only 0.25 mL volume in a special brass–plexiglass cuvette. The cuvette is a dewar filled with liquid nitrogen (a simple glass beaker in polystyrene is also suitable) for approximately 1 min to reach a temperature of −196 °C (boiling ceases at this point) and the measurement can be started. The fluorescence is collected by the second branch of the optical fiber and guided to the emission monochromator. A conventional fluorimeter is easily modified for this purpose. The usual low temperature set-ups in commercial fluorimeters are difficult to handle and typically only accept clear, glassy samples. Depending on the solubility properties of the fluorophore, a variety of solvents of different polarities are chosen; e. g., a mixture of ether–isopentane–ethanol (5:5:2) remains in the glass-form when frozen. Cracks in the glass should be avoided as they introduce scattering artifacts, which is only feasible in practice with samples thinner than 1 mm.

If it is only the kinetic properties of spectrally well-defined fluorophores that are of interest, both monochromators could be replaced by colored filters with much higher light throughput (interference, glass or plastic filters, see Section 3.7). The transmission maxima are chosen to coincide with the excitation and emission peak of the fluorophore. In addition, with the filter arrangement the photodetector, in particular a photomultiplier with a large end-on photocathode, can be placed in close juxtaposition to the sample. In this manner the acceptance cone, and thus the sensitivity, is enlarged by several orders (cf. absorption measurement of turbid samples Section 5.2.8).

An advantageous light source for fluorescence excitation in the UV range is the 254 nm line of a mercury lamp that is easily separated by a special Hg filter. Small, light-weight and low-priced pencil beam lamps that are typically used for calibration tasks, are excellent UV light sources for fluorescence spectroscopy.

6.3.5
Further Sources of Error

In practice, there are often unexpected features in fluorescence spectra, even after appropriate correction has been carried out. These unexpected bands, peaks, shoulders, and even "negative" fluorescence signals, can be caused by many reasons, some of which are given below.

- A lack of filters for the supression of higher scattering orders.
- The influence of Raman bands of the solvent (Section 8.3) increases dramatically with decreasing sample concentration as a consequence of the high sensitivity of fluorimetry. For example, the wavelength of the Raman band of water is always at an energy of 3600 cm^{-1} below the wavelength of the excitation light; i. e., at

762 nm for λ_{ex} = 600 nm and 215.5 nm for λ_{ex} = 200 nm. Characteristic of this type of error is the variation in profile and intensity of the emission spectrum on changing the excitation wavelength. Therefore, this test is mainly useful if fluorescence emission peaks are extraordinarily sharp and superimposed on the basic fluorescence spectrum. On changing the excitation wavelength only the "height", not the shape or the peak maximum of the emission spectrum should undergo change if Raman effects are absent.

- The purity of the solvents is of great importance in fluorescence (and phosphorescence) spectroscopy. Contaminants often exhibit broad band, featureless fluorescence bands. In case of doubt the purification of even "analytically pure reagents" is advised.
- Many optical components (including contaminating fingerprints!) such as *cut-off* filters or dichroic mirrors exhibit prompt and delayed fluorescence. Tests with "clean", empty cuvettes or filters can often lead to surprising emission measurements.
- The strong Rayleigh peak (elastic scattering of the excitation light) including its tail should be suppressed or avoided, since it may cause irreversible damage to the photomultiplier. Therefore, emission signals should be measured a sufficient distance away from the excitation wavelength, which depends on the slit widths.

6.4
Polarization and Anisotropy

From the physical point of view, fluorescence light is a transverse, so-called dipole radiation that possesses a well-defined direction with respect to the structure of the emitting molecule. (transition dipole moment, Section 2.3.3.4). If polarized light is incident on a sample containing dissolved and isotropically distributed fluorophores, the emitted fluorescence light will also be more or less polarized: a very important finding for analytical purposes.

6.4.1
Definitions

Assuming first that the electrical field vector **E** of the exciting light is oriented in the z-direction and further that the transition dipole moments of absorption and emission are parallel (enclosed angle β = 0) and enclose a common angle Θ with the z-axis, the projection onto the x–y plane corresponds to the azimuth Φ (Figure 6.16). The probability of absorption is proportional to $\cos^2 \Theta$. The intensity I_{vv} of fluorescence light emitted parallel to the z-axis is also proportional to $\cos^2 \Theta$, thus,

$$I_{vv} \propto \cos^2\Theta \tag{6.13}$$

the light polarized orthogonal to this direction is

$$I_{vh} \propto \sin^2\Theta \sin^2\Phi \tag{6.14}$$

If all angle-dependencies of excitation and emission are integrated across the whole space for plane polarized light we obtain a value of polarization that is defined by

$$p_0 = \frac{I_{vv} - I_{vh}}{I_{vv} + I_{vh}} \tag{6.15}$$

of 1/2. However, if the dipole moments of excitation and emission are orthogonal to each other ($\beta = 90°$), we obtain a minimum value of $-1/3$. For any value of β, polarization values vary between these limits

$$-\frac{1}{3} \le p_0 \le +\frac{1}{2} \tag{6.16}$$

as demonstrated for the idealized polarization spectrum $p_0(\lambda)$ of a hypothetical molecule, Figure 6.17.

Unfortunately, for physical reasons, all grating monochromators exhibit a strong intrinsic polarization rendering the immediate application of Eq. (6.15) worthless. In order to make this formula useful the apparatus-related polarization must be eliminated. This is achieved by a straightforward but somewhat cumber-

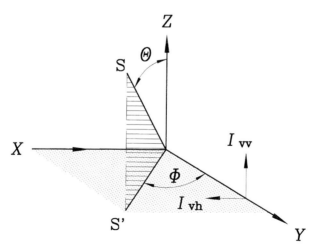

Figure 6.16 Schematic representation of the dependencies of the two orthogonally oriented components of fluorescent light I_{vh} and I_{vv} in the y-direction, with perpendicularly polarized exciting light in the x-direction (**E**-vector in the z-direction). For simplicity we assume that the transition dipole moments of absorption and emission are oriented parallel in the **S** direction (enclosed angle $\beta = 0$). The projection of vector **S** onto the x–y plane corresponds to the azimuth Φ.

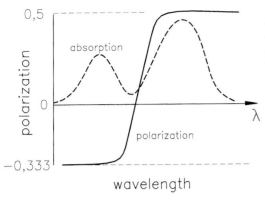

Figure 6.17 Idealized polarization spectrum $p_0(\lambda)$ of a hypothetical molecule, which theoretically can vary between the extreme values of $-1/3$ and $+0.5$.

some measurement procedure requiring the acquisition of four individual spectra with different positions of the polarization filters. However, this is easily realized using a suitable computer-controlled approach (the first index refers to the exciting, the second to the emitted light):

$$p(\lambda) = \frac{I_{vv} - I_{vh}\left(\frac{I_{hv}}{I_{hh}}\right)}{I_{vv} + I_{vh}\left(\frac{I_{hv}}{I_{hh}}\right)} \tag{6.17}$$

In spite of these problems the polarization p, as defined in Eq. (6.15), is still widely used in today's scientific work with filter monochromators. Thus, the total polarization value p of a mixture of fluorophores with individual intensities F_i and individual polarization values p_i yields

$$\left(\frac{1}{p} - \frac{1}{3}\right)^{-1} = \sum_i \frac{F_i}{\frac{1}{p_i} - \frac{1}{3}} \tag{6.18}$$

The so-called anisotropy r, as defined by

$$r_0 = \frac{I_{vv} - I_{vh}}{I_{vv} + 2\,I_{vh}} \tag{6.19}$$

was introduced in place of the polarization p in the first half of the 1960s, and differs simply by the factor 2 in the denominator from the definition of polarization as given in Eq. (6.15). Namely, as seen in Figure 6.18 the dipole radiation is distributed symmetrically around the z-axis, yielding a total light energy of $I_w + 2I_{vh}$, representing a much better reference than $I_{vv} + I_{vh}$. Consequently, instead of Eq. (6.18) the anisotropy of a mixture is obtained by the linear relationship

$$r = \sum_i F_i f_i \tag{6.20}$$

In addition, in contrast to polarization, the decay kinetics of the anisotropy of a single fluorophore induced by an exciting light pulse is strictly exponential. In a mixture of fluorophores it is equal to the sum of the exponentials of the indivi-

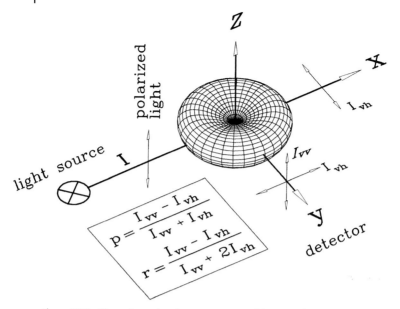

Figure 6.18 Three-dimensional representation of the spatial distribution of fluorescent light with excitation by perpendicularly polarized light in the x-direction, including a definition of polarization and anisotropy.

dual components. Therefore, kinetic studies of complex mixtures use anisotropy rather than polarization.

In accordance with Eqs. (6.16) and (6.20) the anisotropy and polarization are easily interconverted: $p = 3r/(2 + r)$ or $r = 2p/(3 - p)$. These formulae yield limiting values of anisotropy of between -0.2 and $+0.4$.

6.4.2
Energy Transfer

In order to understand the theoretical basis of depolarization which is discussed in Chapter 7, we first have to distinguish between the following mechanisms for energy transfer from one electronically excited molecule to another:

- For the electron exchange mechanism the orbitals of the excited donor molecule and the acceptor molecule in its ground state should overlap in space significantly. The excitation energy is transferred through mutual electron exchange. The range for this interaction is small and requires immediate "contact" of the molecules (collision); the distances should not extend more than 10 Å. The spectral properties of both molecules are irrelevant in this mechanism.

- A second, "trivial" process is energy transfer by emission and subsequent absorption of light quanta: the excited molecule emits a photon that is absorbed by the second molecule thereby transferring it into an excited state. This, of course, requires a transparent solvent. This process, however, is of little analytical interest but might interfere with actual measurements.
- The third, so-called Förster-transfer mechanism (resonance transfer, dipole–dipole transfer) is of great analytical importance. It depends on the fact that excitation energy is transferred over large distances without the mediation of light quanta. The energy transferred is directly proportional to the overlap of the emission spectrum of the donor and the absorption spectrum of the acceptor (in wavenumber, i. e., equivalent to energy presentation). In addition, it is inversely proportional to the sixth power of the distance R and is quantified by the parameter R_c, the experimental determination of which is discussed in Chapter 7.

6.4.3
Depolarization

Only in the ideal case are we able to measure the "real" intrinsic polarization $p_0(\lambda)$ of molecules. In practice, for two reasons, smaller values are always measured, i. e., a graduated depolarization takes place.

- Excitation energy will be transferred from the fluorophore to other neighboring molecules of the same type. Their transition dipole moments do not necessarily have to be parallel before the energy accepting molecules emit light. This energy transfer process is enhanced with increasing fluorophore concentration, leading to a gradual depolarization of the fluorescence emission. This allows the determination of the critical distance R_c, i. e., by definition, the distance at which the transition probability for excitation energy transfer between two molecules with parallel transition dipole moments equals the emission probability $1/\tau$.
- Secondly, owing to rotational relaxation processes, a fluorophore will rotate in the time between the processes of excitation and emission. This process is quantified by the so-called rotational relaxation time ϱ, which describes the rotatory mobility of a molecule which can be determined, for example, by the viscosity of its microenvironment ("microviscosity").

In his original work of 1926 Perrin gave the following relationship:

$$\frac{1}{p} - \frac{1}{3} = \left(\frac{1}{p_0} - \frac{1}{3}\right)\left(1 + \Phi\frac{3T\varkappa\tau}{4\pi a^3}\right)$$

(6.21)

where p and p_0 describe the apparent and intrinsic ("natural") polarization, respectively, $\varkappa$ is the Boltzmann constant, T is the absolute temperature, τ is the fluorescence lifetime, a is the average Einstein radius and Φ is the fluidity (measured in the unit *rhe*) of the solvent (not to be mistaken for the quantum yield which has the same symbol Φ).

$1/\Phi = \eta$ = viscosity; at 20 °C water has the fluidity of 100 rhe and the viscosity $0.01p = 1$ *cp*, centipoise; again, not to be mistaken for the polarization p.

For example, if we measure the various polarization values p of flavin at various solvent fluidities Φ (adjusted by well defined solvent mixtures of glycerol and water), the plot of $1/p$ versus Φ yields a straight line. By extrapolation of this line towards $\Phi = 0$ we obtain the intrinsic polarization $p_0 = 0.3$ (Figure 6.19). Taking this value of p_0 and extrapolating the line in Figure 6.19 to the fluidity of water at $\Phi = 100$, for lumiflavin we obtain an effective Einstein radius of $a = 6.2$ Å. Finally, with a and Φ and using the Einstein formula the rotational relaxation time ϱ can be determined.

$$\varrho = \frac{4\,a\,\pi}{\varkappa\,T\,\Phi} \tag{6.22}$$

Thus, for lumiflavin in water at room temperature we obtain a rotational relaxation time of $\varrho = 0.38$ ns, i.e., approximately 10 times shorter than the fluorescence lifetime of 4.5 ns. Consequently, lumiflavin will not exhibit any apparent fluorescence polarization in aqueous solution.

However, if the flavin molecule is anchored within a membrane by a long aliphatic chain (amphiflavin, e.g., in position 7, cf. Figure 6.7) (Figure 6.20), its rotatory mobility is strongly hindered and fluorescence polarization will be observed depending on the temperature, i.e., the phase transition state of the membrane. Polarization spectra show the spectral properties of polarization (Figure 6.21) and

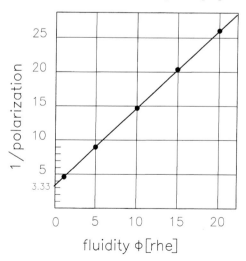

Figure 6.19 Linear dependency of the reciprocal value of polarization p of lumiflavin as a function of fluidity Φ of the solvent (adjusted by various glycerol and water mixtures).

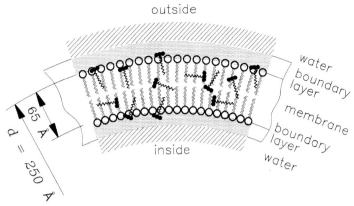

outside

water
boundary
layer

membrane
boundary
layer

water

inside

65 Å

d = 250 Å

Figure 6.20 Representation (to scale) of vesicle bound amphiflavin (section enlargement). There is an extremely strong polarity gradient across biological membranes with an internal dielectric constant ε of 4 up to approximately 80 in the *interface*. This gradient leads to strong fluorescence changes on delocalization of the chromophore within the membrane.

allow electronic states (see Figure 6.17) to be distinguished, a feature not observable in typical absorption spectra. The conclusion derived from kinetic observations (Figure 6.7) is, thus, spectrally verified. With increasing temperature the chromophore undergoes a reorientation into hydrophobic, inner membrane domains as indicated by a hypsochromic shift of the emission spectrum and a concomitant increase in quantum yield of the fluorescence.

Combining Eqs. (6.21) and (6.22), the rotational relaxation time ϱ can be determined on the basis of the temperature dependence of the fluorescence polarization (Figure 6.21). This was observed using vesicle membranes made from three different lipids (dimyristoyl-, dipalmitoyl- and distearoyl-lecithin) with phase transitions at 23, 41 and 59 °C (Figure 6.22). On the basis of Eq. (6.23) (and with $1/\Phi = \eta$) the microviscosities experienced by the fluorophores in the crystalline (125 cp) and liquid crystalline (25 cp) state of the membrane can be determined.

Fluorescence (de) polarization measurements can easily and quickly show whether a (fluorescing) coenzyme such as NADH (nicotinamide adenine dinucleotide) or flavin (vitamin B2) is bound to a protein or a membrane, or whether it is solubilized in its free form. Rotational relaxation times of proteins are in the millisecond range. For example, for rhodopsin, a large membrane protein of approximately 50 Å (Figure 6.23), a value of $\varrho = 20$ ms was determined (Cone, R., *Nature* 1972, *236*, 1234–1244). During its fluorescence lifetime, in the nanosecond time scale, a protein molecule is virtually motionless. In an individual case, spectral properties might support the assumption concerning the binding.

On the basis of the Perrin relationship (Eq. 6.21) Weber (1966) deduced the following equation for a three-dimensional, homogeneous distribution of fluorophores.

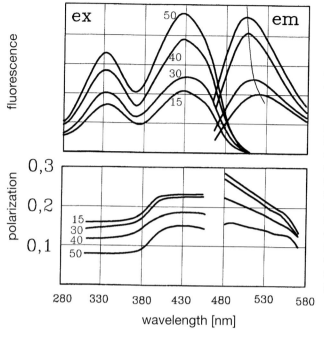

Figure 6.21 Top: If the flavin fluorophore is anchored with a long aliphatic chain (amphiflavin, in position 3, see Figure 6.7) within a membrane (Figure 6.22), the rotational mobility is strongly inhibited leading essentially to a fluorescence polarization, the extent of which depends strongly on temperature, i. e., the transition phase crystalline → liquid crystalline (dipalmitoyl lecithin: 41 °C). Bottom: The spectral properties of polarization.

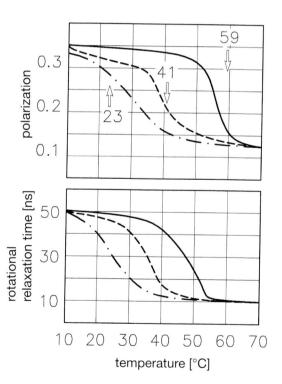

Figure 6.22 The known temperature dependency of fluorescence polarization (Figure 6.23) immediately allows the rotational relaxation time ϱ (ns) to be calculated. This was performed for vesicle membranes made from three different lipids (dimyristoyl-, dipalmitoyl- and distearoyl-lecithin) with phase transitions at 23, 41 and 59 °C.

Gregorio Weber (1916–1997)

$$\frac{1}{p} - \frac{1}{3} = \left(\frac{1}{p_0} - \frac{1}{3}\right)\left[1 + c\frac{(4\pi NR_c^6)}{15(2a)^3}\right] \tag{6.23}$$

where N is the Loschmidt number, $2a$ the effective molecular diameter and c the concentration of the fluorophore. Solving for R_c a straight line results with slope $s = pc^{-1}$:

$$R_c = (2a)^{1/2}\left[\frac{15s}{4N\pi(1/p_0 - 1/3)}\right]^{1/6} \tag{6.24}$$

Measuring the polarization of lumiflavin in solvents of various viscosities as adjusted by different glycerol–water mixtures, we obtain a typical constant of R_c = 29 Å. Thus, on the basis of known values of R_c the measurement of fluorescence efficiency allows the measurement of submicroscopic distances. Using this fluorimetric method, in 1972 Wu and Stryer were able to determine the structure and dimensions of the rhodopsin molecule, which is very important in the vision process. In addition to the natural chromophore, retinal, they bound three further fluorophores to the protein in a well-defined spatial manner (Figure 6.23).

The two-dimensional analog to Eq. (6.24) yields

$$R_c = (2a)^{2/3}\left[\frac{15s}{\pi(1/p_0 - 1/3)}\right]^{1/6} \tag{6.25}$$

If we vary the area-concentration of the membrane bound flavin, according to Eq. (6.25), we again obtain concentration-dependent polarization spectra from which we can deduce a critical distance of R_c = 17 Å. The slope is proportional to the sixth power of R_c, and can therefore be determined very precisely, as shown by the comparison of slopes with slightly different R_c values (Figure 6.24). Thus, within the membrane the energy transfer of flavin is much more efficient compared with freely solubilized flavin. This example underlines the important difference between isotropic and anisotropic photochemistry. Facts deduced on the

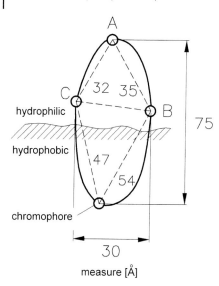

A

C

hydrophilic

B

75

hydrophobic

32 35

47

54

chromophore

30

measure [Å]

Figure 6.23 Fluorimetric determination of size of the rhodopsin molecule, which carries fluorophores of varying spectral behavior at three different positions (A, iodoacetamide; B, fluorescent disulfides; C, acridine). From the known excitation and emission spectra as well as from the known values of R_c the various distances can be determined (from Wu and Stryer, *Proc. Natl. Acad. Sci., USA* 1972, *69*, 1104–1108).

basis of "test tube experiments" cannot simply be transferred to membrane bound components, a highly relevant fact for biological mechanisms.

As also deduced by Perrin in 1929, polarization spectra can be converted into so-called angle spectra

$$\beta(\lambda) = \arccos \sqrt{\frac{(1 + 3p_0)}{(3 - p_0)}}$$

(6.26)

which immediately shows the angle $\beta(\lambda)$ between the corresponding transition dipole moments (Figure 6.25). In the example given the angle between the transition dipole moments is $(S_0 \rightarrow S_1)/(S_0 \rightarrow S_2) = 40° - 31° = 9°$. Theoretical calculations as performed by Song (*Int. J. Quantum Chem.* 1969, *3*, 303,) yield the transition dipoles in Figure 6.25, top. These are spectrally identified in Figure 6.25, bottom.

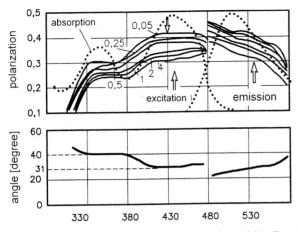

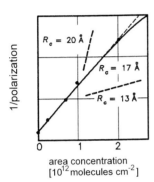

Figure 6.24 Top: Change in polarization spectra of amphiflavin (aliphatic chain at position 3, Figure 6.8) in lipid vesicles made from dipalmitoyl-lecithin as function of area density (given in units of 10^{12} molecules per cm^2; $\lambda_{em} = 540$ nm, $\lambda_{ex} = 440$ nm). Bottom: The polarization values (430 nm) allow calculation of a critical distance of $R_c = 17$ Å, based on the Weber equation [Eq. (6.26)]. The slope is proportional to the sixth (!) power of R_c, and thus R_c can be determined very precisely, as shown by comparison of the slopes with slightly different R_c values. Middle: Recalculation of the polarization spectrum yields the so-called angle spectrum; this immediately allows the angle $\beta(\lambda)$ between transition dipole moments to be determined $(S_0 \rightarrow S_1)/(S_0 \rightarrow S_2) = 40° - 31° = 9°)$.

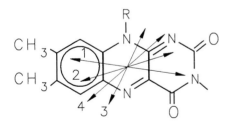

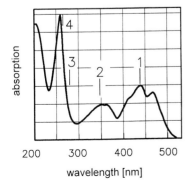

Figure 6.25 Top: Vector representation of the transition dipole moments of different absorption bands of the flavin molecule. Bottom: Spectral identification of the transition dipole moments drawn above (from calculations by Song, *Int. J. Quantum Chem.* 1969, *3*, 303, modified).

6.5
Fluorescence Lifetime

6.5.1
Definitions

As mentioned earlier, fluorescence is normally emitted from the vibrational ground state of the first electronically excited state S_1 of a molecule, which has a lifetime of approximately 10^{-9} s. Further decay mechanisms such as *electron transfer (ET, photochemistry)*, *intersystem crossing* and *internal conversion (IC)* compete with the "fluorescence channel". Only the *apparent* lifetime τ of the first excited singlet state ($S1$) is measurable. Owing to various quenching processes it is always smaller than the intrinsic (natural or "radiative") lifetime τ_0. The decrease in the number of excited molecules n with time is described by a simple differential equation

$$dn = -k\,n\,dt \tag{6.27}$$

where the total rate constant (k), describes the time dependence of a transition from the excited to the ground state. k is the sum of the individual rate constants for fluorescence (k_F), photochemical processes (k_{ET}, energy transfer), radiationless transitions (k_{IC}) and intersystem crossing (k_{ISC}):

$$k = k_F + k_{IC} + k_{ISC} + k_{ET} \tag{6.28}$$

Integration of Eq. (6.28) yields the solution

$$n = n_0\,e^{-kt} \tag{6.29}$$

The time $t = \tau$, for which $k = 1/t$, is the lifetime of the excited state. After this time, $1/e \approx 37\%$ of the molecules are still in the excited state. In addition, the half-life $t_{1/2}$ defines the time after which just half of the molecules have decayed to the ground state S_0. The relationship between both parameters is simple:

$$t_{1/2} = \frac{\ln 2}{k} = \ln 2\ \tau = 0.693\ \tau \tag{6.30}$$

6.5.2
Experimental Determination of Fluorescence Lifetime

Basically, there are two experimental approaches to determining the fluorescence lifetime τ, one indirect and one direct. The first one, rather inappropriately called flash photolysis, is intimately related to the development of short duration flash lamps and lasers as pulsed light sources for excitation. This aspect can be further subdivided in terms of the excitation pulse. In the simplest case, the so-called single pulse method, the chromophore is excited by an individual, intense light flash of short duration. Subsequently the fluorescence decay is monitored (Eq. 6.30) in real time.

With multiple pulse excitation (*pulse sampling*) the sample is irradiated with a high frequency sequence of weaker intensity pulses. Each flash provides one measurement point (the delay between excitation and emission of a single photon) and the full decay kinetics is built up as a histogram of single events (*single photon counting*, Figure 6.26). Thus, although the fluorescence lifetime is very short (ns), with the latter approach its measurement can take many minutes. Flash lamp pulse durations are only slightly shorter or comparable to the fluorescence lifetime τ to be measured. Therefore, the measured decay kinetics are "stretched out" by the finite duration of the excitation pulse, in mathematical terms "folded" or convolved with the decay kinetics of the flash light. Utilizing a mathematical process called *deconvolution* (see Section 5.4.4.4) the true kinetics of the chromophore can be determined, if the temporal profile of the flash lamp is known, and if the lifetime τ to be measured is not considerably shorter than the pulse. Fluorescence lifetimes as short as 100 ps are measurable.

Single photon counting depends on the fact that fluorescence emission is a statistical process similar to radioactive decay: within the duration of the flash the molecules of the sample are excited and a "stopwatch" is started. The first detected photon stops the watch and the time delay between the flash and detection is measured and recorded. In accordance with the kinetics of fluorescence emission, the probability of a photon being emitted by an excited molecule decreases with time. As a rule of thumb, the measurement of complete decay kinetics requires the measurement of 10^5 individual signals for statistical accuracy. The pulse intensity must be weak in order to excite the emission and detection of a single event at approximately every 20th flash. Thus, it is also possible to cover longer time ranges for the kinetics (Figure 6.26).

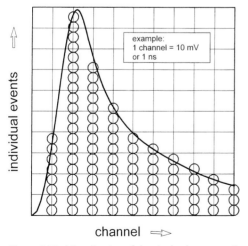

Figure 6.26 Visualization of the *single photon counting* method. Each circle represents an individual measurement (for details see text).

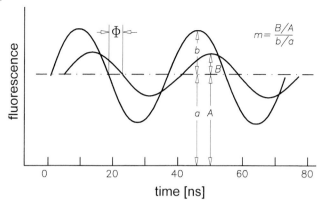

Figure 6.27 Determination of the fluorescence lifetime τ on the basis of modulated excitation light (large amplitude b, offset a). Owing to the intrinsic lifetime τ_0 of the excited state the emission light is phase shifted (Φ) and emitted with a smaller amplitude (B, offset A). The degree of demodulation m as well as the phase angle Φ, which are independent of each other, allow the lifetime τ to be determined.

A second, indirect harmonic method for the determination of fluorescence lifetimes is based on the relationship of amplitude and phase between the exciting and emitted light. The essential component of a phase fluorimeter is the gas-filled discharge lamp that emits a rapid sequence of light pulses of a few nanoseconds duration at a repetition rate of approximately every millisecond. Alternatively, continuous light sources are used where an amplitude variation is generated by an electro-optical modulator (Figure 3.28). These modulated signals are much more reproducible than a flash, in terms of shape and intensity. The modulation of the excitation light is not followed exactly in emission, due to the finite lifetime of the fluorescence, and the emitted light appears after a certain time delay. In addition, the degree of modulation m of the emitted light is also decreased, because not all excited molecules decay to the ground state between two subsequent flashes. The time delay expressed by the phase angle, Φ, as well as the degree of demodulation, m, allow (Figure 6.27) the determination of the lifetime independently of each other.

$$\tau(\Phi) = \frac{1}{\omega} \times \tan \Phi, \quad \text{or} \quad \tau(m) = \frac{1}{\omega} \sqrt{\left(\frac{1}{m^2}\right) - 1}, \quad \text{with } \omega = 2\pi f \qquad (6.31)$$

where f is the modulation frequency and ω the angular frequency. The principle is explained by the example of a sinusoidal modulation.

Unfortunately, various parameters in science and technology are described by identical symbols. For example, "Φ" stands for phase angle, fluidity, quantum yield and azimuth angle, leading to possible misunderstandings.

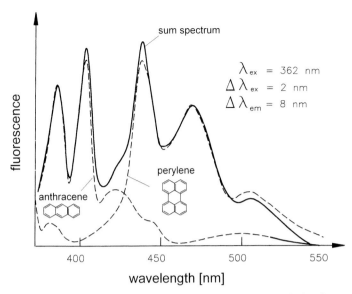

$$\lambda_{ex} = 362\ nm$$
$$\Delta\lambda_{ex} = 2\ nm$$
$$\Delta\lambda_{em} = 8\ nm$$

Figure 6.28 Time resolved fluorescence spectroscopy allows the selective spectral determination of various components in a mixture with different fluorescence lifetimes. In this case a 50:50 mixture of anthracene and perylene in methanol was analyzed using the modulation method. Lifetimes of 4.5 and 6.1 ns were determined. The difference of only 1600 ps allows the clear separation of both components – providing the phase-setting of the *lock-in* amplifier is properly adjusted.

While the direct method measures the complete behavior of the kinetics in their often complex shape, the indirect, harmonic method only allows one individual value to be determined (m or Φ). However, for the latter there is a good test to indicate more complex decay kinetics (superposition of several fluorophores): for a single fluorescing species, we expect equality $\tau(m) \equiv \tau(\Phi)$. However, if the decay is more complex, i.e., several species are involved, then,

$$\tau\ (m) > \tau\ (\Phi) \tag{6.32}$$

However, the lifetime of several components of a mixture can be concomitantly determined in the phase method by changing the modulation frequency, f. This is particularly important in protein-biochemistry and has been discussed in more detail by Lacowicz (1999, Selected Further Reading).

Taking advantage of different fluorescence lifetimes, time resolved fluorescence spectroscopy also allows the spectral resolution of various components coexisting in a mixture. Figure 6.28 demonstrates the technical possibilities of this time-resolved analytical method on a model mixture of anthracene and perylene, which differ only slightly in their fluorescence lifetimes.

6.5.3
Determination of Quantum Yield

The fluorescence quantum yield Φ_F is determined not only by intrinsic molecular properties but by external factors in particular. It is the measure of the probability that the energy of an absorbed quantum is emitted as a light quantum. Eq. (6.1) can be rewritten in form of kinetic parameters:

$$\Phi_F = \frac{k_F}{k_F + k_{IC} + k_{ISC} + k_{ET}} = k_F \times \tau \tag{6.33}$$

where the maximum value of Φ_F is "1" if no further relaxation channels are operating. The fluorescence quantum yield Φ_F can be determined experimentally by various techniques.

- In principle the absorbed and emitted quanta can be determined directly as absolute quantities. However, the expense of the experiments involved is unacceptable for moderately equipped laboratories.
- The second possibility is through the comparison of the intrinsic (τ_0) and apparent lifetime τ ($\Phi = \tau/\tau_0$).
- The simplest, and therefore most commonly applied method, relies on the relative comparison of emission spectra of the compound in question, molecule x, with that of a reference molecule such as quinine sulfate, for which the absolute quantum yield Φ_F is known from the literature (see Eaton, D. F. *J. Photochem. Photobiol.* 1988, *2*, 523–531). Firstly, we determine the interception of the two corresponding absorption spectra: irradiation at this wavelength guarantees identical absorption (however their addition absorptions should be less than $A = 0.05$). The ratio of the areas beneath both emission spectra (corrected and plotted on a wavenumber = energy scale) immediately delivers the quantum yield Φ_F.

Table 6.1 lists the quantum efficiencies of some typical fluorophores with optimum excitation and emission wavelengths (for more information see Wolfbeis, O. S. *The Fluorescence of Organic Natural Products*, in Schulman, S. G., ed. *Molecular Fluorescence Spectroscopy*, Wiley, New York, 1985).

Table 6.1 Absorption and emission wavelengths and fluorescence quantum efficiencies of some representative molecules.

Molecule/ label	Solvent, comments	Absorption-maximum (nm)	Emission-maximum (nm)	Fluorescence-quantum yield (%)
Tryptophan	pH 7, H2O	280	348	14
Tyrosine	pH 7, H2O	274	303	21
Phenyl-alanine	pH 7, H2O	257	282	4
Lumiflavin	pH 7, H2O	450	540	34
Dansyl chloride	Covalent protein bonding	330	510	10
Ethidium bromide	Nucleic acid bonding	515	600	ca. 100
Fluorescein	0.1 N NaOH	490	515	85
Quinine disulfate	0.1 N H2SO4	250	461	55
Rhodamine B	Ethanol	544	571	73
Anthracene	Ethanol	252	400	30
Eosin	0.1 N NaOH	518	540	23
Chlorophyll a	Ethanol	666	675	23
Thionine	0.1 N NaOH	598	621	2,5

6.5.4
Fluorescence Quenching

Radiationless relaxation of the S_1 state can be induced by various mechanisms that lead to fluorescence quenching (see Figure 6.1). This phenomenon is often utilized in the analysis of fluorescence. A great number of specific quenchers can be used, as not all fluorophores are influenced in the same way by different factors. For example, the accessibility of a certain fluorophore can be monitored by individual selective quenchers. This poses many questions. Is the fluorophore embedded in the inner part of a macromolecule? Where exactly is it localized in a membrane? What is the magnitude of the permeability of a membrane to the quencher? What is the diffusion coefficient D of a quencher or fluorophore? Molecular fluorescence is generally quenched by molecular oxygen, xenon, iodide or heavy metal atoms, as these induce an $S_1 \rightarrow T_1$ transition (k_{ISC}) by spin-orbit coupling (see Figure 2.12). For example, the apparent fluorescence quantum yield Φ_F, being controlled by the quenching processes, serves as an inverse measure for other deactivation pathways. The larger the k_{ET}, the smaller the fluorescence quantum yield is [however, the inverse conclusion is not necessarily true, Eq. (6.33)].

From Eq. (6.35) the quantum yield of fluorescence is defined by

$$\Phi_F = \frac{k_F}{\sum k_i} \tag{6.34}$$

In the presence of a quencher Q, which reacts with a rate constant k the quantum yield is decreased:

$$\Phi_Q = \frac{k_F}{\sum k_i + k \ [Q]} \tag{6.35}$$

With $t = 1/Sk_i$ the quotient of Eqs. (6.34) and (6.35) yields

$$\frac{\Phi_F}{\Phi_Q} = \frac{\sum k_i + k \ [Q]}{\sum k_i} = 1 + k \times [Q] \times \tau \tag{6.36}$$

For simplicity, we do not measure the quantum efficiencies themselves, but rather the fluorescence intensities with (I_F) and without the quencher (I_Q). Defining the so-called dynamic quenching constant $K \times [Q] = k \times [Q] \times \tau$ we obtain the so-called Stern–Volmer equation

$$\frac{I_F}{I_Q} = 1 + K \times [Q] \tag{6.37}$$

Plotting $[(I_F/I_Q) - 1]$ versus $[Q]$, we obtain a straight line with the slope K. For example, fluorescence of lumiflavin is efficiently quenched by iodide. The Stern–Volmer plot clearly yields the slope $K = 110 \ M^{-1}$ (Figure 6.29). Using the well known fluorescence lifetime of $\tau = 5 \times 10^{-9}$ s, we obtain $k = 22 \times 10^9$, indicative of a diffusion controlled fluorescence quenching. With the analog measurement, for membrane bound flavin (Figure 6.20) we obtain a quenching constant which is 500 times smaller: the membrane effectively protects it from external influences.

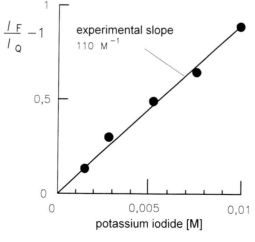

Figure 6.29 If according to the Stern–Volmer equation $(I_F/I_Q - 1)$ is plotted against $[Q]$ for the fluorescence quenching of flavin by iodide, a straight line through the origin with the slope $K = 110 \ M^{-1}$ is obtained.

6.6
Selected Topics

6.6.1
Lanthanides and Actinides

The fluorimetric analysis of inorganic substances is rather limited. The only species which fluoresce in solution are the ions of rare earth elements such as cerium, praseodymium, neodymium up to lutetium. However, commercially much more interesting are the actinides such as, thorium, protactinium or uranium. The superficial explanation of their fluorescence is that the radiationless deactivation is inhibited. But why? If we look at the electronic arrangement of these elements it is obvious that the f-shells are successively filled, as the xenon backbone with 5s- and 5p-electrons is already complete (see Figure 2.9 and Appendix A4) which presents a type of "shield" against exogenous influences. The very narrow bands and lines of the fluorescence spectra of the chelates, the extremely long lifetime of up to 1 μs and the absence of a mirror character of the fluorescence and absorption spectra are particularly interesting. This is all explained by low lying f-atomic levels, which are filled by *intersystem crossing* from higher lying molecular triplet states.

6.6.2
Proteins

Apart from the chromophore groups of many proteins (e. g., heme, flavin, retinal, phytochrome) only the aromatic amino acids phenylanaline, tyrosine and tryptophan fluoresce (see Table 6.1). Their absorption and emission spectra are shown

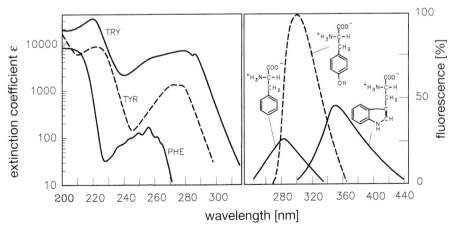

Figure 6.30 Absorption spectra (left, logarithmic plot) and fluorescence spectra (right) of the three fluorescing amino acids tryptophan, tyrosin and phenylalanin (from Wetlaufer, *Advan. Protein Chem.* 1962, *17*, 303 and Teale, Weber, *Biochem. J.* 1957, *65*, 476, modified).

in Figure 6.30. Phenylalanine exhibits an absorption maximum at 250 nm, but has a very small absorption coefficient and a small quantum yield on fluorescence and is, therefore, of less significance analytically. The typical absorption and fluorescence of proteins is basically dominated by tyrosine and, particularly, tryptophan. An analytically important fact is that effects due to tryptophan in particular are observed at excitation wavelengths above 295 nm (tyrosine barely absorbs at this wavelength).

6.6.3
Labels and Probes

The intrinsic fluorescence of cells and tissues is often weak, complex and fairly unspecific. It arises mainly from nucleic acids, proteins or metabolic products such as porphyrins and also from pigments and coenzymes. The latter are of great importance as they are involved in certain metabolic processes which somehow mirror their functionality. For example, the autofluorescence of many cells upon excitation with blue or close to ultraviolet light is mainly caused by nicotinamide adenine dinucleotide (NADH), flavins and pterins, depending on their oxidation state. Metabolic disorders, e. g., within the respiratory chain of mitochondria, are reflected by a lower oxidation of NADH and thus a higher fluorescence activity.

The so-called fluorescence markers that are used to label certain cellular locations or organelles in a highly specific way are important. There are, e. g., acridine orange derivates to label cell nuclei and lysosomes, rhodamine 123 and carbocyanines for mitochondria or 1-aniline naphthalene sulfonate (ANS) to investigate membranes. In immuno-fluorescence, the fluorescein and rhodamine dyes (FITC = fluorescein isothiocyanate, TRITC = tetramethylrhodamine isothiocyanate) are preferred. These are covalently bound to antibodies or antigens, and allow immediate monitoring of highly specific antigen–antibody reactions. Typical fluorophores of these reactions are fluorescein, rhodamine or naphthol, which, in favored cases, can be detected at concentrations as low as 10^{-18} M. For the specific, quantitative detection of enzymes certain fluorophores, such as 4-methylumbelliferon (4-MU), are coupled to the corresponding substrate. In this form they are non-fluorescent as they have not been converted from their esters into the strongly fluorescent phenolates (i. e., the dissociated phenol) by adiabatic (photo)dissociation. This fluorescence quenching is eliminated by enzymatic de-coupling of, e. g., 4-MU, and the fluorescence increases in proportion to the corresponding enzyme. Commercial labeling kits for the detection of a large number of important enzymes are available.

6.6.4
Chelates

Metals are often detected fluorimetrically by means of chelating agents. Typically, these possess two functional groups, one carrying a labile hydrogen atom, the other a lone pair of electrons. The groups of such a reagent must have the correct

Figure 6.31 8-Hydroxyquinoline (oxine) is used in various fields of analytical chemistry such as gravimetry or colorimetry. It forms complexes with a range of metals such as magnesium (planar) or aluminum ions (octahedral). Through ring closure, rigidity and size of the chromophore and thus its absorption coefficient ε are increased. On the other hand a lone electron pair is accepted by the metal, destroying the possibility of a low lying $n \rightarrow \pi^*$ transition: thus the complex fluoresces.

steric arrangement in order to form a five- or six-membered ring with the metal. There are thousands of such complex forming agents, and many more of still greater specificity for particular metals and certain physico-chemical properties are being developed.

Classical examples are morin for beryllium and aluminum ($\lambda_{ex} = 420$, $\lambda_{em} = 525$ nm), or 8-hydroxyquinoline (oxine), which complexes a large variety of metals. The complex with magnesium (planar) or aluminum ions (octahedral) is uncharged (Figure 6.31). For better solubility in water the sulfur group (SO_3H) is incorporated at position 5. Through ring formation the rigidity and size of the chromophore are increased, thereby increasing the absorption coefficient ε. On the other hand, an electron lone pair is accepted by the metal thereby eliminating the possibility of a low lying $n \rightarrow \pi^*$ transition and the complex is fluorescent.

6.6.5
Determination of Calcium

A series of markers have been developed for the determination of intracellular calcium (Ca^{2+}) in its role as a *messenger*. The most important of these is FURA II. This molecule combines the properties of tetracarboxylated chelators such as EDTA (ethylenediaminetetraacetic acid) with the strong fluorescence of stilbenes (Figure 6.10). Upon binding of the calcium the absorption maximum is hypsochromically shifted from $\lambda = 362$ to 335 nm (Figure 6.32), while the emission spectrum is only shifted by 7 nm from $\lambda = 512$ to 505 nm. The specificity for Ca^{2+} – compared with Mg^{2+} – is very high. The dissociation constant at ionic strengths of 0.1 is approximately 6 mM for magnesium, but only approximately

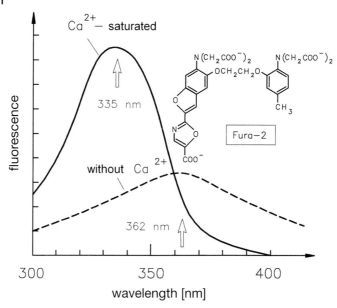

Figure 6.32 Determination of intracellular calcium (Ca^{2+}) by FURA II. Upon complexing the calcium, the fluorescence excitation spectrum is shifted and changed in a characteristic manner from λ = 362 to 335 nm. This allows the quantitative measurement of calcium.

200 nM for Ca^{2+}. For the determination of $[Ca^{2+}]$ the fluorescence signal is measured at λ = 510 nm with alternating excitation at 340 and 380 nm. The ratio of the corresponding fluorescence signals together with the known absorption coefficients and fluorescence quantum yields (0.33 for the bound complex and 0.27 for the free molecule) allow the calcium concentration to be determined. We will not go into the details here but refer the reader to the original work (e. g., Grynkiewicz et al. *J. Biol. Chem.* 1966, *6*, 3440).

6.6.6
Total Fluorimetry

In fluorescence spectroscopy typically excitation or emission wavelengths are varied; and the corresponding spectra are termed excitation and emission spectra (or fluorescence spectra for short), respectively. In order to obtain the complete spectral information of a mixture of many fluorescent components the excitation wavelength must be scanned stepwise; and for each single step the corresponding emission spectrum must be obtained. All excitation and emission wavelengths are then plotted in an orthogonal x–y coordinate system. The emission signal is either plotted as contour lines or in a 3D-manner (Figure 6.33). In this way we obtain a characteristic spectral "landscape" for each sample.

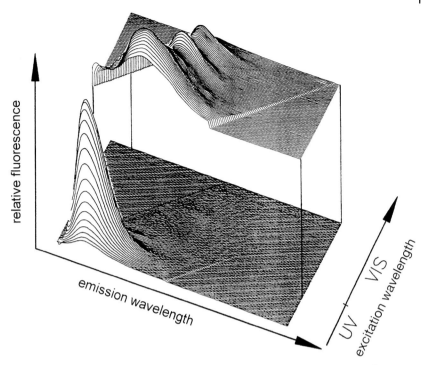

Figure 6.33 Complete fluorescence spectra of human serum in phosphate buffer, visualized in a three-dimensional projection. The fluorescence intensities in the visible wavelength range are plotted in a 800-times magnification in the upper part of the image (6.7 mmol L^{-1}, pH 7.4, modified from Leiner, M. J.P. and Wolfbeis, O. S., *Time-resolved Laser Spectroscopy in Biochemistry*, SPIE, International Society for Optical Engineering, Bellingham, WA., 1988, 909).

For routine measurements of this type, most typical fluorimeters are not practical, as such a measurement would last several hours. Currently spectral resolution of such total spectra is less significant and we use special spectrometer systems based on a simple principle. Firstly, we generate a spectrally resolved "fan with a prism" (see Figure 1.2), which is directed as a "multi-spectral" light source onto a special cuvette containing the sample. A second prism at right angles to the first prism disperses the fluorescent light emitted from the various spots of the cuvette (various excitation wavelengths). In this way we obtain the required two-dimensional spectral distribution, where each point contains information on the "fluorescence strength". Finally, a space-resolving, image-processing detector is necessary, for example, a typical CCD-camera, and we obtain the required three-dimensional presentation (*excitation–emission matrix*, EEM). Required corrections of the non-linear dispersion of the prisms and the spectral sensitivity of the lamp and camera are performed by the on-line computer. Figure 6.33 shows the total fluorescence spectrum of human serum in phosphate buffer in a three-dimensional plot (see Leiner et al., *Total Luminescence Spectrometry and*

Its Application in the Biomedical Sciences in Baeyens et al. eds., *Luminescence Techniques in Chemical and Biochemical Analysis*).

6.6.7
Fluorescence Sensors

Chemical sensors allow the continuous and automatic measurement of chemical parameters in *real time* in research, industry and medicine. In an ideal case sampling and sample preparation are not required. Amongst the classical electrodes for the measurement of various components such as oxygen, pH or different metal ions, so-called optodes (sometimes optrodes) are increasingly gaining importance. In contrast to the former, the measurements are not electric but – corresponding to their name – of an optical nature. Owing to the great sensitivity, fluorescence measurements are much more important than absorption measurements. We can distinguish two types of optodes. The first is simply a light fiber with appropriate specifications (length, diameter, material). With this type, e.g., fluorescing uranium compounds in dangerous, remote and inaccessible places can be measured, the infrared absorption of methane in caves or the NADH (nicotinamide adenine dinucleotide) fluorescence in bioreactors. The second type of optodes depend on a specific indicator or an enzyme which is attached (immobilized) at one end. Specific reactions taking place there can be monitored optically.

Unfortunately, many interesting substances do not fluoresce or absorb in the visible wavelength range (which is easy to handle). In these cases we can utilize well established colorimetry (indicator-chemistry). For nearly all molecules and ions of interest, e.g., in biochemistry or environmental analysis, we know the reagents with which they form reversible, specifically colored and/or fluorescing products, which lead to quantitative identification (cf. photometric analysis by Lange–Vejdêlek, *Verlag Chemie*, 1980). A typical example is the oxygen-optode, which depends on fluorescence quenching of fluorophores by molecular oxygen. The indicator is dissolved in a polymer such as a silicone rubber, and attached to one end of a glass fiber through which the excitation as well as the emission light is guided (Figure 6.34, here very thin), e.g., for oxygen measurements in tiny

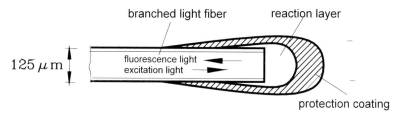

Figure 6.34 Oxygen optode, which relies on fluorescence quenching of many fluorophores by molecular oxygen. The indicator is dissolved in a polymer such as a silicone rubber and applied to the end of a (here very thin) glass fiber through which excitation as well as emission light is guided. In addition, a specific thin layer serves for mechanical protection (from Wolfbeis, O. S. in *Optical Fiber Sensors*, Springer, Berlin, 1989, modified).

Table 6.2 Typical optodes, their detection principle and ranges (from Wolfbeis).

Detected molecule	Detection principle	Range
Oxygen	Fluorescence quenching	0.5–300 Torr
	Phosphorescence quenching	0.0005–0.1 Torr
CO_2	By pH measurement in the internal buffer	0.5–100 Torr
SO_2	Fluorescence quenching	70 ppm–5 %
Toxic gases	Absorption in the visible	0.01–5 %
Sodium and potassium ions	Chromo-ionophores	1–100 mmol L^{-1}
Halogens	Fluorescence quenching	20–200 mmol L^{-1}
Nitrate	Ion coextraction	5–100 mg L^{-1}
Glucose	Enzymatically via oxygen consumption	1–30 mmol L^{-1}
Hydrogen	By interferometry	5–2000 ppm

blood vessels. Various examples of detection ranges are listed in Table 6.2 (from Wolfbeis, in *Optical Fiber Sensors*, Springer, 1989).

6.6.8
The Pulse Amplitude Modulation Fluorimeter (PAM)

Photosynthesis, i. e., the synthesis of carbohydrates from water and CO_2 through light by plants, is the key process to life. Generations of biologists have studied photosynthesis using Warburg manometry and later the oxygen electrode. The concomitant chlorophyll fluorescence originates largely in photosystem II (PSII) rather than photosystem I (PSI) [Figure 6.35(b)] and was viewed originally as ancillary. However, theory and measurement of chlorophyll fluorescence have progressed considerably during recent decades and – concomitant with simple sample preparation – the conventional gas exchange measurements have largely been replaced. The now worldwide dominant so-called **p**ulse **a**mplitude **m**odulation (PAM) fluorimeter was developed by Schreiber of the University of Würzburg (Germany) and is manufactured and marketed by Walz in Effeltrich (Germany). There is a simple and clear relationship between chlorophyll fluorescence and the efficiency of photosynthetic energy conversion:

100 % of the impinging light energy = fluorescence + photochemistry + heat

or in short,

$E = F + P + D = 1$ (or 100 %, "D" is dissipation or heat)

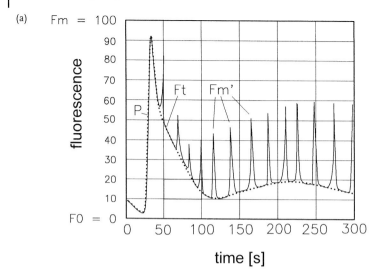

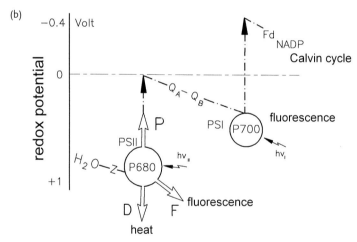

Figure 6.35 (a) Modified fluorescence induction kinetics (Kautsky effect) as monitored by the PAM fluorimeter. F_0 Initial small fluorescence of dark adapted specimen; P fluorescence maximum during fluorescence induction; F_t fluorescence as function of time; F_m maximum fluorescence as caused by saturating red light pulses; and F_f final asymptotic fluorescence (for details see text).
(b) Typical Z-scheme of photosynthesis showing the electron path through the two coupled photosystems and energetic relationships. Chlorophyll fluorescence *in vivo* is mainly generated by photosystem II (PSII).

This is an equation with three unknowns, where we are ultimately interested in the photochemistry P, i.e., the conversion of light energy in biochemical/biophysical equivalents. If we could determine the parameter "fluorescence" F and "heat" D, we would obtain the third parameter: "photochemistry" P, and this can be done with a PAM fluorimeter.

If we irradiate (as with a classical chlorophyll fluorimeter) a dark adapted, photosynthetically active plant (e. g., its leaves), we observe the fascinating phenomenon of "fluorescence induction" [the Kautsky effect, Figure 6.35(a)]. With the onset of a continuous, photosynthetically active red light the fluorescence F increases within 1–2 s from low values (F_0) up to a maximum P and, thereafter it decreases again in a characteristic manner, in a time scale of minutes, down to a low level (F_t): the photosynthetic apparatus is only partially active in darkness. Therefore, electrons can not be transferred as quickly as they are set free from the water by PSII. When 1–2 s after the onset of light the dark reactions become active, electrons are again increasingly withdrawn from PSII, and the fluorescence slowly decreases to a low, stable level. Essentially, the Kautsky kinetics mirror the build up of the proton gradient across the thylakoid membrane and the concomitantly dissipated energy. If now – the ingenious idea of Schreiber – in addition to the continuous light, separated, e. g., by 20 s, very short, saturating (3500–10 000 μmol m^{-2} s^{-1}) red light pulses are applied ($P = 0$), we ensure the maximum output of fluorescence and heat dissipation. From these data, finally, we are able to determine the photochemical quantum yield P of PSII utilizing the following formula (we have omitted its simple derivation):

$$P = 1 - F - D = 1 - F - F\,(1 - F_\mathrm{m})/F_\mathrm{m} = \Delta F/F_\mathrm{m}.$$

6.7
Phosphorescence

In principle, fluorescence and phosphorescence yield analytically comparable results. However, because of the spin-forbidden transition $T_1 \rightarrow S_0$ the so-called intrinsic lifetime (i. e., the lifetime that is not reduced by external influences such as quenchers) of the T_1-state is in the order of milliseconds to seconds, most molecules do not phosphoresce at room temperature and in solution, i. e., they relax radiationless on impact.

Various methods have been developed which inhibit a radiationless, premature relaxation of the first triplet state:
- Cooling of the sample (usually down to approximately −196 C°, liquid nitrogen).
- Incorporation of the phosphorescing molecule into micelles or cage structures such as cyclodextran
- Immobilization within a fixed matrix such as sodium acetate, polyacrylic acid or simply paper.

The simplest and most commonly utilized method for inducing phosphorescence of ionic, organic substances at room temperature is, indeed, the adsorption onto filter paper. For this method the hydrogen bonding of the hydroxyl groups of the

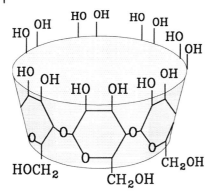

Figure 6.36 Cage structure of cyclodextran, in which the phosphorescing molecule is enclosed and immobilized in order to allow the observation of phosphorescence even at room temperature.

paper play a decisive role: on destruction of the hydroxyl groups by silination the phosphorescence decreases by more than 90%.

Phosphorescence is measured with so-called phosphoroscopes. The sample is irradiated by a series of light pulses, and the sample is measured in the dark time gaps in between. In the classical phosphoroscope of Becquerel this is done using two mechanically coupled rotating disks (Figure 6.37). A typical modification is the rotating-can phosphoroscope where the sample is placed within a cylinder which rotates along its axis. The excitation lamp and detector are mounted orthogonal to each other and radial to the rotation axis. A hole in the cylinder mantle opens the light path from the lamp to the sample and from the sample to the photodetector alternately. The rotational speed has to be adapted to the phosphorescence lifetime. Today a more elegant solution is to use a pulsed light source covering a large frequency range, presumably by a laser or a light emitting diode (LED) of suitable wavelength. The measurement takes place dur-

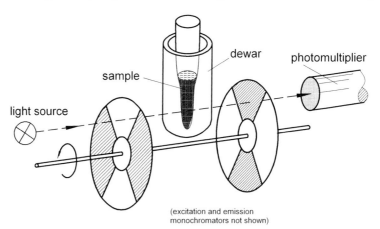

(excitation and emission monochromators not shown)

Figure 6.37 Classical phosphoroscope according to Becquerel. Using coupled rotating shutters a sample placed in a dewar is sequentially excited to phosphoresce. Phosphorescence is measured in the dark periods in between.

ing the dark periods thus avoiding any mobile mechanical parts. The photodetector has to be protected from the excitation light (note: most blocking filters exhibit intrinsic phosphorescence!).

6.8
Chemo- and Photobioluminescence

6.8.1
Chemoluminescence

In chemiluminescence excited molecules, which are also good fluorophores, are generated via strongly exothermic reactions (see Section 6.2). The decisive property of all luminescence methods is their extreme sensitivity in comparison with other common analytical procedures: thus, the detection limits for ATP and NADH are in the femtomole, and of H_2O_2 in the nanomole range. On the one hand this is feasible through the use of photomultipliers, but on the other it depends on the high fluorescence quantum yield of the fluorophores. Three basic mechanisms are discernible: (1) oxidation of a substrate with molecular oxygen, (2) electron transfer and (3) fragmentation. The most powerful chemiluminescence reactions depend on the breakdown of peroxides. The reaction of luminol with hydrogen peroxide is particularly important in analysis:

$$\text{Luminol} + 2H_2O_2 + OH^- \xrightarrow[\text{(Microperoxidase)}]{\text{Catalyst}} \text{Aminophatalat} + N_2 + 3\,H_2O + h\upsilon$$

$$(6.38)$$

where either luminol (in an appropriately chemically modified form) or the peroxidase could serve as the *tracer*.

Many of these processes include the exothermic degradation of intermediary dioxetanes (so-called McCapra mechanisms). On heating, tetramethyl dioxetane only exhibits weak chemiluminescence which, however, strongly increases through energy transfer to a fluorophore (*enhancer*). There is a series of well studied and analytically important examples of chemiluminescence, where an intermediary dioxetane is involved. A model reaction is the thermolysis of tetramethyl-1,2-dioxetane (TMD), which degrades with an efficiency of 0.5 % to singlet acetone and with 50 % to triplet acetone and thereby both fluoresce as well as phosphoresce:

$$\text{TDM} \rightarrow 2(CH_3)_2\,CO + \text{phorescence (430 nm)} + \text{fluorescence (400 nm)} \quad (6.39)$$

$$
\begin{array}{c}
\quad\quad\quad CH_3 \\
\quad\quad\quad\quad \backslash \\
H_3C - C - O \\
\quad\quad | \quad\quad | \\
H_3C - C - O \\
\quad\quad / \\
\quad\quad CH_3
\end{array}
$$

6.8.2
Bioluminescence

From the mechanistic point of view, bioluminescence is a type of chemilumines-cence. The natural bioluminescence of certain glow-worms and bacteria is utilized for analysis in biochemistry, pharmacy and most of all biotechnology. We will dis-cuss the most important applications. In all cases this is a so-called luciferase (= enzyme) catalyzed oxidation of luciferine (= substrate) by molecular oxygen (an energy and light delivering reaction).

6.8.2.1 Bioluminescence of Glow-worms

The determination of ATP/ATPase is performed routinely in a concentration range between 10^{-6} and 10^{-13} M. For the detection of extraterrestrial life, NASA reported a sensitivity limit of as low as 2×10^{-17} M, in order to detect contamina-tion by bacteria and thus the slightest traces of life. The basis for this is the sys-tem luciferin (LH2)/luciferase (E) of the American glow-worm *Photinus pyralis*. A reaction chain with ATP leads to the emission of light:

$$E + LH_2 + ATP \overset{Mg^{2+}}{\leftrightarrow} E - LH_2 - AMP + PPi$$
$$E - LH_2 - AMP + O_2 \rightarrow \text{oxyluciferin} + AMP + CO_2 \qquad (6.40)$$
$$+ \text{ light } (\lambda_{max} = 560 \text{ nm})$$

The intensity of light is directly proportional to the ATP concentration. The half-life of these reactions is only a few seconds. The relevant commercial success was created by modification of this basic reaction to slow down the decrease in light intensity in order to simplify the measurements. Thus, various luciferin/luci-ferase "kits" are available today.

6.8.2.2 Bacterial Bioluminescence

Another luciferin/luciferase system is obtained from certain oceanographic bac-teria. In principle the mechanisms involved and the generation of light are always the same. Thus, in the following coupled reaction, NAD(P)H can be determined in the concentration range between 10^{-7} and at least 10^{-11} M, and on the basis of a nitrogen laser even lower. Therefore, the fluorimetric determination of NAD(P)H is at least 1000- to 10000-times more sensitive than the photometric method.

The flavin mononucleotide (FMN) is reduced first:

$$NAD(P)H + FMN \overset{\text{FMN reductase}}{\rightarrow} NAD(P) + FMNH_2 \qquad (6.41)$$

and subsequently via an aliphatic, long-chain aldehyde with at least eight carbon atoms under light emission is then reoxidized

$$FMNH_2 + RCHO + O_2 \rightarrow FMN + RCOOH + H_2O + h\upsilon \text{ (490 nm)} \qquad (6.42)$$

It is with this reaction that fat becoming rancid is associated. This process is always connected with the generation of long chain aldehydes, which are stochiometrically required for the above reaction.

6.8.2.3 Enzyme-catalyzed Systems that Produce Hydrogen Peroxide

Various oxidase-catalyzed reactions generate hydrogen peroxide:

$$\beta\text{-D-Glucose} + O_2 \xrightarrow{\text{Glucose oxidase}} \Delta\text{-Glucono-1,5-lactone} + H_2O_2$$

$$\text{Cholesterol} + O_2 \xrightarrow{\text{Cholesterol oxidase}} \text{Cholest-4-ene-3-one} + H_2O_2 \tag{6.43}$$

$$\text{D-Glutamate} + O_2 \xrightarrow{\text{D-Glutamate oxidase}} \text{2-Oxoglutarate} + NH_3 + H_2O_2$$

$$\text{L-Lactate} + O_2 \xrightarrow{\text{Lactate oxidase}} \text{Pyruvate} + H_2O_2$$

$$\text{Pyruvate} + HPO_4^{2-} \xrightarrow{\text{Pyruvate oxidase}} \text{Acetyl phosphate} + CO_2 + H_2O_2$$

H_2O_2, in turn, can be determined as discussed in Section 6.8.1.

6.8.2.4 Instrumentation

Various more or less sophisticated "luminometers" have been developed over the years. There are very simple, battery powered instruments that are used in the field, in addition to rather complex and thermostatically controlled set-ups with automatic sample injection, flow cuvettes and a microcomputer interface. However, even the most simple apparatus allows two modes of measurement: (1) of maximum light emission and (2) of integrated light emission. In the latter mode the emitted light is typically integrated between approximately 5 and 60 s. Sometimes also the rate of the initial increase in light emission can be determined quantitatively (Figure 6.38). The absolute detection limit for chemiluminescence is determined by the dark current of photomultipliers, as these are still the most sensitive photodetectors. It can lowered still further by cooling the photomultiplier. The response near to the detection limit is strictly linear.

6.8.2.5 Ultraweak Luminescence

Chemo- and bioluminescence is called *ultraweak* if there are only a few photons per second emitted. In this case the most sensitive *single photon counting* method is applied (cf. Figure 6.26). It is of no surprise that at the detection limit of only a few photons per second virtually all organic samples such as cells, tissues, blood suspensions, etc., exhibit a weak light emission modified by various factors (e. g., cancer, diabetes). Naturally, the interpretation is difficult, even if these signals might carry valuable information (Popp, F.A. *Biophotonen*, Dr. E. Fischer Publisher, Heidelberg, 1976).

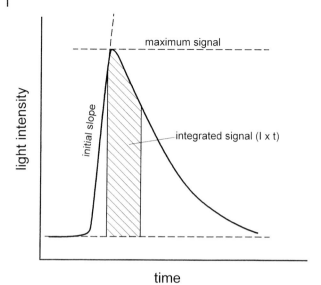

Figure 6.38 Luminometers allow two measuring modes: (1) the measurement of maximum light emission and (2) of the integrated light emission. In the latter mode the emitted light is typically integrated between 5 and 60 s ($I \times t$). Sometimes the initial increase in light emission is also determined.

6.9
Delayed Luminescence

6.9.1
Fundamental Principles

Delayed luminescence is defined as light emission of excited molecules that lasts longer than the lifetime τ_0 of prompt/pulse fluorescence and which sometimes can be observed for minutes and even hours. The photochemist distinguishes between two types of delayed luminescence (DF). In E-type DF (after eosin, a molecule where this type was discovered in 1930), energy flows from the excited T_1-state, concomitant with reception of activation energy $E^{\neq}$, uphill back to the S_1-state (Figure 6.39) and is observed for molecules with activation energies between 20 and 90 kJ mol^{-1}. The lifetime corresponds to that of phosphorescence, however the spectrum to that of fluorescence. Because the process is monomolecular, the signal is proportional to the light energy.

In a second process molecules in their $T1$-state react to an excimer (singlet), from which an S_0-(ground) and a S_1- (excited) state is generated from where fluorescence light with the expected spectrum is emitted:

$$T_1 + T_1 \rightarrow (2T) \rightarrow S_1 + S_0 \tag{6.44}$$

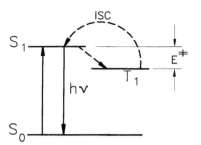

Figure 6.39 After reception of the difference (and activation) energy $E^{\neq}$, in E-type DF the electron jumps from the T_1-state (excited triplet) back to the S_1-state (singlet), which finally emits typical fluorescence light as the corresponding emission spectra demonstrate.

The decay constant is twice that of phosphorescence, according to a bimolecular reaction. As expected, the signal is proportional to I^2 and depends strongly on the concentration as well as on the viscosity and temperature of the solvent. Because this type of luminescence was observed for the first time with the solvent pyrene, Parker and Hatchard classified it as P-type DF (*Trans. Faraday Soc.*, 1963, 59, 284).

Of particular interest biologically is a more complex type of DF, the so-called recombination luminescence which is generated by whole molecular systems. Three criteria have to be met which we clarify on the basis of a *potential surface* (Figure 6.40):

- ΔG_0 has to be equal or larger then hc/λ_m, where ΔG_0 describes the free energy and λ_m the long wavelength limit for the excitation of the product molecule.

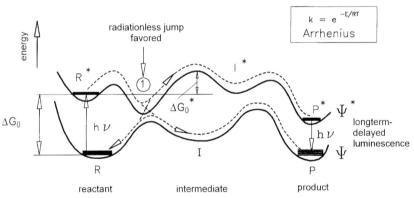

Figure 6.40 Schematic two-dimensional potential diagram of a complex system consisting of a ground state surface and a further surface in the first excited state, both being defined fully by the wave functions ψ and ψ^*. There are two stable states R (reactant) and P (product), which can be converted into each other by light. The light energy yields the required reaction enthalpy (G_0) for the transition $R \rightarrow R^*$. Acceptance of further activation energy (ΔG_0) from the environment converts the system via various intermediates I* into the excited product state P*, from where it finally, with the emission of normal fluorescence light, relaxes to the ground state P. The delay of light emission is determined by the Arrhenius equation which includes ΔG_0. The quantum yield of luminescence is defined by radiationless transitions taking place particularly in cases where the potential surfaces ψ and ψ^* come close to each other (1).

- There must be an effective reaction path from the originally excited molecule *R* (reactant) to the molecule *P* (product) in order not to loose excitation energy by radiationless de-excitation.
- The fluorescence quantum yield Φ of the product *P* has to be adequate, i. e., the distance between both potential curves has to be sufficiently large.

If the last requirement is not fulfilled, the possibility remains of generating luminescence by a strongly luminescent molecule (*enhancer*). Luciferin, 6-hydroxybenzothiazol or *p*-iodophenol are often used as enhancers. We are frequently dealing with delayed fluorescence where the light is emitted from the first excited singlet state S_1 of a product molecule *P*, even if phosphorescence (emission from T_1) sometime takes place.

Recombination luminescence is observed, e. g., in photosynthetic, dark adapted organisms. The normal electron transport takes place – when light quanta of the two photosystems PSII and PSI connected in series are received – from the electropositive site (H_2O) to the electronegative site (NADP) of the electron transport chain. Concomitantly a large electrical and a large proton gradient is built up across the thylakoid membrane (several millions of volt V cm^{-1} and at least 3 pH units, inside acid and positive). This results in a (very small) reverse leakage current (not "required" by nature). Even though this cannot of course be seen by the naked eye, it can be measured with sensitive photomultipliers and monitored kinetically. During the initial seconds the kinetics exhibit a highly structured starting phase which reflects the time course of the electrical potential gradient Ψ and the subsequent proton gradient ΔpH across the thylakoid membrane (Figure 6.41). The emission spectrum shown here differs markedly from the spectra of prompt fluorescence excluding a simple P- or E-type DF (Schmidt, W., Senger, H. *Biochim. Biophys. Acta, 1987, 890,* 15). The kinetics allow, similar to the induction kinetics of prompt/pulse fluorescence (PAM fluorimeter, Section 6.6.8) the electron transport chain and the vitality of the particular plant to be determined. Therefore, this type of kinetics has been successfully utilized as an analytical tool in forest decline research (Schmidt, W., Schneckenburger, H. *Photochem. Photobiol. 1995, 62, 4,* 745–750).

6.9.2
Technical Requirements

One of the reasons for the low popularity of delayed luminescence in environmental analysis is probably its complexity as well as its low intensity. Currents of light quanta as low as 10^{-16} E cm^{-2} s^{-1} are typical (cf. daylight of approximately 10^{-10} E cm^{-2} s^{-1}). This requires specific measuring set-ups. Normal fluorimeters are not sensitive enough to detect 10^7 quanta cm^{-2} s^{-1} (sun light is 10^{16} quanta cm^{-2} s^{-1}). We can distinguish two types of spectrophotometers for measuring delayed luminescence. These differ markedly in sensitivity, price and specification. Bio- and chemiluminescence are easily monitored with low cost, simple spectro-

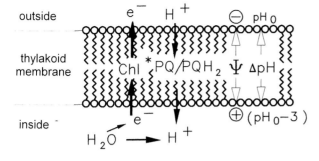

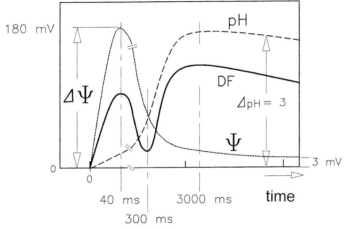

Figure 6.41 Top: The light-driven electron and proton transportation across the thylakoid membrane of photosynthetic organisms is mediated by chlorophyll (Chl). It leads to an electric (ψ) and concomitantly a proton gradient ΔpH (Chl = chlorophyll, PQ = plastoquinone). Bottom: The complex time course of light-induced delayed luminescence (DF) of photosynthetic and dark adjusted organisms is explained by the very small, but significant and reverse electron current across the thylakoid membrane – with concomitant light emission. These DF kinetics were measured in dark adjusted green mature leafs (most of them show similar results) with a small laboratory-made, computer-controlled fluorimeter. Irradiation was performed in the millisecond time scale with red rectangular light pulses, measurements taken between the dark periods by a red-sensitive end-on photomultiplier.

photometers using dc techniques. These set-ups are offered specifically for this purpose by various manufacturers (detection limit of several 10^5 quanta s^{-1}). Even if these simple set-ups are not capable of measuring single photons, they are sufficiently sensitive, without compensation for the background radiation. For this purpose the sample to be analyzed is in close proximity to the photocathode of the photomultiplier and – after adequate magnification – the photocurrent is measured. If the emitted signal has to be measured within a wavelength range, a broad band (good transmission) filter is placed between the sample and the photomultiplier. Owing to the low light intensities, the use of wavelength scanning monochromators is not feasible (except in Fourier spectrometers).

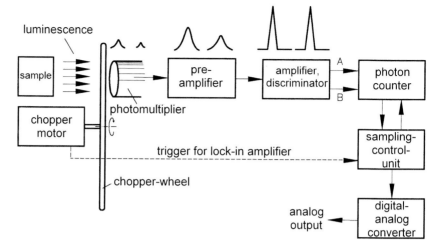

Figure 6.42 Set-up for measuring extremely small light quantum fluxes (*single photon counting, ultraweak luminescence*). The sequence of light pulses emanating from the sample is sequentially covered by a mechanical rotating chopper and detected by an end-on photomultiplier. After amplification and discri- mination ("unit pulse") the individual signals are further processed and counted. During the period when the sample is shadowed the non- specific background is measured and finally subtracted from the total signal (background + sample signal).

For ultraweak light signals (*ultraweak luminescence*) we use the technique of *single photon counting*. With biological/medical samples, it is usually the background radiation not originating in the sample itself that disturbs the measurement. This is largely excluded by use of a reference measurement. A rotating chopper se- quentially closes the light path between the sample and photodetector thus allow- ing the combined measurement of sample *and* background, or just background. By subtraction of both signals we obtain the corrected sample signal (Figure 6.42). The photomultiplier pulses are initially amplified, normalized in a discriminator and finally amplified and electronically counted. With this method very low quan- tum fluxes as low as a few photons per second can be measured.

6.9.3
Thermoluminescence

The induction of delayed luminescence by heat is called *thermoluminescence* (see Figure 1.1). So-called *glow curves* represent the thermoluminescence of photosyn- thetic organisms (cf. Prafullachandra and Rutherford, 1986). (1) Photosyntheti- cally active organisms/samples are frozen to the liquid nitrogen temperature of $-196\,°C$ while irradiated with white light. (2) On subsequent heating in darkness up to $60\,°C$ they exhibit *thermoluminescence* at specific temperatures, a character- istic light emission of specific wavelengths. The analysis of these glow curves yields valuable information about the photosynthetic electron transport chain.

Thermoluminescence has been used for years with great success for dating archeologically interesting ceramics and minerals. On this basis, countless artifacts have been unmasked as fakes. The time range covered is large. Pottery from the Viking era that was 900 to 1000 years old has been dated precisely. However, thermoluminescence is of particular interest for time scales of more than 50000 years where the more usual radiocarbon method (^{14}C) fails.

The basics of thermoluminescence analysis is simple. All minerals represent a specific crystal structure. By natural radioactive irradiation they become intrinsically ionized to a certain extent and electrons are permanently stored in so-called traps within the lattice. Only by heating to several hundred degrees Celsius is this structure loosened with respect to the vibration. The electrons become mobile as free electrons within the lattice (more precisely conduction bands) and are finally trapped again by so-called *luminescence centers*, with or without emitting light. Luminescence centers are represented by metals such as Ag^{2+} or Mn^{2+} which determine the *thermoluminescence*; however the method requires a lot of technical experience.

Finally, the successful application of thermoluminescence in food control needs to be mentioned. In many countries food is conserved by radioactive irradiation. However, it is often forbidden by law (e. g., in Germany). Thermoluminescence offers a safe and fast analytical method of detecting radioactive conservation of food.

6.10
Selected Further Reading

Aitken, M. J. *An Introduction to Optical Dating: The Dating of Quaternary Sediments by the Use of Photon-Stimulated Luminescence*, Oxford University Press, Oxford, **1998**.

Bach, P. H., Reynolds, C. H., Clark, J. M. *Biotechnology Applications of Microinjection, Microscopic Imaging and Fluorescence*, Plenum, New York, **1993**.

Baeyens, W. R.G., de Keukaleire, D., Korkidis, K. eds., Marcel Decker, New York, **1991**.

Barnes, R. M., ed. *Emission Spectroscopy*, Dowden, Hutchinson & Ross, **1976** (a historically oriented, general application book of optical spectrocopic emission methods).

Boetter-Jensen, L. *Optically Stimulated Luminescence Dosimetry: Survey of Theory and Applications*, Elsevier Science, Amsterdam, **2003**.

Brand, L., ed., *Fluorescence Spectroscopy, Methods in Enzymology*, Vol. 278, Academic Press, New York, London, **1997**.

Brandl. H., Albrecht, S. *Experimentelle Chemolumineszenz und Biolumineszenz*, Hütling Verlag, Heidelberg, **1998** (in German).

Cundall, R. B., Dale, R. E. *Time Resolved Fluorescence Spectroscopy in Biochemistry and Biology*, Plenum, New York, London, **1983**.

Deell, J. R., Toivonen, M. A. *Practical Applications of Chlorophyll Fluorescence in Plant Biology*, Kluwer Academic Publishers, **2003**.

Demas, J. N. *Excited State Lifetime Measurements*, Academic Press, New York, London, **1983**.

Demtröder, W. *Laser Spectroscopy*, Springer Verlag, Berlin, , **1998** (general and comprehensive textbook, in German).

Dewey, T. G. *Biophysical and Biochemical Aspects of Fluorescence Spectroscopy*, Plenum, New York, **1991**.

Dick, J. R. *Fluorescence*, University of Georgia Press, Atlanta, **2004**.

van Dyke, K. *Luminescence Biotechnology*, CRC Press, Boca Raton, FL, **2001**.

Garcia-Campana, A. M., Baeyens, W. R.G. *Chemiluminescence in Analytical Chemistry*, Marcel Dekker, New York, **2001**.

Guilbault, G. G. *Practical Fluorescence*, 2nd revised edn., Marcel Dekker, New York, **1990**.

Geddes, C. D. *Who's Who in Fluorescence 2004*, Plenum, New York, **2004**.

Grynkiewicz, G., Poenie, M., Tsien, R. Y. *J. Biol. Chem.*, **1966**, 6, 3440–3450.

Haugland, R. P. *Handbook of Fluorescence Probes and Research Products*, 9th edn., Molecular Probes Inc. **2004** (a valuable book, includes theory and practice).

Hurtubise, R. *Phosphorimetry: Theory, Instrumentation, and Applications*, Verlag Chemie (VCH), Weinheim, **1990** (one of the rare books on phosphorimetry).

Jameson, D. M., Reinhart, G. D., eds. *Fluorescent Biomolecules: Methodologies and Applications*, Plenum, New York, London, **1989**.

Jursinic, P. A. *Delayed Fluorescence: Current Concepts and Status*, in Govindjee et al., eds. *Light Emission by Plants and Bacteria*, Academic Press, New York, London, **1986**, pp. 291–223.

Lakowicz, J. R., ed. *Topics in Fluorescence Spectroscopy: Techniques*, Plenum, New York, London, **1991**.

Lakowicz, J. R., ed. *Topics in Fluorescence Spectroscopy: Biochemical Applications*, Plenum, New York, London, **1992**.

Lakowicz, J. R., ed. *Topics in Fluorescence Spectroscopy: Nonlinear and Two-photon-induced Fluorescence*, Plenum, New York, London, **1992**.

Lakowicz, J. R., ed. *Topics in Fluorescence Spectroscopy: Principles*, Plenum, New York, London, **1992**.

Lakowicz, J. R., ed. *Topics in Fluorescence Spectroscopy: Probe Design and Chemical Sensing*, Plenum, New York, London, **1992**.

Lakowicz, J. R., Lakowicz. J. *Principles of Fluorescence Spectroscopy*, Plenum, New York, London, **1999**.

Lakowicz, J. R., Thompson R. B., eds. *Advances in Fluorescence Sensing Technology V*, Progress in Biomedical Optics and Imaging, SPIE-International Society for Optical Engineering, Bellingham, WA, **2001**.

Lakowicz, J. R. *Topics in Fluorescence Spectroscopy*, Kluwer Academic Publishers, **2003**.

Lichtenthaler, H. K., ed. *Applications of Chlorophyll Fluorescence in Photosynthesis Research, Stress Physiology, Hydrobiology and Remote Sensing*, Kluwer Academic Publishers, **2002**.

Mycek, M.-A., Pogue, B. W. *Handbook of Biochemical Fluorescence*, Marcel Decker, New York, **2003**.

Nalwa, H. S., Rohwer, L. S. *Handbook of Luminescene, Display Materials, and Devices*, American Scientific Publishers, Stevenson Ranch, California, **2003**.

Niemantsverdriet, J. W. *Spectroscopy in Catalysis, Handbook*, Wiley, New York, **2000**.

Parker, C. A. *Photoluminescence of Solutions*, Elsevier, Amsterdam, **1968**.

Rendell, D. *Fluorescence and Phosphorescence*. John Wiley, New York, **1987**.

Rettig, W., Strehmel, B., Schrader, S., Seifert, N., eds. *Applied Fluorescence in Chemistry, Biology and Medicine*, Springer, Berlin, **1998**.

Rost, M., Karge, E., Klinger, W. *J. Biolumin. Chemilumin.* **1998**, *13(6)*, 355–63.

Schmidt, W., Senger H. *Biochim. Biophys. Acta* **1987**, *890*, 15–22.

Schulman, S., Sharma, A., Schulman, S. G. *An Introduction to Fluorescence Spectroscopy*, Techniques in Analytical Chemistry, John Wiley, New York, **1999**.

Sharma, A., Schulman, S. G. *Introduction to Fluorescence Spectroscopy*, Wiley-Interscience, New York, **1999**.

Slavik, J., ed. *Fluorescence Microscopy and Fluorescent Probes*, Plenum, New York, **1998**.

Szöllösi, J., Damjanovich, S. Mulhern, S. A., Tron, L. *Prog. Biophys. Mol. Biol.* **1987**, *49*, 65–87.

Takahashi, A., Camacho, P., Lechleiter, J. D., Herman, B. *Measurement of Intracellular Calcium*, Physiological Reviews, American Physiological Society, **1999**, Vol. 79, No. 4, pp. 1089–1125.

Taylor D. L., Wang, Y-L, eds. *Fluorescence Microscopy of Living Cells in Culture Part B: Quantitative Fluorescence Microscopy – Imaging and Spectroscopy*, Academic Press, New York, London, **1990**.

Tsien, R. Y. *Trends Neur. Sci.* **1988**, *11*, 419–424.

Wang, X. F., Herman, B. *Fluorescence Imaging Spectroscopy and Microscopy*, Chemical Analysis, Vol. 137, John Wiley, New York, **1996**.

Wilde, E., *Geschichte der Optik 1 + 2*, **1838– 1843**, Sändig Reprint Verlag, H. R. Wohlwend, reprinted unchanged, **1967** (a valuable book on the history of optics; in German).

Wolfbeis, O. S., ed. *Fiber Optic Chemical Sensors and Biosensors*, Vols. 1 and 2, CRC Press, Boca Raton, FL, **1991**.

7
Photoacoustic Spectroscopy

> Longum iter est per praecepta, breve efficax per exempla
> (Seneca)
>
> *Teaching can be awful, but examples make it much more*
> *enjoyable*

7.1
Introduction

Absorption and luminescence spectroscopy, as the main representatives of opto-spectral analysis, have proved to be indispensable in science and industry. Even though it is possible to analyze strongly absorbing and/or scattering samples by absorption (Sections 5.2.8 and 5.5) and fluorescence measurements (Section 6.3.4), there are definite limits. Another measuring technique is required that is not, or is only slightly, limited by the optical consistency of the sample material. Without doubt photoacoustic spectroscopy (PAS), as a complementary method to the frequently used absorption and fluorescence spectroscopy, meets these requirements. Most of all, PAS allows the "optical analysis" and measurement of parameters which are not, or are barely, accessible by other methods using a combination of spectrophotometric and calorimetric methods.

The basic concept of PAS was developed by scientists with great names such as Bell (1881), Rayleigh (1881), Röntgen (1881), Preece (1881) or Tyndall (1888). However, only within the past 30 years has PAS been championed under the leadership of Rosencwaig, Adams and Kirkbright. The potential analytical benefits in chemistry, surface physics, medicine and biosciences were clearly recognized. Nevertheless, in spite of the availability of photoacoustic spectrometers from various manufacturers, PAS is still perceived more as an exotic method rather than an established routine procedure.

Optical Spectroscopy in Chemistry and Life Sciences. W. Schmidt
Copyright © 2005 WILEY-VCH Verlag GmbH & Co. KGaA, Weinheim
ISBN 3-527-29911-4

Alexander Graham Bell (1847–1922)

7.2
Basic Principle of Photoacoustic Spectroscopy (PAS)

The basic idea of PAS is surprisingly simple and is illustrated by Figure 7.1. The molecules of the sample to be investigated are excited by a light beam that is modulated with the frequency f. For a short period of time, the *lifetime* of the excited molecule(s), the excitation energy is stored and on subsequent relaxation is partly or completely released in the form of heat (cf. Figure 6.1). Because the exciting light is intensity modulated, the subsequent heat production will also be

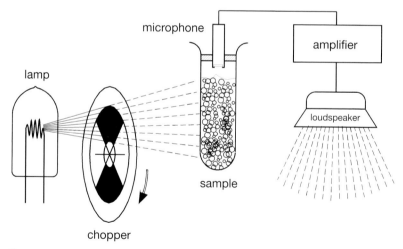

Figure 7.1 Experimental set-up for demonstration of the basic principle of the photoacoustic effect in photoacoustic spectroscopy (*PAS*). A sample is irradiated with a light source. The light beam is interrupted by a rotating chopper. The sample absorbs part of the incident light, and subsequently the absorbed energy released again in the form of heat. Depending on the modulation frequency f, heat emission takes place with corresponding pressure and volume changes, i. e., sound. The latter is converted into an electrical signal by a microphone, which in turn – after appropriate amplification – can be made audible with a loudspeaker.

modulated. The periodic heating of the sample, in turn, leads to a periodic heat flow to the mediating gas above the sample. This leads to periodic volume changes and pressure waves, i.e., sound waves being generated. These are registered by a sensitive microphone, converted into an electrical signal, amplified, measured – or for the purpose of demonstration as shown in Figure 7.1 – converted into a tone by a loud speaker.

During his extensive experiments while working towards the invention of the telephone, around 1880 Sir Alexander Graham Bell, using his *Photophon*, performed light-acoustic experiments hoping to convert acoustic vibrations into electrical signals, to conduct them to a distant receiver and finally to reconvert them into an acoustic signal. In his first trial – initially without any involvement of electricity – he conveyed a light beam that was modulated by a 1000 Hz chopper onto a thin solid diaphragm of any material. With a connecting funnel and without any amplification this signal could be recognized acoustically.

The method is extremely sensitive; temperature changes of as little as $10^{-6}\,°C$ and changes in power density of less than 10^{-6} J cm^{-3} s^{-1} are measurable. Figure 7.2 shows a "PAS cuvette", consisting of a massive aluminum cell and including the sample chamber itself. The relatively large *thermal mass* is important, i.e., the mass compared to the size of the sample chamber. The chamber shown here is tunable to a specific resonance frequency f by a variable cavity. At this frequency it is particularly sensitive exhibiting an extraordinary high signal to noise ratio (SNR).

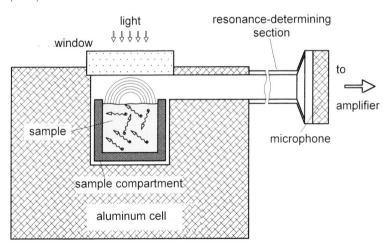

Figure 7.2 Photoacoustic sample chamber with adjustable resonance frequency f. By correct choice of the resonance-determining section, the chamber can be adjusted to the required resonance frequency. At this specific frequency the signal to noise ratio (SNR) is particularly enhanced (Rosencwaig, 1977).

7.3
Theory of Photoacoustic Spectroscopy

7.3.1
Overview

The measured photoacoustic signal is directly proportional to the absorbed energy. On the one hand it depends on the sample depth within which the light absorption takes place, and on the other on the efficiency of heat transport through the sample, from the sample surface to the boundary layer, and from here to the mediating gas and finally to the microphone.

The *boundary layer* is a thin layer of the order of 1 mm and of great theoretical and practical significance in PAS, because it is here that the essential conversion of heat into acoustic waves takes place. In the Anglo Saxon technical literature, where the plane sound wave is generated is termed a *piston*.

Therefore, the measuring signal is a function of all individual steps involved, particularly of the total optical absorption coefficient $\beta(l)$ and the thermal diffusion constant S, where β and S are defined as proportionality constants by

$$\mathrm{d}I = -\beta\,I\,\mathrm{d}x$$

$$\mathrm{d}\delta = S\,\Delta\delta\,\mathrm{d}x$$

(7.1)

Here $\mathrm{d}I$ describes the change in light intensity I within the thickness $\mathrm{d}x$ and $\mathrm{d}\delta$, the diffusion related decrease of the heat gradient $\Delta\delta$ in $\mathrm{d}x$. Both constants β ($\beta = \ln 10 \times \varepsilon \times c = 2.3026 \times \varepsilon \times c$, with ε = molar extinction coefficient and c = concentration) and S have the same unit: cm^{-1} [Eq. (7.1)]. Their reciprocal values define the distance at which the energy of the incident light (*optical absorption length* $\mu_\beta = 1/\beta$) or the amplitude of the heat waves (*thermal diffusion length* within the sample $\mu_S = 1/S$) have decreased to $1/e$ of the original value, respectively.

As mentioned, sound waves in the sample chamber originate in the boundary layer between the sample and mediating gas. It is only this thin gas layer from the whole thickness of the *thermal diffusion length* in the piston μ_G that is heated rhythmically by the sample surface, and for air with a modulation frequency of 20 Hz measures approximately 0.6 mm. Of course, a rough surface with a specific structure allows more efficient heat transport than a smooth, polished surface, thereby resulting in a greater PA signal. The thermal diffusion length μ_S for the inner sample corresponds to the distance that is traveled by the heat wave within one period of modulation (Figure 7.3). We obtain following formula (Cahen et al., 1980):

$$\mu_S = \sqrt{\frac{2\varkappa}{\omega \varrho c}}$$

(7.2)

with $\varkappa$ = thermal conductivity (J cm^{-1} s^{-1} grad^{-1}), ϱ = density (g cm^{-3}), c = specific heat (J g^{-1} grad^{-1}), ω = $2\pi f$ = angular frequency and f = modulation frequency (s^{-1}). On this basis, for example, modulation frequencies of between 1 and 1000 Hz yield values for μ_S of between 170 and 4 μm (Figure 7.4), assuming a typical value of $\varkappa$ = 0.001 (J cm^{-1} s^{-1} grad^{-1}). Taking the thermal diffusion length μ_S and sample thickness l (measured in the direction of the optical axis of the incident light beam) we can identify six different situations (Rosencwaig and Gersho, 1975a,b). First of all we can distinguish two different types of samples: optically transparent samples with large optical absorption length μ_β, where the light is absorbed throughout the total sample thickness l. Secondly, there are optically dense/turbid samples with small μ_β, where light is absorbed only within a small layer – compared with the sample thickness l. For each type, in turn, there are different relationships, depending on the thermal diffusion length μ_S (thermally thin, medium and thick). Here we will only take the situation that is most important in practice, that of an optically turbid and thermally thick sample, demonstrating the extraordinary capabilities of PAS where conventional optical spectroscopy runs into difficulties or even fails. Let us assume $\mu_S \ll l$. In the first approach (cf. Section 5.1.1) the light intensity decreases exponentially with the thickness of the sample (Figure 5.1). The distance $\mu_S = \mu_\beta$ separates two regions, which we call the *working region* ($\mu_S < \mu_\beta$) and the *saturation region* ($\mu_S > \mu_\beta$, Figure 7.4).

The theoretical approach to PAS based on the Rosencwaig–Gersho theory is mathematically demanding, however for practical work an exact understanding is not required, as long as we pay attention to the principles behind the theory.

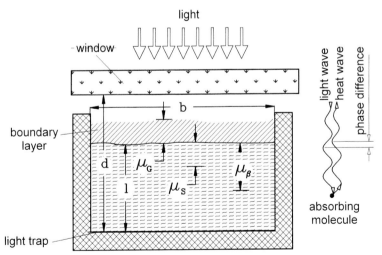

Figure 7.3 Magnified section of the sample chamber in Figure 7.2. In order to minimize the background signal, the inside is polished (e. g., quartz coated) and as – as with a light trap – blackened. The parameters d and b are the crude dimensions of the chamber; l is the sample thickness. μ_G, μ_S and μ_β are discussed in the text. The phase shift between the light-and the heat-wave are indicated (Moore, 1984).

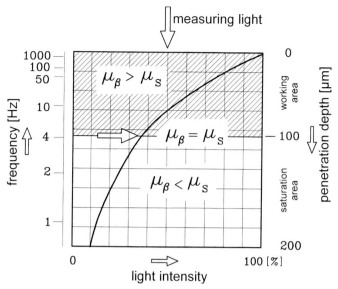

Figure 7.4 Dependence of the intensity of the exciting light and the amplitude of the resulting photoacoustic signal as a function of the sample thickness and angular frequency $\omega = 2\pi f$. The dominant factor is the relationship between the *optical absorption length* μ_β and thermal diffusion length μ_S. The photoacoustic spectrum only represents the absorption spectrum of the sample if μ_S is smaller than μ_β; otherwise saturation effects will disturb the spectrum. The corresponding borderline between the working and saturation range is a $\mu_\beta = \mu_S$, where by definition $I = I_0 \times e^{-\beta x}$.

At a minimum, the theoretical derivation of the PA signals may be roughly traced. Typically we start with a sine-modulated exciting light wave, even if this type of modulation pattern only plays a subordinate role in the practical work. However, on this basis the corresponding differential equations are easily solved with complex numbers. The temporary (t) and spatial (x, usually only one dimension) change of temperature is described by the so-called wave- or transport-equation; mathematically this is a partial differential equation of the second order in x and first order in t. As for all differential equations, a solution requires the definition of specific boundary conditions, which fix the limits of the function values "at the limits". In the following discussion we simply accept the most important results of these calculation. Finally, in a practical situation we have to solve such a wave equation for all "transport media" involved, e. g., for a chloroplast suspension (particles which mediate photosynthesis in higher plants, see Figure 7.13) this implies the chloroplasts themselves, the surrounding liquid medium, the boundary layer and finally for the mediating gas up to the microphone. It should also be mentioned that the conversion of the heat wave into a sound wave is not restricted exclusively to the boundary layer of the sample–air (even if this is the main source of sound). It may have taken place earlier so we also have to determine the sound propagation within the sample itself. Clearly, for practical work only the experiments actually matter, and such theoretical considerations are only of academic interest.

7.3.2
Photoacoustically Detected Absorption Spectra

The PA signal originates in a very thin layer of the sample in the order of μ_S (measured from the locus of the absorption). For $\mu_\beta >> \mu_S$ the total energy of the absorbed light quanta is proportional to β. Consequently, the wave dependency of the PA signal immediately mirrors the absorption spectrum of the sample (cf. Chapter 5). Of course, such a rudimentary spectrum subsequently has to be corrected with respect to the wave dependency of the illuminating system. Figure 7.5 compares the typical absorption spectrum of a dilute solution of cytochrome c (A) with the PA spectrum (B) of cytochrome c in the solid state (Rosencwaig, 1973). This result suggests that the PAS allows absorption spectra of solid biological samples to be obtained that are comparable with typical absorption spectra in solution, which is particularly important if the substance to be investigated is completely insoluble.

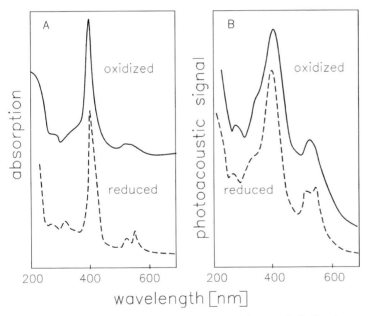

Figure 7.5 Comparison of the absorption spectrum of an aqueous solution of oxidized and reduced cytochrome c (A) with the PA spectrum of solid cytochrome c (powder spectrum, B) (cf. Figure 5.31). The differences between the spectra observed with the two methods – mainly the broadening of the bands in the PA spectrum – are real and are caused by the different aggregation states of the molecule rather than by methodological reasons (Rosencwaig, 1973).

7.3.3
Saturation Properties

When μ_S becomes comparable with μ_β or even if it is larger, the PA signal no longer represents the absorption spectrum of the sample. This is demonstrated by Figure 7.6 (cf. absorption spectrum of blood, Figure 5.32), which displays a tendency towards saturation with decreasing modulation frequency f on the basis of a smear preparation of human blood. Finally, upon further decrease of the frequency f the PA signal is completely saturated, i.e., in a thin surface layer of the sample all absorbed light quanta are "seen". The signal becomes independent of wavelength and simply represents the wave-dependency of the measuring system (*power spectrum*, Rosencwaig, 1978). As shown by Eq. (7.2), μ_S can be in principle be decreased by choosing higher modulation frequencies f thereby avoiding saturation. However, this possibility is limited, because the PA signal becomes smaller with increasing modulation frequency f – a decreasing sample thickness is measured (Figure 7.4) – and simultaneously the SNR decreases (see Section 7.4). The theory also shows that at saturation the PA signal is proportional to ω^{-1}, and at non-saturation proportional to $\omega^{-3/2}$. This property offers a fast and simple test for saturation.

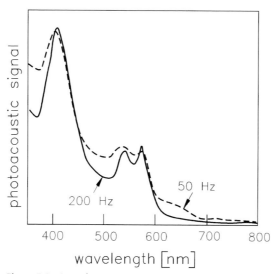

Figure 7.6 According to Figure 7.4, in order to avoid saturation of a PA signal the thermal diffusion length μ_S should be significantly smaller than the optical diffusion length μ_β. The required border frequency can only be found experimentally. For the PA spectra of a smear of human blood a significant saturation is clearly observed at $f = 50$ Hz; $f = 200$ Hz is still located within the working range (from Munroe and Reichert, 1976).

7.3.4
Depth Profiles of PA Spectra

Depth profiling is a specific feature of PAS which normally can only be studied by attenuated total reflection (ATR) spectroscopy (Section 8.4.1). There are two independent procedures. The first one is based on the Rosencwaig–Gersho theory [Rosencwaig and Gersho, 1975a,b, Eq. (7.2)] and depends on the change in μ_S on variation of f, with the precondition that the modulation frequency f is chosen high enough to remain within the working range of PAS.

Figure 7.7 shows an impressive example of the absorption spectra of the peel of a red apple. With $f = 220$ Hz only the typical protein absorption at 280 nm under the colorless thin layer of wax is obtained. With 33 Hz we see an additional absorption of the deeper lying red pigment layer with a maximum in the green spectral range. It is interesting to note that the three layers with different pigments are no longer discernable in a clear-cut way with PAS, not even with the phase method as discussed in the following section.

This was developed by Kirkbright's group (cf. dual wavelength spectrophotometry, Figures 5.44 and 5.46). In principle, the PA signal is an ac (alternating current) signal, e. g., in contrast to the signals for normal fluorescence or single beam absorption spectrophotometers. Thus, particularly for biochemical/biological systems, in addition to amplitude and frequency the relative phase of a PA signal – compared with the exciting light – carries additional information. Under the prerequisite of a thermally thick sample, i. e., $\mu_S \ll l$ (Figure 7.3), the relative

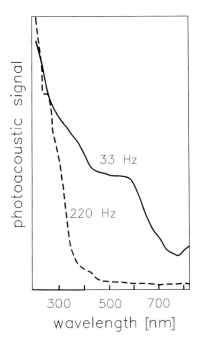

Figure 7.7 Photoacoustic spectra of an intact, red apple peel. According to Eq. (7.2) different layers are monitored with different modulation frequencies f. At $f = 220$ Hz only the protein band near 280 nm (cf. Figure 6.30) is seen, however, at $f = 33$ Hz the absorption band of the red pigment layer is also discernable. In PAS the choice of the modulation frequency f is most decisive (from Rosencwaig, 1978, Academic Press, London).

phase θ_r between the exciting and the measuring light can be determined. Ignoring various, more technological influences, e. g., the phase shift between the mediating gas and the microphone, or within subsequent electronic components, the following relationship is a good first approach (Poulet et al., 1980):

$$\tan \theta_r = \beta \times \mu_S + 1 = \frac{\mu_S}{\mu_\beta} + 1 \qquad (7.3)$$

For $\beta\mu_S \ll 1$ there is virtually no phase shift across the whole absorption band. However, if $\beta\mu_S \geqslant 0.1$ this equation allows β to be determined (l), i. e., the absorption spectrum. For $\mu_S < \mu_\beta$ (Section 7.4, the working range), θ_r will reach its maximum value of $\pi/4$. For any biological sample, the structure of which is neither qualitatively nor quantitatively known, the frequency dependency of the PA signal is complex. Nevertheless, relatively reliable predictions regarding the distribution of chromophores within thin layers of from 1 µm up to 200 µm – as measured from the sample surface – are possible.

The size of the PA signals depends on the phase adjustment of the *lock-in* amplifier (see Figure 7.8). According to the rules of vector analysis the phase of the signal is determined by two orthogonal components. Thus we select those two components which are phase shifted by 90°, which is called the quadrature. For an easier understanding we assume that one of the components is fixed to the PA signal that has its origin close to the distance where $\mu_S = \mu_\beta$ (see Figure 7.4). Owing to the time interval required by heat distribution to the sample surface the PA signals originating in the inner sample only appear with some phase delay, i. e., phase shifted.

Figure 7.8 demonstrates the spectral analysis of a model sample. A plastic film of only 25 µm thickness was painted on the upper side with a blue, and on the lower side with a red color; the optical density was approximately $A = 0.5$ for each. Spectra were recorded for μ_S values larger than 20 µm corresponding to a modulation frequency of $f = 26$ Hz = const. The figure shows that the PA signal is defined completely by only two parameters: either by the resulting component A and the corresponding total phase angle θ_m, or – mathematically equivalent – by the two components $A \cos \theta_m$ and $A \sin \theta_m$. The latter values, and thus the full information, are elegantly measured with a two phase *lock-in*-amplifier (see Section 5.5.2) and a two channel analog–digital converter. The phase diagram in Figure 7.8 shows that measurement under a phase angle of 0° yields the pure signal of the red, and measurement at an angle in the X' direction the pure signal of the blue color component. All other phase angles yield mixed spectra. With a correctly programmed on-line PC the axis X' can be continuously rotated and the corresponding spectra calculated and visualized. The problem with unknown samples is the appropriate choice when determining the "right" phase angle, and thus the corresponding spectra.

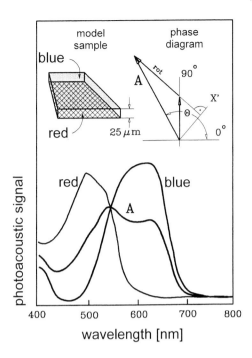

Figure 7.8 Spectral depth analysis of a model sample on the basis of PAS. A plastic film of 25 μm thickness carries on its upper and lower sides a blue and a red pigment layer, respectively, with an absorption of approximately $A = 0.5$ each. By fixing the relative phase of the PA signal with respect to the exciting light, absorption spectra at varying depths are measurable. The total PA signal can be described by the value (length) of vector A and a phase angle θ_m. The phase diagram in the inset shows the composition of the total signal of the individual "blue" and "red" components. The phase diagram and resulting absorption spectra are coordinated (Moore, 1984).

7.3.5
Measurement of Photophysical Parameters with PAS

A particular advantage of PAS is the determination of various photophysical parameters such as the fluorescence quantum yield or the measurement of energies and lifetimes of long lived, metastable quantum states. In Chapter 6 we discussed the fact that the relaxation process of excited molecules is based on different individual mechanisms (Figure 6.1). On the one hand the first excited singlet state will dissipate its energy through fast processes such as fluorescence emission, photochemical reactions, or simply by radiationless transition to the singlet ground state. Only the last process can be followed directly by PAS. On the other hand, the first excited singlet state (S_1) can undergo a change in multiplicity and subsequently relax via internal conversion (see Section 6.2.1) by emission of a small amount of energy to the first excited triplet state (T_1). This triplet state, in turn, will either trigger a photochemical process, or it looses its energy by emission of phosphorescence light (Section 6.7) or, again, by a radiationless transition to the singlet ground state. The last process can also be measured directly by PAS. Owing to the relatively long lifetime of the T_1 state the modulated heat emission will be observed after a given phase delay with respect to the exciting light. This phase delay is essentially determined by the lifetime of the triplet state (see Figure 6.27). Thus, a plot of phase difference as a function of the exciting frequency f allows the determination of the triplet lifetime. In addition, this measurement enables the rate of population of the triplet state to be determined, i. e., the intersystem crossing (Moore, 1984).

In particular PAS allows the measurement of fluorescence quantum yield Φ_F (Section 6.5.3). For this purpose a non-fluorescing reference sample is measured which has the same absorption as the sample at the exciting wavelength. The identical geometry of the sample, the same volume and buffer guarantee a constant correction factor $R(f)$ [see Eq. (7.5)]. On this basis the quantum yield γ for radiationless relaxation can be determined and from this the fluorescence quantum yield Φ_F [Eq. (7.4) and Section 6.5.3]:

$$\phi_F = \frac{\nu_F}{\nu} (1 - \gamma) \tag{7.4}$$

7.4
Experimental Methods

Figure 7.9 shows the typical set-up of a PA spectrometer. Light source 1 generates the required excitation light incident on the grating monochromator, which selects a monochromatic light beam of variable wavelength. This is frequency modulated by a chopper. The optimization of the SNR (see Section 5.2.5) is most important in PAS because the final signal involves the conversion of various signal forms: firstly the light signal (in watts) is converted into sound (pascals) and finally into an electrical signal (volts per amp). The light source shows fluctuations, which are passed on via sound waves to the electrical signal. However, the sample itself will carry a noise signal that is generated by heat noise and stray light (see Section 5.2.7), which is also further processed by a microphone and amplifier. Of course, the preamplifier and microphone are also not noise free. These

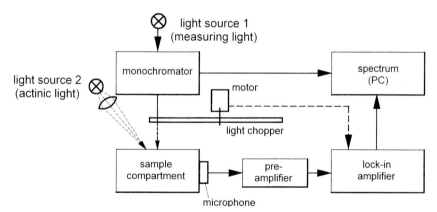

Figure 7.9 Principle set-up of a PA spectrometer. The basic difference compared with a conventional optical single beam spectrometer with a chopper [cf. Figure 5.12(b)] is the detection of the signal being measured by a microphone rather than a photodetector. The subsequent signal processing is identical.

sources yield the so-called asynchronous noise, because it is independent of the modulation frequency f.

Another source of interference is the so-called synchronous noise. This is mainly due to the PA signal of the empty sample cell. A frequent reason is the mechanical coupling of the chopper and microphone. Good damping of the sample chamber and mechanical decoupling, e. g., through light fibers and a stable spectrometer platform, may help.

The sample chamber should be constructed so as to optimize the SNR. Particularly critical, again, with respect to SNR, is the thickness of the boundary layer between the sample and the mediating gas, which reacts sensitively to small temperature changes. This is in the order of the thermal diffusion length μ_G, which for air and at $f = 50$ Hz is approximately 0.34 mm. The difference between the window and the sample surface is chosen to be of this order (Figure 7.3).

Various specific microphones have been developed for PAS, the most common being the condenser or electret microphone with a diameter of between approximately 0.5 and 1 in (1 in = 2.54 cm). Its sensitivity is between 10 and 60 mV Pa^{-1}. However, for photoacoustic flashlight spectroscopy (Section 7.6.2) microphones are not suitable because they only work in the acoustic range. For this purpose fast, piezoelectric crystals are adopted as transducing elements.

For optimization of the SNR, the sample chamber has to be isolated from the "outside world", and the signal for the empty chamber should be as small as possible. The window material, contaminations and even the chamber walls themselves should absorb as little light as possible in order to reduce the background signal to a minimum. For a large SNR it is most important that a "thermal mass" as large as possible is chosen (Figure 7.2), because then it will undergo only a small temperature change ΔT upon heat conveyance and thus generate only a small background PA signal. The temperature changes are calculated by the formula $\Delta T = \Delta W \, m^{-1} \, c_p^{-1}$, where ΔW is the amount of heat released, m the mass and c_p the heat capacity of the sample chamber at constant pressure.

Of course, the PA spectrometer shown in Figure 7.9 only delivers spectra that are not corrected with respect to the wavelength dependency of the whole system. We have been confronted with the same problem when analyzing fluorescence excitation spectra. We previously discussed the correction procedures on the basis of quantum counters (Section 6.3.2). Correction of PA spectra is achieved in an analogous way, i. e., we need a reference sample which absorbs all quanta, independent of the wavelength of the exciting light. Subsequently the absorbed energy should be completely released in the form of heat and finally – in the absence of any chemical processes – appear in full as a PA signal. For measuring PA reference spectra a very fine layer of carbon-black has proved to be the best material so far. In a fairly similar manner to the correction of fluorescence excitation spectra, a PA spectrum is corrected by division by the reference curve. Of course, all analytical procedures described previously (see Section 5.4.4) can be applied to PA spectra.

7.5
Photochemically Active Samples

7.5.1
Modification of PA Signals

Photobiologically active samples are usually also photochemically active; i. e., not all photons initially converted into excitation energy are subsequently immediately detectable in the form of heat. Firstly, a photochemical, stable end-product of higher enthalpy can be formed. Alternatively there are several intermediates – presumably with electron transport chains – connected in series, and the initial state is only reached again with some delay according to their characteristic lifetimes (Figure 7.10). The following general formula for the observed PA signal was developed by Cahen et al. (1980):

$$PAS(f,\nu) = A_\beta(\nu)\ R(f)\left[1 - \frac{\sum_i \Phi_{pi}\ \Delta E_{pi}(f\ )}{N\ h\nu} - \Phi_F\ \frac{\nu_f}{\nu}\right] \tag{7.5}$$

where $A_\beta(\nu)$ describes the total absorbed energy ($\mu_S < \mu_\beta$), $R(f)$ describes a correction factor that depends on the spectrometer used and the thermal properties of the sample and reference (see below). Φ_{pi} is the quantum yield for the formation of the intermediate of a photochemical reaction i and ΔE_{pi} describes the stored energy of the individual step. $Nh\nu$ is the energy of one Einstein (1 mol quanta)

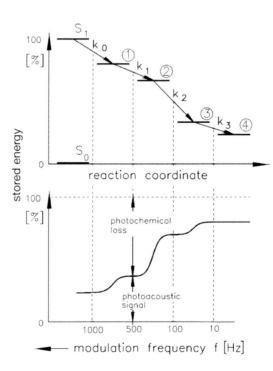

Figure 7.10 Top: Schematic energy diagram of a photochemical reaction chain. The ground state of most organic molecules is a singlet state. Immediately after excitation they relax to the first excited singlet state S_1 by *internal conversion*. From here, by *intersystem crossing* part of the molecule undergoes transition to the first excited triplet state (k_0), which, in turn, will trigger a reaction chain of photochemical/photophysical and/or photobiological reactions (k_1–k_3). Finally, a new equilibrium (4) is reached. Bottom: According to the energies released in the individual steps we measure the PA signal as shown schematically in the lower picture. The difference in the energy originally "pumped in" is stored chemically: we call this the *photochemical loss* (from Cahen et al., 1980).

at the wavelength $\lambda = c\nu^{-1}$; ν is the frequency of the incident light, f the modulation frequency, Φ_F the fluorescence quantum yield and ν_f the frequency of the fluorescing light. The brackets in Eq. (7.5) stand for the so-called photochemical loss of a PA signal, which reduces the maximum possible measuring signal, including any luminescence emission (Cahen et al., 1980).

7.5.2
Frequency Spectrum of PA Signal

Applying Eq. (7.5) we have to take into account that ΔE_{pi} is a function of the modulation frequency f. An intermediate is only observable if its lifetime is longer, and its generation rate, however, smaller than the duration of a complete cycle of the modulation frequency $[\omega^{-1}= 1/(2\pi f)]$.

If the PA signal is monitored at a given excitation wavelength λ as a function of the modulation frequency f, the short lived intermediates will be detected at higher and the longer lived at lower frequencies. Stable end-products are recognized by the fact that even at modulation frequencies approaching zero the whole of the expected PA signal is not measured. As an example, Figure 7.11 shows the frequency dependency of bacteriorhodopsin in the purple membrane of *Halobacterium halobium*, for which a light-induced cycle is formulated (see Haeder, Tevini, 1985).

The curve in Figure 7.11 mirrors at least three independent exothermic processes with lifetimes of 0.5 ms (330 Hz), 0.85 ms (185 Hz) and 1.7 ms (92 Hz), and finally for $f > 20$ Hz an exothermic process of more than 5.3 ms. While the three slowest processes are readily attributed to the intermediates N_{530}, M_{412} and O_{660} known from flash light spectroscopy (see Section 6.5.2), the shoulder at 290 Hz does not correspond to any known intermediate.

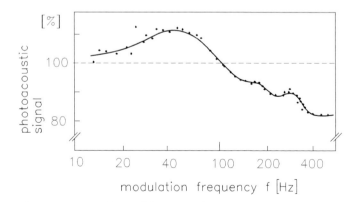

Figure 7.11 Dependence of the PA signal on the modulation frequency f for bacteriorhodopsin in the purple membrane of *Halobacterium halobium* (excitation wavelength in the maximum of the absorption band at 565 nm). Dependent on the modulation frequency f the curve mirrors three exothermal processes with lifetimes of $\tau = 0.5$, 0.85 and 1.7 ms and an endothermic process with a lifetime of 5.3 ms (from Garty and coworkers, 1978).

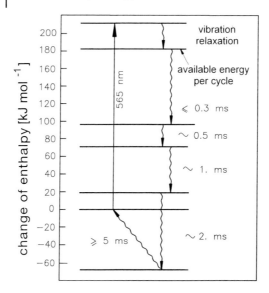

Figure 7.12 The energy scheme for the photo cycle of bacteriorhodopsin as derived from the frequency dependence in Figure 7.11 (from Garty et al., 1978).

The PA measurement provides additional calorimetric data, characterizing the storage capacity of the purple membrane. Firstly, the total energy of the initially absorbed photons of 212 kJ Einstein^{-1} (650 nm) is minimized by 38 kJ Einstein^{-1} by trivial vibrational relaxation leaving 184 kJ for the photo cycle. From the remainder, 100 kJ are stored in a fast process and about 70 and 20 kJ in the two subsequent processes, respectively. Finally, the PA signal at about $f = 50$ Hz exceeds the maximum value expected and corresponds to a decrease in enthalpy of approximately 60 kJ. This band is not found at a low pH (3.8) or at a high ion concentration, which might indicate a conformational change of the protein–lipid complex. Such changes could cause an increase in entropy, and thus a decrease in free enthalpy ($\Delta G = \Delta H - T\Delta S$). This is the prerequisite for a spontaneous reaction. Figure 7.12 shows the energy scheme derived from this experiment (Garty and Kaplan, 1982).

7.5.3
Chloroplasts

As a non-invasive technique, PAS allows tissues and organelles from various sources such as plants, microorganisms and humans to be investigated; for example, leaves, hard tissues such as teeth and bones, or soft tissues such as skin or muscles. In particular, suspensions of any type are easily measured by PAS, which is therefore an ideal supplementary method, and often the only alternative to conventional absorption spectroscopy.

The following is a typical example. The photosynthetic electron transport in chloroplasts of green plants is strongly inhibited by the herbicide DCMU

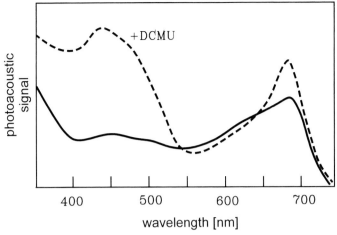

Figure 7.13 PA spectra of the primary leaves of the radish, with and without DCMU [3-(3,4-dichlorphenyl)-1,1-dimethylurea]. DCMU blocks the photosynthetic electron transport and thus the transduction of light energy into biochemical equivalents. As a consequence, in the presence of DCMU more energy is converted into heat, which is evident in an extended and distorted PA spectrum (from Buschmann and Prehn, 1981, Springer, Heidelberg).

[3-(3,4-dichlorphenyl)-1,1-dimethylurea], i.e., the photochemical activity is significantly reduced. Consequently, we can expect an increased PA signal in the presence of DCMU. This can be demonstrated by an experiment with isolated chloroplasts of radish (Figure 7.13). It is interesting to note that not only the height, but also the shape of the photoacoustic spectrum is changed, allowing conclusions about details of the inhibition of the electron transport chain to be drawn (from Buschman and Prehn, 1981).

7.6
Selected PAS Techniques

7.6.1
Photoacoustic Microscopy

PAS has been extended to microscopic dimensions in what is known as **photo-acoustic microscopy** (PAM). A light beam of a few micrometers diameter is focussed onto a microscopically small area of the sample. Thus, in analogy with *laser scan microscopy*, "acoustic pictures" are generated.

While in conventional microscopy larger sample areas are irradiated with light simultaneously, in laser scan microscopy the sample is scanned point by point with two highly sensitive crossed mirror galvanometers in the *x*- and *y*-directions, analogous to a television screen. The measured signal at each individual point is stored for further processing in an electronic memory.

Today, PAM is used routinely, particularly in metallurgy and microelectronics. As in macroscopic PAS, on the basis of a variable modulation frequency *f*, PAM enables microscopic depth monitoring to be performed. The first PAM measurements of biological samples, such as individual chromosomes, soon delivered promising results (Wong et al., 1978).

7.6.2
Laser-induced PAS

In addition to fast, normal spectrophotometric measurements, PAS can also be utilized in short-time spectroscopy, known as laser-induced optoacoustic spectroscopy, LIOAS. LIOAS has been successfully adopted for the measurement of extremely fast photochemical and photophysical mechanisms. Firstly, similar to stationary PAS, the molecules to be investigated are electronically excited by light. On the basis of very short laser pulses, e. g., in the nanosecond time range, of defined wavelengths and intensity the induced PA signal is monitored over a long period of approximately 20 µs. This represents a damped oscillation approaching zero after a few milliseconds. The amplitude of the so called *first acoustic wave, H*, defines the *LIOAS* signal. The following relationship has been derived (Tam and Patel, 1980):

$$H = \frac{\xi E_0 \, A}{\pi c_p \tau_p^2} \sqrt{\frac{v_a \tau}{2\pi f}} = \varkappa \, E_0 \, A \tag{7.6}$$

with ξ = thermal expansion coefficient, c_p = specific heat at constant pressure, A = absorption of light by the sample in the direction of light path, E_0 = energy of the laser pulse, $\tau_p = 2\tau$ duration, v_a = velocity of sound. With the presumption that the absorbed energy during the observation time is fully released as heat, proportionality between H and E_0 is found for several orders of magnitude.

As previously mentioned, a normal microphone is unsuitable for detecting such short time acoustic signals in the microsecond time range. Typically, piezoelectric crystals are used as transducing elements, converting sound waves into electrical signals within short time periods. After suitable amplification the signal is further processed and stored – in a similar manner to other fast processes such as flash light photolysis (cf. Section 6.5.2).

For example, using the laser pulse induced PAS the reverse reaction of the first *Phytochrome* intermediate I_{700} of the $P_r \rightarrow P_{fr}$ phototransformation can be measured; a result barely accessible with any other spectroscopic method (Jabben et al., 1984).

Phytochrome is an important *photochromic* plant pigment, which is mutually converted by light of 660 nm and 730 nm between the two forms P_{fr} and P_r, respectively, thereby undergoing changes in color and absorption maxima (r = red, fr = far-red). Only the P_{fr} form is physiologically active and responsible for a large number of light controlled reactions throughout the plant kingdom. Therefore, analyzing the individual reaction steps of the photoconversion $P_r \leftrightarrow P_{fr}$ it is scientifically highly relevant as it is a key process in plant development.

7.6.3
Photothermal Radiometry

Periodic temperature fluctuations as induced by the absorption of modulated light are not only detectable as PA signals. According to the radiation law of *Stefan–Boltzmann* (Section 3.4.1)

$$W = \varepsilon \; \sigma \; T^4 \tag{7.7}$$

$$\delta W = 4\varepsilon\sigma T^3 \; \delta T$$

small temperature changes δT also generate modulated heat emission of a black body radiator (cf. Section 3.4.1), in this context known as a *photothermal signal*. The σ describes the *Stefan–Boltzmann* constant and ε the emission of the respective sample (not to be confused with the extinction coefficient ε of the *Lambert–Beer* law). *Photothermal radiometry* can be used for remote measurements up to distances of approximately 1 km. Detailed theoretical investigation has shown the proportionality of absorption A_β and W (Tam, 1986). With modern infrared (IR) detectors values far below $W = 10^{-7}$ W are measurable with good SNR. In practice, the method requires no sample preparation and no sample chamber: similar to fluorescence light, IR emission of the sample is focussed onto the detector with suitable IR lenses. While a normal PA signal is mainly generated by the boundary layer between the sample and air, photothermal signals from well within the inner volumes of the sample are detectable, presuming that the sample is transmitting in the infrared. According to the thermal relaxation time (see Figure 6.4) reactions down to time ranges of 10^{-11} to 10^{-10} s are resolved by *photothermal radiometry*, thus offering itself as another flash light method. Moreover, the fast IR detectors available today allow fast kinetics in the ps range to be monitored.

Similar to isotropic fluorescence light in fluorimetry, diffuse IR radiation is not easy to focus. Therefore, an "end-on" arrangement of sample and detector in close proximity are the best choice.

7.6.4
Fourier Transform–Infrared Photoacoustic Spectroscopy (FTIR–PAS)

Compared with conventional optical spectroscopy the *SNR* of *PAS* is relative low. Particularly in the infrared, where signals are smaller than in the visible spectral range, fast Fourier transform (FFT) techniques are today becoming more popular in optical spectroscopy. With respect to interferometry, the same holds true for

PAS (see Figure 5.8). Most manufacturers of optical FTIR spectrometers offer suitable PAS chambers as an optional accessory, making FTIR–PAS possible. With this method, infrared spectra of biological sample that barely transmit in the IR range have been obtained for the first time. Spectra of milligram samples of protoporphyrin, hemoglobin and peroxidase have been measured (Rockley et al., 1980).

7.7
Resume and Outlook

Novel insights regarding the thermal properties of samples can be obtained and specific (bio-) chemical reaction pathways can be derived essentially on the basis of PAS. Absorption spectra of completely turbid samples are measurable, *in vivo* and *in situ* (another alternative to conventional absorption spectroscopy is offered by diffuse reflection, Section 8.5, or by total internal reflection, ATR, Section 8.6). In addition, PAS allows the measurement of absorption spectra of intact samples as a function of depth, largely independent of their optical properties. In contrast to classical optical absorption spectroscopy, in PAS elastically scattered and transmitted photons are automatically excluded from the measurement. In absorption spectroscopy the reference beam I_0 [see Figure 5.12 (c–e)] has to be precisely processed as "ballast" in a linear and fairly noise free manner by the photodetector and the subsequent electronics.

Because PAS is a complementary method to normal fluorescence spectroscopy, luminescence and photochemical processes are accessible in an indirect way. The non-optical nature of the PA signal is its great advantage, avoiding any optical elements between the sample and detector.

The limitations of PAS are caused in particular by the relatively large light intensities required compared with absorption and fluorescence spectroscopy (in the order of 100 W m^{-2}). This can cause some undesirable photolysis, i. e., non-specific photooxidation and destruction of organic molecules. It would appear that interferometry in association with Fourier transform techniques will provide the advances in the future (see Figure 5.8). In addition, the window of the sample chamber (Figures 7.2 and 7.3) has to be fairly transparent to the individual optical radiation source to exclude it as far as a possible as a source of interfering signals. This latter point raises serious problems for many spectral ranges. Therefore, *photothermal spectroscopy,* which in principle does not require a sample chamber at all – and thus no window – may become more important in the future. Even if there are some limitations in principle, we might predict that the PAS will gain more weight as an analytical method, and become complementary to optical spectroscopy at some point. Of course, this is also a challenge for the manufacturers to offer versatile and affordable PAS spectrometers to extend known applications and to open up new ones.

7.8
Selected Further Reading

Adams, M. J., Beadle, B. C., King, A. A., Kirkbright, G. F. *Analyst* **1976**, *101*, 5533.

Adams, M. J., King, A. A., Kirkbright, G. F. *Analyst* **1976**, *101*, 73.

Adams, M. J., Kirkbright, G.F. *Analyst* **1977**, *102*, 281.

Bell, A. G. *Philos. Mag.* **1881**, *11(5)*, 510.

Buschmann, C., Prehn, H. in: *Modern Methods of Plant Analysis*, Linskens, H. F., Jackson, J. F., eds. New Series, Vol 11, Physical Methods in Plant Sciences, Springer-Verlag, Berlin, **1990**, pp. 148–180.

Cahen, D., Bults, G., Garty, H., Malkin, S., *J. Biochem. Biophys. Meth.* 3, **1980**, 293.

Campbell, S. D., Yee, S. S. *IEEE Trans. Biomedical Eng.* **1979**, *26*, 220.

Garty, H., Caplan, S.R. *Biophys. J.* **1982**, *34*, 405.

Garty, H., Cahen, D., Caplan, S. R., in: Caplan, S. R. and Ginzburg, M. (eds.), *Energetics and Structure of Halophilic Microorganisms.* Elsevier/North-Holland Biomedical Press, Amsterdam: **1978**, 253–259.

Häder, D.-P., Terini, M. *Allgemeine Photobiologie*, Georg Thieme Verlag Stuttgart, New York, **1985**.

Jabben, M., Heihoff, K., Braslavsky, S. E., Schaffner, K., *Photochem. Photobiol.* 40, **1984**, 361.

Kanstad, S. O., Nordal, P.-E., Hellgren, L., Vincent, J. *Naturwissenschaften* **1981**, *68*, 47.

Michaelian, K. H., Wineforder, J. D. *Photoacoustic Infrared Spectroscopy*, Wiley Interscience, New York, **2003**.

Moore, T. A., in: *Photoacoustic Spectroscopy and Related Techniques Applied to Biological Materials*, Smith, K. C., ed., Photochemical and Photobiological Reviews, Vol. 7, Plenum, New York, **1984**, pp. 187–221.

Munroe, D. M., Reichard, H. S., *PAR Application Note 147*, **1976**.

Pfund, A. H. *Science* **1939**, *90*, 326.

Poulet, P., Chambron, J., Unterreiner, P. *J. Appl. Phys.* **1980**, *51*, 1738.

Preece, W.H., *Proc. R. Soc. London* **1881**, *31*, 506.

Rayleigh (Lord), *Nature (London)* **1881**, *23*, 274.

Rockley, M. G., Davis D. M., Richardson, H. H., *Science* 210, **1980**, 218.

Röntgen, W. C. *Philos. Mag.* 11, **1881**, *(5)*, 308.

Rosencwaig, A., in: *Adv. Electronics and Electron Physics*, Vol 46, L. Marton (ed.): New York Academic Press, **1978**.

Rosencwaig, A., *Science* **1973**, *181*, 657.

Rosencwaig, A., *Rev. Sci. Instrum.* **1977**, *48*, 1133.

Rosencwaig, A. *Photoacoustics and Photoacoustic Spectroscopy (Chemical Analysis)*, John Wiley, New York, **1990**.

Rosencwaig, A., Gersho, A. *J. Appl. Phys.* **1975a**, *47*, 64.

Rosencwaig, A., Gersho, A. *Science* **1975b**, *190*, 556.

Tam, A. C. *Rev. Modern Phys.* **1986**, *58*, 81.

Tam, A. C. Patel, C. K.N. *Optics Lett.* **1980**, *5*, 27.

Tyndall, J. *Proc. R. Soc. London* **1881**, *31*, 307.

Viengerov, M. L. *Dokl. Akad. Nauk SSSR* **1938**, *19*, 687.

Wong, Y., Thomas, R. L., Hawkins, G. F., *Appl. Phy. Lett.* 32, **1978**, 538.

8
Scattering, Diffraction and Reflection

Natura expelles furca, tamen usque recurret
(Horace)

Even if you drive out nature with a pitchfork,
she will always return

8.1
Introduction

When analyzing the light absorbing properties of a sample ($A = \log I_0 - \log I$), part of the incident light that does not reach the detector is not absorbed but has been lost by scattering (Figure 5.18). The best known example of this behavior is the Tyndall effect (1869), which occurs with colloidally distributed particles: if a white light beam is incident at a cuvette containing clean water that is mixed, for example, with a single drop of milk, a bluish shine is observed orthogonal to the light beam, whereas the transmitted light appears reddish. Analogous to the Lambert–Beer law (Section 5.1.1) the total power of the scattered light is described by the formula

$$I(x) = I_0 \, e^{-S' \, x} \tag{8.1}$$

where I_0 is the intensity of the incident light beam, $I(x)$ is the intensity at depth x lost by scattering and S' is the scattering coefficient, also known as turbidity, attenuation constant or apparent extinction coefficient. For dilute solutions the scattering cross section σ_s is deduced directly by conventional absorption measurements from the turbidity of the sample:

$$S' = N \times \sigma_s \tag{8.2}$$

where N is the particle density. As implied by the name, σ_s defines the effective cross section of the scattering particles where incident light quanta are deflected. The term *scattering* has already been used in previous chapters without a full definition. It is a collective term for various physical phenomena such as refraction,

Optical Spectroscopy in Chemistry and Life Sciences. W. Schmidt
Copyright © 2005 WILEY-VCH Verlag GmbH & Co. KGaA, Weinheim
ISBN 3-527-29911-4

diffraction, reflection and others, which distort a light beam in its geometrically correct light path, and possibly change its wavelength. We can refer to *inelastic scattering* or if there is no wavelength change involved it is termed *elastic scattering*. The most important representative of inelastic scattering due to interaction with vibrational quanta of scattering molecules is *Raman scattering* with frequency shifts in the order of 10^{11} to 10^{13} Hz (visible light is approximately 10^{14} Hz).

Sound waves are also capable of causing scattering, as was shown for the first time by the French physicist Brillouin in 1915. These are induced by typical thermal molecular vibrations (cf. Section 2.3.3.3) and – in contrast to *PAS*, where the exciting light is slowly frequency modulated (Chapter 7) – of high frequency ranging between 10^8 and 10^{10} Hz. As a result of *Doppler shifts*, even smaller frequency shifts are observed by interaction of light with translatory or rotatory motions of molecules and particles: $10 - 10^6$ Hz. This is termed *quasi elastic scattering*. By measuring the *temporal average* intensity of elastic scattering, quasi elastic scattering yields the *temporal variations* in intensity, which is thus termed *dynamic light scattering*.

Scattering phenomena cannot really be ignored in optical spectroscopy, in a positive as well as in a negative respect. On the one hand they can give enormous false photometric measurements, and on the other they open up new fields of analysis and thus allow measurements to be made that would otherwise be impossible. The scattering theory is highly versatile, and at times complex and demanding. Thus, in this chapter we will become acquainted with at least the main aspects of scattering theories and – as in former discussions – we will adopt and visualize the essential results without explicit derivations. Firstly we will discuss the elastic, then the quasi elastic and finally the inelastic scattering (Raman scattering).

Scattering experiments are not just restricted to the visible wavelength range on which we are focusing in this chapter. The molar mass of even the smallest molecule can be determined using visible light (see Figure 8.11), however structures and sizes below approximately $d = 200$ Å are not resolvable, because the resolution is inversely proportional to λ^2. Moreover, all molecules, particularly water and air, exhibit strong absorbance at these wavelengths and only become transparent again in the X-ray range – e.g., at 1.54 Å of the often utilized Cu K_α-line. However, in this range the extreme forward scattering is dominant with deviations of only a few degrees from the direction of the incident light beam. On the basis of the so-called *X-ray small angle scattering* the sizes of various small molecules such as ribonuclease (responsible for hydrolysis of ribonucleic acid, size approximately 10 Å) or the size of the rhodopsin molecules (the light absorbing pigment in the visual process, Figure 6.23) can be determined, which however is not part of the present considerations.

Light scattering depends primarily on the ratio of the wavelength of light λ and particle size d, the specific type and shape of the scattering particle is only of secondary importance. For example, telescopes designed for astronomical observations of the night sky are occasionally used for monitoring eclipses of the sun in the infrared. To protect them from light they are thus covered with an opaque plastic foil that is highly scattering in the visible wavelength range. However, this does not interfere with the light path in the infrared and allows excellent pictures to be taken. For the present discussion we will assume that the concentration and/or the sample thickness are so small that only a single scattering process is allowed. The medium scattering length – i.e., the average distance between two scattering processes – should be significantly larger than the thickness of the layer investigated. This is the case when the product of the scattering coefficient S' and layer thickness x is smaller than 1, that is if we exclude multiple scattering processes, for which to date no standard theory exists (see Section 8.5.1). Depending on the particle size d – relative to the wavelength λ, e. g., 633 nm for the HeNe lasers; see Section 3.4.5 – we can in theory and by practical analysis distinguish various scattering ranges: *Rayleigh scattering* between approximately $d = 1$ nm and $0.1 \lambda \approx 60$ nm, *Rayleigh–Gans–Debye scattering* between 60 and 200 nm, *Mie scattering* between 200 nm and 5 mm, and finally *Fraunhofer scattering* from 5 mm up to the mm range.

Many samples, particularly in biological and medical research, are extremely opaque, and consequently cause multiple scattering. For example, this is the case when measuring absorption spectra of tissues such as flowers, suspensions of algae or blood smears. Particularly interesting and revealing are *in vivo* measurements of living tissues. Thus the effect of herbicides on foliage, the influence of environmentally relevant substances on microorganisms, or the oxygen content of human blood as a function of respiration can be measured directly in a non-invasive way. Then absorption spectra and (diffuse) reflection spectra become almost indiscernible. Therefore, the latter are highly important for the measurement of optically opaque, highly scattering samples. A suitable quantitative description of these phenomena is given by the Kubelka–Munk theory, which we will discuss in more detail at the end of this chapter.

8.2
Elastic Scattering

8.2.1
Derivation of Rayleigh's Equation

In the classical description of scattering theory the electric vector E of a (in the z-direction) linearly polarized electromagnetic light wave (Figure 2.1) induces a forced sine vibration of the "electron cloud" of the scatterer. The ease or extent with which this displacement takes place is described by the term polarizability α, discussed earlier. The result of the displacement is an electric dipole moment μ. In various directions α can have different values (mathematically α is a tensor): we call this *anisotropy* (see Section 6.4). We assume that the wavelength of the exciting light is spectrally a long way from the absorption band, and that there is no interference on the absorption (scattering with concomitant absorption will be discussed later). According to the laws of electrodynamics, such an oscillating charge represents – similar to a television antenna – a tiny Hertz dipole with a characteristic emission pattern. The incoming electric light wave E is described by

$$E = E_0 \, \cos 2\pi\nu \left(t - \frac{x}{c} \right) \tag{8.3}$$

with an orthogonal magnetic field

$$H = H_0 \, \cos 2\pi\nu \left(t - \frac{x}{c} \right) \tag{8.4}$$

For the following considerations only the change in the electric field E at the *scattering center* $(x = 0)$ with respect to time is important:

$$E = E_0 \, \cos 2\pi\nu t \tag{8.5}$$

As a consequence, the electrical scattering field E induces a correspondingly oscillating dipole

$$\mu = \alpha E = \alpha E_0 \, \cos 2\pi\nu t \tag{8.6}$$

For isotropic molecules this oscillates in the orthogonal direction of the z-axis. In the following we will first discuss small scattering particles with dimensions that

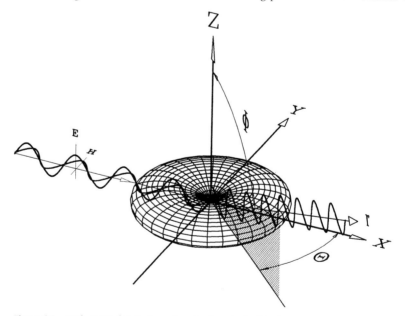

Figure 8.1 With monochromatic and vertically polarized incident light the distribution of Rayleigh scattering is independent of the azimuth angle θ. Electric (E) and magnetic field components (H) are indicated.

Heinrich Rudolf Hertz (1857–1894)   John William Strutt, Lord Rayleigh (1842–1919)

are significantly smaller than the wavelength (e. g., $\lambda/20 = 25$ nm). According to Maxwell's theory for the electric field strength E_r of the emitted (scattered) light at distance r (i. e., in Figure 8.1 a defined point r' on the axis r) and an angle Φ to the electric dipole μ we obtain

$$E_r = \left(\frac{\alpha E_0 4\pi^2 \sin \Phi}{r\lambda^2} \right) \cos 2\pi\nu(t - r/c) \tag{8.7}$$

This first pair of parentheses describes the amplitude of the scattered light. According to the rules of physics, the radiation intensity is proportional to the square of the amplitude. In an analogy with the definition of transmission [Eq. (5.6)] the scattering intensity is usually related to the intensity of the incident light I_0 and we obtain the famous Rayleigh equation (Lord Rayleigh, 1871):

$$\frac{I_s}{I_0} = \frac{E_r}{E_0} = \frac{\left(\frac{\alpha E_0 4\pi^2}{r\lambda^2} \sin \Phi \right)^2}{E_0^2} = \frac{16\pi^4 \alpha^2 \sin^2 \Phi}{r^2 \lambda^4} \tag{8.8}$$

This equation yields important results:
- The scattering intensity decreases – as expected – with the square of the distance.
- The scattering intensity only depends on $\sin^2 \Phi$, not on the azimuth angle θ. This is visualized by a rotating body with corresponding cross section around the z-axis (Figure 8.1, surface of constant scattering intensity).
- The scattering intensity decreases with the fourth power of the wavelength. This finding is particularly important in *UV* spectroscopy. For example, the scattering of light with a wavelength of $\lambda = 250$ nm is approximately 100 times stronger than that of light at $\lambda = 800$ nm.

The most frequently mentioned example of Rayleigh scattering is the blue color of a clear sky, particularly when the sun is high. In his more detailed theory of scattering for the scattering coefficient S' of gas Lord Rayleigh obtained

$$S' = \frac{8\pi^3}{3N\lambda^4} (n^2 - 1)$$

where N is the number of molecules cm^{-3} and n is the refractive index. The lower the wavelength, the greater the scattering coefficient. Evening and morning glows are explained by the fact that the sun light has to traverse a very thick layer of gas where red light is predominantly transmitted. However, the shortest visible wavelengths correspond to the color violet and not blue! This apparent discrepancy is explained by the fact that the physiological recognition of color is the result of the combination of both the sensitivity curve of the human eye and the physical stimulus (Schmidt, 1999, Section 3.4; mathematically speaking this is a *convolution*, see Section 5.4.4), hence yielding the resulting "blue".

When performing scattering experiments we generally use non-polarized incident light. Because this is a superposition of light waves from all polarization directions we also have to rotate the rotating body (Figure 8.1) around the x-axis, converting the Φ dependency into a θ dependency. The exact calculation, which we will not perform here, yields, as a modification of Eq. (8.8), a replacement of $\sin^2 \Phi$ by $(1 + \cos^2 \theta)/2$:

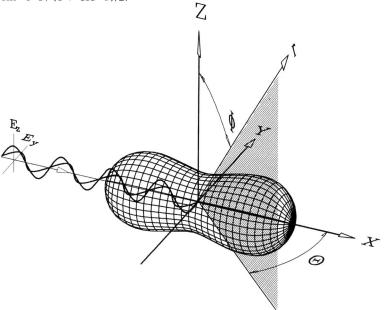

Figure 8.2 With incident monochromatic and non-polarized light the distribution of Rayleigh scattering depends on the azimuth angle θ as well on Φ. Electric (**E**) and magnetic field components (**H**) are randomly distributed around the x-axis.

$$\frac{I_s}{I_0} = \frac{8\pi^4 \alpha^2}{r^2 \lambda^4} \ (1 + \cos^2 \Theta) \tag{8.9}$$

The corresponding scattering body is shown in Figure 8.2 and exhibits mirror symmetry with respect to the y–z-plane.

8.2.2
Determination of Molar Mass (Molecular Weight)

Firstly we take into account the light scattering of many identical "small" (with respect to the wavelength λ, point-like) particles with no interactions, e. g., macromolecules in a solvent of much smaller molecules. The scattering waves emanating from the individual scattering centers (e. g., atoms) are in practice always in-phase, interfering constructively, i. e., they amplify each other. On the basis of the law proposed by Clausius–Mosotti, the polarizability α of independent particles of an ideal solution is related to the refractive index n:

$$n^2 - 1 = 4\pi N\alpha \tag{8.10}$$

where N describes the particle density.

According to Eq. (8.7) the scattering amplitude E_r is proportional to the polarizability α. If we irradiate a thin and clear sample (e. g., water) with a layer thickness of Δx with monochromatic light

$$E = E_0 \ \cos 2\pi n\gamma(t - x/c)$$

most of the light will be transmitted. Only a very small portion is scattered in the direction of the transmission:

$$E_s = E_0 \frac{4\pi^2 A\alpha}{\lambda} \sin 2\pi n\gamma(t - x/c)$$

After mathematical transformations, the addition of transmitted and scattered light waves results in a sine vibration E_r which, however, is delayed by a small phase angle τ compared with the incident light wave:

$$E_r = E_0 \cos 2\pi n\gamma(t + \tau - x/c)$$

As a result, this leads to a lower velocity of light, $c_{H_2O} = c_0/n$ in water, with the refractive index n finally yielding Eq. (8.10).

For dissolved scattering macromolecules for a polarizability α with respect to the solvent (n_L) we obtain:

$$n^2 - n_L^2 = 4N\pi\alpha \tag{8.11}$$

and can write:

$$(n - n_L)(n + n_L) = 4\pi N\alpha \tag{8.12}$$

We then solve the equation for α and – with a little trick – expand the fraction with the concentration of the dissolved substance C (in g cm^{-3}):

$$\alpha = \frac{(n + n_0)}{4\pi} \frac{(n - n_0)}{C} \frac{C}{N} \tag{8.13}$$

The value $(n - n_0)/C$ is the specific increase in the refractive index n of the solute, which in infinitesimal or calculus notation is written as dn/dC. If N describes the total number of dissolved molecules, M the molar mass and A Avogadro's number, then $C/N = M/A$. As a good approach, for diluted solvents we can write $n + n_0 = 2n_0$ and obtain

$$\alpha = \frac{n_0}{2\pi} \left(\frac{dn}{dC}\right) \frac{M}{A} \tag{8.14}$$

Replacing this expression in Eq. (8.8), we obtain the scattering contribution from one particle:

$$\frac{I_s}{I_0} = \frac{2\pi^2 n_0^2 (dn/dC)^2 M^2}{\lambda^4 r^2 A}(1 + \cos^2\Theta) \tag{8.15}$$

For N particles the scattering intensity is increased $N = CA/M$ times:

$$\frac{I_s}{I_0} = \frac{2\pi^2 n_0^2 (dn/dC)^2 (1 + \cos^2\Theta)}{\lambda^4 r^2 A} CM \tag{8.16}$$

Now we will define the so-called Rayleigh ratio, R_Θ which comprises all measuring parameters:

$$R_\Theta = \frac{I_s}{I_0} - \frac{r^2}{1 + \cos^2\Theta} \tag{8.17}$$

Using the abbreviation

$$K = \frac{2\pi^2 n_0^2 (dn/dC)^2}{A\lambda^4} \tag{8.18}$$

we can collect all optical constants together and for an ideal solution (where the scattering molecules do not show any interaction) finally obtain:

$$R_\Theta = \frac{2\pi^2 n_0^2 (dn/dC)^2}{A\lambda^4} = KCM \tag{8.19}$$

Clearly this equation can be used to determine the molar mass M, and it can be written in its simple form as

$$\frac{KC}{R_\Theta} = \frac{1}{M} \tag{8.20}$$

However, in practical work we are dealing with real solutions and we have to use a more generalized equation:

$$\frac{KC}{R_\Theta} = \frac{1}{M} + 2BC + \dots \qquad (8.21)$$

in which – additionally – the so-called second (and even higher) virial coefficient B is introduced taking the interaction of particles into account (including the fact that several particles cannot be placed in the same space at the same time). To determine the molar mass M we have to monitor the scattering as a function of concentration and finally to extrapolate for $C \rightarrow 0$ (Figure 8.3). The slope gives the virial coefficient B directly. Figure 8.4 shows a typical measuring set-up, a so-called goniometer. While in earlier experiments white light sources with corresponding filters and point apertures of 0.1 to 2 mm had to be used for collimation, today a laser is the ideal light source, because it delivers a well collimated and monochromatic light beam. A photomultiplier, which is preferred to a solid state detector, is included in a light-tight box and can be rotated on a carriage around the sample, where the azimuth angle θ is adjusted precisely in relation to the direction of the incident beam.

For the determination of the molar mass M, according to Eq. (8.21) we have to determine the Rayleigh ratio R_θ for a series of concentrations C and, e. g., for $\theta = 90$, and to extrapolate the expression KC/R_θ for $C \rightarrow 0$. Moreover, we also require the value of dn/dC, which can be assumed to be constant. Because the difference in the refractive indices of the solvent with and without scattering molecules is very small [of the order of 10^{-5}, based on the approach used in Eq. (8.14)], for this purpose we use a so-called differential refractometer because it is the most sensitive.

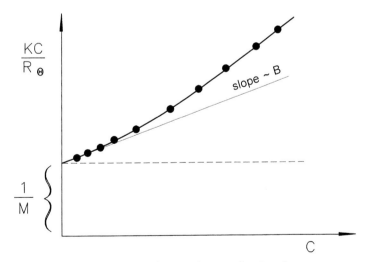

Figure 8.3 Measuring a series of KC/R_θ values as a function of the concentration C enables, on the basis of this plot, the second virial coefficient as well as the molar mass M to be determined.

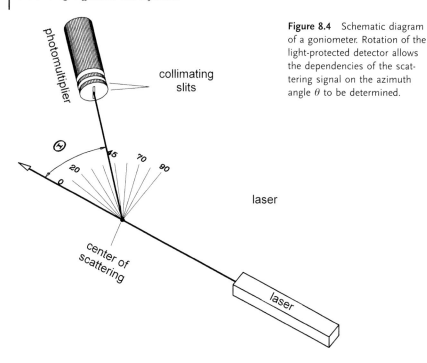

Figure 8.4 Schematic diagram of a goniometer. Rotation of the light-protected detector allows the dependencies of the scattering signal on the azimuth angle θ to be determined.

8.2.3
Scattering by Large Particles

So far we have discussed the scattering properties of small particles that are much smaller than the wavelength of light. In this case the scattering intensities of the individual particles I_s can simply be added together [Eq. (8.16)]. If the size of the scattering particles becomes comparable to the wavelength of the light λ, in addition to the refractive indices of particles and the medium $(n_T - n_M)$ we also have to take into account the size of the particles, and we have to sum the amplitudes rather than the intensities of the scattered light waves. Interference between scattering centers becomes important. In contrast to Rayleigh scattering the so-called Rayleigh–Gans–Debye approximation is more complex and more difficult to interpret, however, it is more informative: in addition to the molar mass M we also obtain information about the molecular size and shape. Three parameters are involved: the wavelength λ, the particle size d and the distance r between the particle and detector. For typical scattering experiments the so-called far-field approach is used $(r > d^2/\lambda)$. Here again we simply adopt the most important results of the theory without derivation.

At the beginning of the 20th century, the German scientist Mie obtained the absolute solution to Maxwell's differential equations, the general far-field approach for spherical particles [taking the boundary conditions into account, the Maxwell equations describe the spatial and temporal interaction of electric (E) and magnetic field strengths (H)]. In this situation the

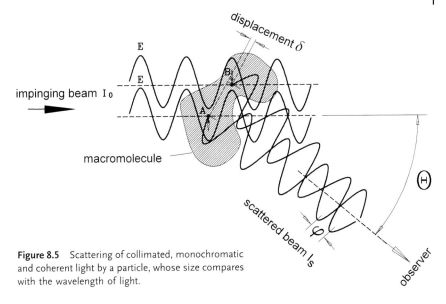

Figure 8.5 Scattering of collimated, monochromatic and coherent light by a particle, whose size compares with the wavelength of light.

scattering intensity is a complex function of the scattering angle θ, the ratio λ/d, the direction of polarization and the (complex) refractive indices n_T and n_M of the particle and solvent. Calculations based on the exact Mie theory are laborious but are achievable using fast computers. Unfortunately, within the "Mie range" (Figure 8.11) the scattering cross section undergoes strong variations ("oscillations") making the analysis difficult.

Let us assume that a collimated monochromatic light beam (laser) hits a scattering particle of molar mass M (Figure 8.5). Two scattering centers A and B (e. g., individual atoms of a molecule) become excited with a particular phase shift, and consequently emit light with a similar phase shift.

The image of discrete light waves shown here only serves as a visualization, however, it is a crude simplification because of the dual nature of light (Section 2.1): the light field is homogeneous. Even with respect to the "particle picture" a typical laser beam of 0.1 mW, $\lambda = 600$ nm and 1 mm^2 cross section represents approximately 10^{11} quanta s^{-1} = wavelets on the scattering particle of 1 mm^2.

In addition, for observations below an angle θ both scattering centers have different distances (δ). Depending on the scattering particles, refractive indices and direction of observation, interferences will be positive or negative. We will now define the so-called form factor $P(\theta)$, taking the relevant interference effects into account. With this factor the interference-free Rayleigh scattering of a small particle (compared with the wavelength λ it is an infinitely small molecule) of the same molar mass M has to be corrected in order to obtain the current scattering intensity as a function of θ:

$$R_\Theta \text{ (real particle)} = P(\theta) \times R_\theta \text{ (pure Rayleigh scattering)} \tag{8.22}$$

According to the theory the form factor $P(\theta)$ in its general state is a double sum:

$$P(\theta) = \frac{1}{N^2} \Sigma_{i=1}^{N} \Sigma_{j=1}^{N} \frac{\sin hR_{ij}}{hR_{ij}} \tag{8.23}$$

where N stands for the number of scattering centers within the molecule (in Figure 8.5 only two: A and B) and R_{ij} for the distance between any two scattering centers A and B. The parameter h (sometimes abbreviated to Q) is defined by

$$h = \frac{4\pi}{\lambda} \sin\frac{\theta}{2} \tag{8.24}$$

λ, again, is the wavelength of light within the medium concerned. $\lambda = \lambda_0/n$, with λ_0 being the vacuum wavelength and n the refractive index. Expansion of Eq. (8.23) in a Taylor series yields

$$\frac{\sin hR_{ij}}{hR_{ij}} = 1 - \frac{(hR_{ij})^2}{6} + \frac{(hR_{ij})^4}{120} + \dots \tag{8.25}$$

and for most scattering molecules we obtain a satisfactory approach for the form factor if we only consider the first two terms of the row:

$$P(\theta) \cong 1 - \frac{h^2}{6N^2} \Sigma_{i=1}^{N} \Sigma_{j=1}^{N} R_{ij}^2 \tag{8.26}$$

For $h \rightarrow 0$, $R_{ij} \rightarrow 0$ and $P(\theta) = 1$, i. e., at very small scattering angles and very small particles, respectively, (compared with λ), we obtain Rayleigh scattering (Figure 8.7). By definition, $P(\theta)$ is always ≤ 1. With increasing particle size the scattering

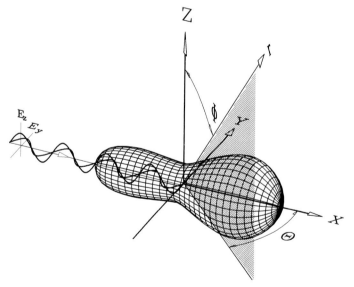

Figure 8.6 With greater particle sizes the scattering intensity I_s increases rapidly, and at the same time the distribution of the scattered light is changed: forward scattering.

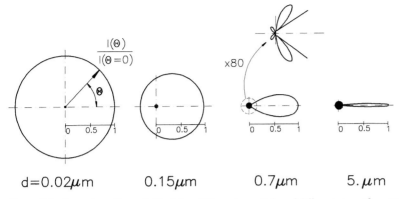

$$\frac{I(\Theta)}{I(\Theta=0)}$$

×80

d=0.02μm	0.15μm	0.7μm	5. μm

Figure 8.7 Scattering of laser light of λ_0 = 633 nm by particles of different sizes (from Wiese, modified; GIT Fachz. Lab., 1992).

intensity I_s increases greatly and at the same time the distribution of scattered light changes: multiplication of Eq. (8.9) by the corresponding form factor $P(\theta)$ < 1 results in the forward scattering as shown in Figure 8.6. This phenomenon is shown in Figure 8.7 for various particle sizes and incident laser light of λ_0 = 633 nm.

As expected, for various molecular shapes but with identical sizes, $P(\theta)$ will have different values, because the conditions for interference are different. To evaluate the molecular size we use the so-called gyration radius R_G, which is defined by the distribution of mass, i. e., the moment of inertia of the scattering particle: for a particle consisting of i elements of mass m_i and the distance r_i from the center of gravity we obtain

$$R_G^2 = \frac{\Sigma_i m_i r^2}{\Sigma_i m_i} = \frac{1}{2N^2} \Sigma_{i=1}^{N} \Sigma_{i=1}^{N} R_{ij}^2 \tag{8.27}$$

For example, for a sphere with the radius r it follows that $R_G = (3/5)^{1/2} r = 0.775r$, for a rod of length L with R_G approximately $(12)^{-2} L = 0.289L$.

For small angles θ the form factor $P(\theta)$ provides a useful formula, even if it is only crudely adapted to the molecular volume of the medium:

$$P(\theta) \cong e^{\frac{-h^2 R_G^2}{3}} \tag{8.28}$$

On this basis, for example, for the tobacco mosaic virus (rod-like, 300 nm long, 18 nm diameter) a gyration radius of R_G = 92.4 nm and a molar mass of M = 39×10^6, and for myosin (which is required in the building of the thick filaments of muscle fibers) R_G = 46.8 nm and $M = 49 \times 10^4$ can be calculated.

In the general case shown in Eq. (8.21) we have to include the *correction factor* $P(\theta)$:

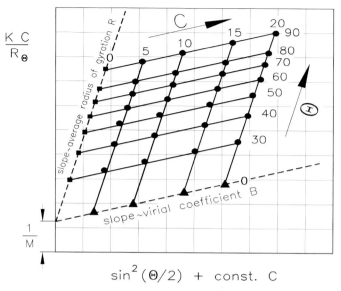

Figure 8.8 On the basis of the so-called Zimm plot the parameters R_0, R_G, B and M can be determined simultaneously.

$$\frac{KC}{R_\Theta} = \frac{1}{P(\theta)}\left(\frac{1}{M} + 2BC\right)$$ (8.29)

Taking the approach of the formula given in Eq. (8.26) into account we finally obtain

$$\frac{KC}{R_\Theta} = \left(1 + \frac{16\pi^2 R_G^2}{3\lambda_2}\sin^2\frac{\theta}{2}\right)\left(\frac{1}{M} + 2BC\right)$$ (8.30)

In order to obtain $1/M$ and thus finally M, we have to extrapolate $C \to 0$ as well as $\theta \to 0$. Following the elegant method of Zimm, this is done by a single plot. Each measurement of R_θ yields defined values of C and θ which we plot in Figure 8.8, where either C or θ is kept constant. In this way we generate a grid of elements which allow – as indicated – the determination of all three significant values of R_G, B and M. The constant of $C = 2000$ has been chosen deliberately and simply allows an appropriate plot to be made. According to Eq. (8.30) we can obtain the required curves for $C = 0$ and $\theta = 0$ simply by setting C and θ to 0, respectively.

8.2.4
Fraunhofer Scattering

Fraunhofer scattering is derived from the Mie theory for scattering of large spherical particles and non-polarized light. It is also the far-field approach of the classical *Fresnel–Kirchhoff* theory of light scattering (diffraction at slits, see Section
3.6.3) and for spherical particles yields a series of diffraction rings, which rapidly
decrease in intensity with increasing azimuth angle θ (Figure 8.9). The scattering
intensity I_s is described by the so-called *Airy function*, where J_1 is the *Bessel function* (cylinder function) in the first order. It is the solution of the so-called *Bessel
differential equation*, with the parameter n, that defines the "order" and which here
is an integer; in the following we only indicate the argument $[Kd \sin (\theta)/2]$ of the
function:

$$I_s = \text{const.} \times d^4 \times K^2 \left[\frac{2J_1(Kd\times \sin\theta/2)}{Kd\times \sin\theta/2} \right]^2 \tag{8.31}$$

The so-called wave vector is described by K. Depending on the refractive indices of
the particles and the medium, particles with sizes starting from $d = 5\lambda$ can be
determined. As a consequence of Eq. (8.31) smaller particles will deflect the scattered light towards larger angles (Figure 8.10). For example, for $\lambda_0 = 633$ nm and
with a diameter of the particle of $d = 4$ mm we obtain the first scattering maximum on the surface of a cone with an acceptance angle of $\theta = 9°$. A particular
feature of the scattering device in Figure 8.9 is the Fourier lens, which focuses
the scattering light of a specific particle size onto one of the up to 30 detector

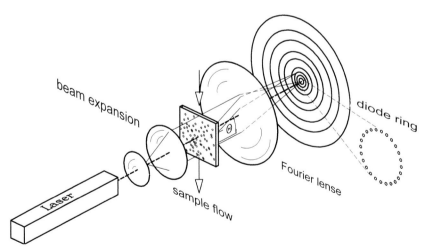

Figure 8.9 For larger particles the Fraunhofer
scattering yields a sequence of refraction rings,
the intensities of which decrease rapidly with
increasing azimuth angle θ. Typical set-ups for
measuring Fraunhofer scattering use up to 30
detector rings (or arcs of rings). By means of a
so-called Fourier lens the scattered light is focused onto one and the same ring (from Wiese, modified, *GIT Fachz. Lab.*, 1992).

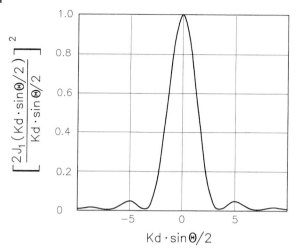

Figure 8.10 The intensity I_s of Fraunhofer scattering as a function of the azimuth angle θ is described by the so-called Airy function. Note: J_1 is the so-called Bessel function (cylinder function) in the first order, where the bracket in the numerator is an assumption about this function and cannot be divided by the denominator where the assumption is a real number (from Wiese, modified, *GIT Fachz. Lab.*, 1992).

diode rings, independent of its position and velocity. The integration effect significantly increases the signal to noise ratio (SNR). Again, the presumptions are absence of absorption and exclusion of multiple scattering, i. e., a high dilution of the scatterer (e. g., 1 mg L^{-1}). The determination is fast and relatively simple and the measuring time short (approximately 1 min per analysis). As with the dynamic light scattering, only a little information on the sample is required.

8.2.5
Scattering Intensity I_s and Particle Size

Figure 8.11 shows the normalized dependence of the scattering intensity I_s and the scattering cross section s_s as a function of the particle size d for the range between 10^{-3} and 200 mm, which is covered by visible light. The scattering intensity increases over the whole measurable range by as much as 20 orders, while the particle diameter only changes by 5 orders. For this demonstration spherical particles with a typical refractive index of $n_T = 1.59$ were suspended in a medium of $n_M = 1.33$ and irradiated with light of wavelength of $\lambda_0 = 633$ nm (HeNe laser). The scattering was observed under azimuth angles of $\theta = 5°$ and $90°$, respectively, and, for easier evaluation, plotted with the scattering cross section in a double logarithmic plot. According to the various theoretical treatments following the theories of Rayleigh, Rayleigh–Gans–Debye (RGD), Mie and Fraunhofer, we have to distinguish between different ranges for the particle sizes.

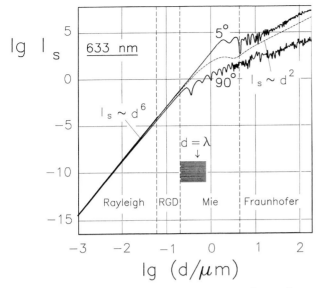

Figure 8.11 Normalized scattering intensity I_s and scattering cross section σ_s as function of particle size d within the whole range of sizes between 10^{-3} and 200 μm. The scattering intensity embraces as much as 20 orders (!) while the particle diameters only increase by five orders. With respect to the various theoretical treatments by Rayleigh, Rayleigh–Gans–Debye (RGD), Mie and Fraunhofer, we can distinguish different ranges of particle size. The visible range is indicated (from Wiese, modified, *GIT Fachz. Lab.*, 1992).

The "smooth" dependence in the range for Rayleigh and RGD scattering is striking, independent of the scattering angle θ, where the scattering intensity increases to the sixth (!) order of the particle diameter: if the particle diameter is small compared with the wavelength, all scattering centers P of the particle scatter "in phase" (see Figure 8.5). In the range of coherent scattering the scattering field E_s of a particle is proportional to the number of scattering centers P and the scattering intensity I_s is

$$I_s = E_g^2 = (P \times E)^2 = P^2 \times E_s^2 \tag{8.32}$$

Clearly the number of scattering centers per particle P is proportional to the particle volume, i. e., proportional to d^3 and therefore $I_s \approx d^6$. Correspondingly, even the smallest contaminations such as fingerprints or small floating particles are capable of giving rise to false measurements. Thus, proper cleaning of the sample is an indispensable prerequisite to a reliable scattering measurement.

The extreme dependence of the scattering cross section on the particle size allows highly sensitive measurements to be made. Example: on anaerobic photoreduction, 3-methyl flavin (see Figure 6.7) aggregates spontaneously with EDTA (ethylendiaminetetraacetate), a chelating agent (see Section 6.6.4) yielding a weakly bound complex. Surprisingly the altered scattering cross section is observed simply using a typical fluorimeter (Section 6.3.1), where a comparably small excitation wavelength of 350 nm is chosen (scattering $\propto \lambda^{-4}$). Scanning the emission wavelength between 335 and 365 nm (Figure 8.12) enables the Rayleigh peak to be measured (90° scattering, cf. Figure 6.11). If the cuvette is subsequently opened in order for the sample to come into contact with oxygen, the flavin molecule is spontaneously and reversibly re-oxidized, as observed using fluorescence (only the oxidized form of flavin shows significant fluorescence). The complex decomposes into the individual components. Plotting the state of the photoreduction as a function of scattering we obtain the dependence shown in Figure 8.13. Obviously, the aggregation is not a simple or trivial linear effect but exhibits significant cooperation (a phenomenon which is essential to the oxygen load of blood) with a so-called critical micelle concentration (*cmc*), where aggregation sets in. Scattering measurements as a function of temperature yield a free energy for the formation of aggregates of $\Delta G = 48$ kJ mol^{-1}, suggesting the presence of 5–10 hydrogen bonds per aggregate. Simple scattering measurements particularly in (bio-) chemical analysis are extremely valuable for the detection of very small changes in the size d and/or the polarizability α of a molecular complex or a scattering particle: e. g., phase transitions of membranes can be monitored simply, without using a fluorescence label that could cause interference (cf. Section 6.6.3).

If the particle size d becomes comparable to the wavelength of light λ, the situation becomes more complex, particularly at large azimuth angles θ (note the *logarithmic* plot in Figure 8.11). Because of interference, the forward scattering now becomes dominant. Strong and rapid variations in the particle cross section as a function of particle size take place in the Mie range, making the scattering analysis difficult. Interestingly, the curve in Figure 8.11 shows a significant inflexion

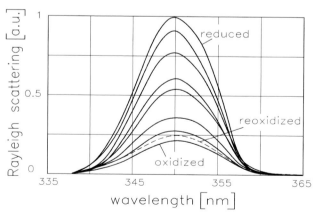

Figure 8.12 Dependence of Rayleigh scattering on the photoreduction of 3-methyl flavin with EDTA (ethylenediaminetetraacetate), a chelating agent. The reaction is fairly reversible (from Schmidt and Wenzel, *Biophys. Struct. Mechanisms* 1983, *10*, 61–69).

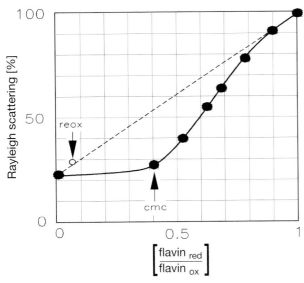

Figure 8.13 Dependence of Rayleigh scattering as a function of a newly formed complex of 3-methyl flavin with EDTA (ethylenediaminetetraacetate) (plotted from Figure 8.12).

in the visible wavelength range and the scattering then depends approximately only on the second power of the diameter, i. e., on the cross section of the particle. In general, measurements are more sensitive and reliable if the particle size d is small compared with the wavelength λ.

8.2.6
Dynamic Scattering

When measuring dynamic or quasi elastic light scattering we have not been interested – in the discussions so far – in the static, average temporal value of scattered light. Measuring temporal fluctuations of the scattering intensity or (small!) wavelength shifts we gain information on the dynamics of the systems as well as on the distributions of the particle size. Again, the presumptions are incoherent scattering, i. e., mutual independence of the particles, as well as highly coherent light (laser). The half-width of laser light of $\Delta\lambda = 0.01$ Å corresponding to $\Delta\nu = 50$ Hz is extended to approximately $\Delta\nu = 20$ kHz by a Doppler shift (Figure 8.14). The scattering intensity is described by a spectral Lorentzian distribution:

$$I_s(\nu) \cong \frac{DK^2/2\pi}{(\nu - \nu_0)^2 + \left[DK^2/(2\pi)\right]^2} \tag{8.33}$$

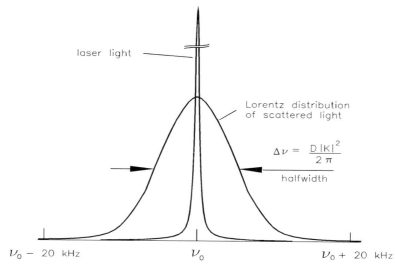

laser light

Lorentz distribution
of scattered light

$$\Delta \nu = \frac{D\,|K|^2}{2\,\pi}$$

halfwidth

$\nu_0 - 20$ kHz ν_0 $\nu_0 + 20$ kHz

Figure 8.14 Owing to the Doppler shift, the half-width of laser light of approximately $\Delta\lambda = $ 0.01 Å corresponding to $\Delta\nu = 50$ Hz is exten- ded by scattering by motile particles to the order of $\Delta\nu = 20$ kHz. The scattering intensity follows a spectral Lorentzian distribution.

where ν_0 is the frequency of the laser light, D the so-called diffusion constant and

$$|K| = \frac{4\pi}{\lambda}\,\sin\frac{\theta}{2} \tag{8.34}$$

the modulus of the wave vector of the scattered light.

Scattering phenomena are effectively described by the vector notation because "light beams" are directed entities. The incident light vector is described by $k_0 = 2\pi/\lambda \times e_0$, where e_0 is the unit vector and $2\pi/\lambda$ its amplitude. Correspondingly, the vector in the direction θ of the scattered light having, by definition, the same amplitude and the unit vector e_s, is termed k_s. The wave vector is defined as the vector difference $K = k_0 - k_s$, yielding the modulus in Eq. (8.34) purely on the basis of a geometric argument.

According to classical kinetic gas theory the diffusion length l as covered by a particle in the time t is proportional to the square root of the time range. The constant of proportionality is D and the medium length l traveled per time unit is defined by $D = l^2/\tau$. The diffusion is, in contrast to our macroscopic assessment, on the microscopic scale an extremely fast process. Following the rules of thermodynamics the medium velocity of a gas particle of mass m is

$$\bar{u} = \sqrt{\frac{3}{m} \times \varkappa \times T} \tag{8.35}$$

where $T = $ absolute temperature and $\varkappa = $ the Boltzmann constant. Thus, at thermal equilibrium for particles of 0.5 µm diameter and density $\varrho = 1$ g cm^{-3} at room temperature we obtain a velocity of 5000 μm s^{-1} (corresponding to 10000 particle diameters). For elastic scattering, theory yields a Doppler shift (energy shift → quasi elastic scattering) of $\Delta v = DK^2/2\pi$ allowing D to be determined experimentally. However, as mentioned previously, the expected frequency shifts are extremely small.

Even in the 100th scattering order a dispersion grating of 10 cm width and 1200 grooves mm^{-1} shows a resolution of "only" $R = 10^6$ [Eq. (3.34)], still a long way off the required $R = v_0/v = 10^{14}/50$. Therefore, other methods have to be applied. The first one is only mentioned briefly: on the basis of an "optoelectronic trick (*self beating spectroscopy*) the offset v_0 of the scattered light is eliminated allowing the easy measurement of the spectral distribution of the scattered light as shown in Figure 8.14. For typical macromolecules such as proteins in an aqueous phase, D values of between 10^{-6} and 10^{-8} (cm^2 s^{-1}) can be determined.

The second method is based on statically fluctuating scattering signals, similar to electronic noise (Section 5.2.5). This time dependent scattering signal depends directly on the thermal motion of samples (*Brownian* molecular motion) and shows variations depending on the particular interference conditions. Small particles move faster than larger ones, which is reflected by the bandwidth of the "noise signal" (Figure 8.15). The medium relaxation rate of translation is 10^2–10^5 Hz. As discussed previously, the technique of autocorrelation is a highly versatile tool for signal analysis (Section 5.4.4.1). Thus, the autocorrelation of a rectangle yields a triangle (Figure 5.43). Correspondingly, the autocorrelation of a noise signal yields an exponential function with the correlation time τ

$$g(\tau) = 1 + \zeta \times e^{-2DK^2 \tau} \tag{8.36}$$

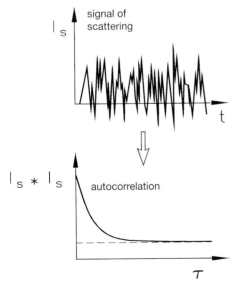

Figure 8.15 Top: Time dependent scattering signal I_s. Bottom: Autocorrelation $I_s * I_s$ of the scattering signal above yields an exponential function with correlation time τ (from Wiese, modified, *GIT Fachz. Lab.*, 1992).

where $0 < z < 1$ describes a constant that is typical for the apparatus used. The factor $2DK^2$ in the exponent of Eq. (8.36) is the medium relaxation rate, which allows the diffusion constant D to be extracted directly and which, in turn, according to the Stoke–Einstein relationship

$$D = \frac{\varkappa\, T}{3\pi\eta d} \tag{8.37}$$

where $\varkappa$ = Boltzmann's constant and η = viscosity of the dispersion medium, allows the determination of the particle diameter d. The dynamic light scattering enables the diffusion coefficients in the range between 10^{-9} and 10^{-6} cm^{-2} s^{-1} to be determined, corresponding to particle sizes of from 5 nm up to 5 μm.

8.3
Raman Scattering Versus Infrared Spectroscopy

In contrast to the elastic scattering discussed so far, *Raman scattering* (Raman, 1928) of molecules changes the wavelength of the light. The experimental set-up is similar to Figure 8.4, typically with a fixed observation angle of $\theta = 90°$. The wavelength change of the scattered light results from an interaction of the incident light with vibration quanta of the scattering molecule. In elastic scattering we are dealing with forced vibrations of the scattering particle where the induced (radiating) dipole is proportional to the so-called polarizability α, which is supposed to be constant for a given molecule. However, we can drop this grave restriction and assume in a more realistic manner that at the frequency of the molecular vibration ν_0 its polarizability α will also change.

   This is exemplified well using the linear CO_2 molecule. Similarly to Section 2.3.3.3, let us consider the relevant vibration modi, Figure 8.16. In principle we can distinguish symmetrical and asymmetrical vibrations (cf. Figure 10.2). With symmetrical vibrations the "electron cloud" changes its size and "softness", i. e., its polarizability α, with the frequency of the exciting wave ν_0:

Sir Chandrasekhara Venkata Raman (1888–1970)

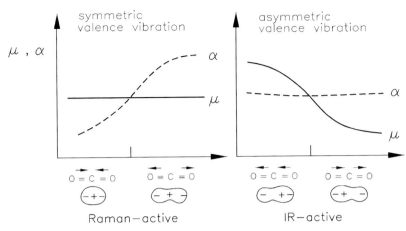

Figure 8.16 Representation of Raman or IR active vibration modes of the CO_2 molecule. With Raman active vibrations the polarizability α is changed, and with IR active vibrations it is the dipole moment μ that changes.

$$\alpha = \alpha_0 + \alpha_M \; \cos 2\pi\nu_M t \tag{8.38}$$

with an amplitude α_M of the change in polarizability. After rearrangement, Eq. (8.6) yields

$$\mu = \alpha_0 \; \mathbf{E}_0 \; \cos 2\pi\nu_0 + \tag{8.39}$$

$$+ \; 1/2 \; \alpha_M \mathbf{E}_0 \left[\cos 2\pi(\nu_0 + \nu_M)t + \cos 2\pi(\nu_0 - \nu_M)t \right]$$

The first term corresponds to the normal Hertz dipole, i. e., Rayleigh scattering with a frequency ν_0. The second term represents the dipole that vibrates with the frequency changed by $\pm\nu_M$. To continue with the analogy to an antenna as used earlier, by modulation of the incident light wave of frequency ν_0 in addition to the Rayleigh peak, we obtain a so-called upper and a lower side band (Figure 8.17). For the upper side band, so to speak, the scattered wave has taken up a vibration quantum from the scattering molecule, and for the lower side band has given one up. We are referring to *anti-Stokes* and *Stokes Raman* lines, respectively. Of course, a vibration quantum only can be taken up, if it actually exists, if the scattering molecule is already in an excited vibrational state. However, according to the Boltzmann distribution, at room temperature this probability p is very small:

$$p = e^{\frac{h\nu_M}{\varkappa T}} \tag{8.40}$$

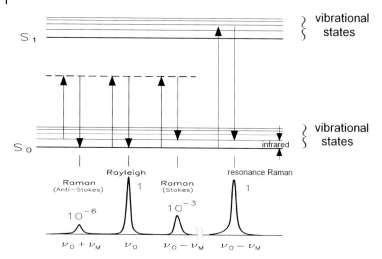

Figure 8.17 Representation of Rayleigh peaks at the frequency ν_0 and two side bands using the concept of an energy diagram with "virtual levels". The relative intensities are indicated and do not correspond proportionally to the drawing. The lower side band corresponds to the Stokes, the upper one to the anti-Stokes Raman lines. Resonance Raman lines are observed if the scattered light induces only an electronic transition (absorption).

For a typical vibration band, at room temperature this probability is close to $p = 0.001$. Consequently, the anti-Stokes Raman scattering is incredibly small and thus for practical work we exclusively utilize Stokes Raman scattering. However, this is also due to the small excitation probabilities with approximately $I_s = 10^{-6}I_0$ being very small. This requires both very intense and at the same time monochromatic light sources which only became available with modern lasers. The resulting strong Rayleigh scattering has to be suppressed using narrow and effective *Notch*-filters (see Section 3.6).

In the case of asymmetric vibrations (Figure 8.16, right) we observe exactly the contrary: the polarizability α hardly changes, in contrast to the electric dipole moment μ of the molecule. The change to the electric dipole moment has already been discussed in Section 2.3.3.4 as being an indispensable prerequisite to any absorption; this also holds true for vibration transitions. To summarize:

- Asymmetric vibrations cause changes to the electric dipole moment μ and thus are IR active; they do not take place in Raman spectra.
- Symmetric vibrations lead to changes in polarizability α and therefore are Raman active. They are not seen in IR spectra.

In both cases we are measuring the vibration of molecules, however this is based on very different physical phenomena, which, in an ideal way, largely complement each other. Sometimes peaks appear in both types of spectra, but usually with very different intensities: as is often the case in quantum mechanics, the

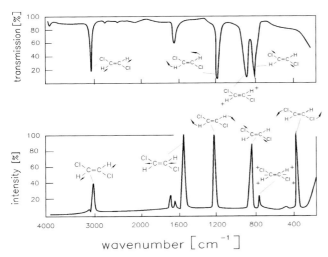

Figure 8.18 Comparison of infrared and Raman spectra of *trans*-dichlorethane. The most important peaks are attributed to the individual vibrational modes. According to convention, infrared spectra are plotted "downwards" and Raman spectra in a linear manner "upwards" (from Hesse et al, modified, with permission, Thieme Verlag).

symmetric–asymmetric classification is fairly gradual and we are dealing with mixtures of both types.

In Figure 8.18 the infrared and Raman spectra of *trans*-dichlorethane are compared. The most important peaks are attributed to individual vibrational modes. By convention, infrared spectra are typically plotted "downwards" (often in logarithmic form) and Raman spectra upwards in a linear format. While the wavelength scale of IR spectra are plotted as absolute values, Raman spectra are always plotted against relative frequencies with respect to the exciting wavelength.

Because the scattering intensity depends on the polarizability α, compared with the exciting light, the scattered light will experience some depolarization. If the incident light is, for example, vertically polarized, then the depolarization of the scattered light is

$$\varrho = \frac{I_\perp}{I_\parallel} \tag{8.41}$$

where – similar to the measurement of fluorescence polarization – the measurement takes place orthogonal to the exciting light with corresponding polarization direction. If we turn the polarization of the exciting light into the horizontal position the scattering intensity approaches 0 (cf. Figure 8.1). Now it is possible to show that for isotropic polarizability α the extent of depolarization ϱ ranges from 0 to a maximum value of 3/4. With anisotropic polarizability, i.e., without any symmetrical features of the molecules, we obtain $\varrho = 3/4 = $ const. Determination of the extent of the depolarization allows individual Raman peaks to be attributed to certain elements (polarization Raman spectroscopy).

For the Raman scattering discussed so far the absence of absorbance is an essential prerequisite. The scattering is the result of displacement of charge within "electron clouds": the scattering intensity is proportional to α^2. However, when the frequency of the exciting light induces an additional electronic transition (Figure 8.17, absorption), the charge displacement is dramatically increased and the Raman signal often becomes comparable to the Rayleigh scattering (*resonance enhancement*) or to that in *UV* absorption spectroscopy. For chromo proteins in particular we obtain far less complex Raman spectra on excitation in the absorption peak of the chromophore: it is just this that essentially contributes to the Raman spectrum of the whole molecule, while the protein part remains "invisible". It is anticipated that resonance Raman spectroscopy will turn out to be a powerful tool not only in the field of research, but also in analytical and clinical studies.

8.4
Rainbows

The first documented spectrum can be found in the bible: in Chapter 9, verse 13, *Genesis*, when God speaks to Noah:

I do set my bow in the cloud, and it shall be for a token of a covenant between me and the earth

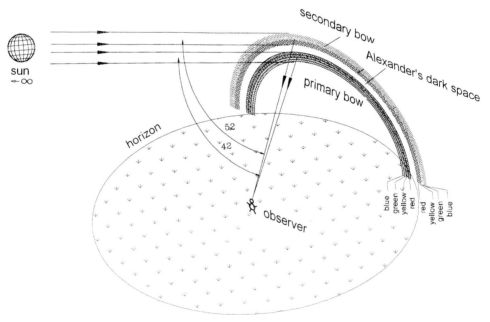

Figure 8.19 Diagram of a rainbow with a secondary bow above, including approximate angles. The position of the observer is indicated. The secondary bow is much weaker than the primary one. The sun, the observer and the center of the bows are always aligned.

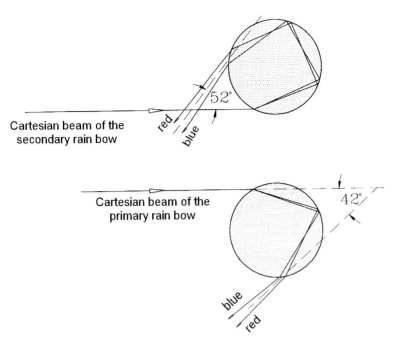

Figure 8.20 Light paths within a drop of water leading to the phenomenon of rainbows. Only so-called *Cartesian rays* hitting the drop at a distance of approximately 7/8th (primary bow, below) 19/20th (secondary bow, top), respectively, are capable of generating spectral phenomena.

A rainbow is known to be the visible spectrum of sunlight, which is caused by the scattering of light by small water droplets. However, why are rainbows so rare even though both factors are very often present, even if to different extents? In addition to the primary bow we often find a weaker and somewhat broader secondary bow with the order of colors reversed: red on the bottom and blue on the top (Figure 8.19). It is much easier to observe a rainbow under the shower or when using a lawn sprinkler under direct sun light. What conditions are required for a rainbow to occur? The final answer to this surprisingly complex phenomenon was obtained by laboratory experiments with lasers and large, well defined droplets.

If a light beam hits a spherical drop of water with a diameter of between 0.3 and 20 mm, part of the light is refracted into the drop, the rest is reflected. Only so-called *Cartesian rays* (after the French mathematician René Descartes) hitting the drop at a distance of 7/8 (primary bow) or 19/29 (secondary bow) of the radius of the middle line will cause spectral phenomena. For the primary bow we have one and for the secondary bow two internal reflections (Figure 8.20). All other rays divert too far and are unable to show any amplifying interference. Owing to the relatively small dispersion of water with a refractive index of only $n = 1.333$ (Figure 3.15) the colored angle range of rainbows is rather restricted – and to some extent is dependent on the diameter of the drop and the number of internal reflections (Table 8.1).

Table 8.1 Details of dispersion angles in a rainbow.

Diameter	1 mm	20 mm
Primary bow	40.5 ± 2	41.5 ± 1
Alexander's dark space	46 ± 3	46 ± 3
Secondary bow	53 ± 4	52 ± 2

The smaller the drops, the more "washed out" the rainbow is. If the drops have diameters of less than approximately 0.3 mm, the overlap of the individual spectral colors is so strong (i.e., the spectral resolution so bad), that they complement each other leading to white (white rainbows), and finally the rainbow disappears completely. According to Fresnel formulae, the colored light of a rainbow is highly polarized (tangential to the bow, **E**-vector): orthogonal to the scattering area polarized light is predominantly reflected inside the drop (cf. Figure 3.16), while the remainder leaves the drop and is visible.

8.5
Reflection Spectroscopy

8.5.1
Theoretical Considerations

Reflection spectroscopy is used to investigate the spectral properties of light that is emitted from solid samples. We can distinguish between three types of reflection: the regular reflection of mirrors (*specular reflection*), internal reflection and the diffuse reflection of rough surfaces. In practice all three types of reflection will be present at the same time, although with different contributions. Fairly independent of the particle diameter, part of the incident radiation is reflected at surface particles (micro mirrors), part is conveyed to the inner sample from where – after numerous refractions, diffractions, reflections and also absorption processes – it is re-emitted again somewhere from the surface.

Although a rigorous theory of multiple scattering does not exist, the general and widely accepted theory of the diffuse reflection and transmission of optically opaque samples is the so-called two-component theory, which was developed in the 1830s and 1840s by Kubelka and Munk in Germany. In the field of reflection spectroscopy this is of similar importance to the Bouguer–Lambert–Beer law in absorption spectroscopy of clear samples (cf. Section 5.1). It has been shown to be an excellent tool in practical work. The theory presumes that the reflected radiation is isotropic, i.e., not dependent on the direction (Lambert's cosine law, Figure 3.3), and that the irradiating light is monochromatic.

Originally Kubelka and Munk were working on the development of paper for newspapers which was both thin, i. e. light in weight, but nevertheless as opaque as possible: a classical spin-off effect of optical spectroscopy.

In order to describe the transport of radiation that is both strongly scattering and absorbing, the light flux is divided into two opposite flows (Figure 8.21). The incoming flow I is from the $+x$ direction from the top into the sample (possibly even down to the reflecting bottom surface: R_g), the scattered part of the flow J is from the $-x$ direction, upwards and out of the sample (note the convention used for the direction). The sample is not limited at the sides [thus the boundary conditions for solving the differential equations are inapplicable, Eq. (8.43), below]. Of course, even if the layer is filled in all directions by radiation, this approach offers – analogous to the kinetic theory of gases – a reasonable solution. It leads to the important statement that on multiple scattering and with completely diffuse radiation the medium wavelength dx inside the sample is identical to that of parallel irradiation under $60°$ – or in other words– exactly double that usually obtained with collimated vertical irradiation:

$$\overline{d\xi_I} = \overline{d\xi_J} = 2 \times dx \tag{8.42}$$

According to the theory of Kubelka–Munk, the absorption coefficient is termed K and the scattering coefficient S [cf. Eq. (8.2)] and we obtain two coupled differential equations:

$$- dI = - KIdx - SIdx + SJdx$$
$$dJ = - KJdx - SJdx + SIdx \tag{8.43}$$

With the boundary conditions

$$x = 0: (J/I)_{x = 0} = R_g = \text{reflection off the bottom}$$
$$x = d: (J/I)_{x = d} = R = \text{reflection off the sample}$$
$$x = \infty: (J/I)_{x = \infty} = R_\infty = \text{reflection off an indefinitely thick sample } (R_g = 0)$$

and without further assumptions the system of equations can be solved. As a result, the diffuse reflection R_∞ of a sample depends exclusively on the ratio K/S, however not individually on the scattering or absorption:

$$\frac{K}{S} = \frac{(1 - R_\infty)^2}{2 R_\infty} = F(R_\infty) \tag{8.44}$$

$F(R_\infty)$ is termed the Kubelka–Munk function, which is listed in tables (Kortüm, 1969). Similarly to absorption, and particularly emission, spectroscopy the wavelength-selective response of apparatus has to be eliminated. This is done (1) by measuring reflectance of the sample itself and (2) by measuring a perfect reflecting surface [$R_{Ref}(\lambda)$], e. g., a surface covered with a thin layer of barium sulfate,

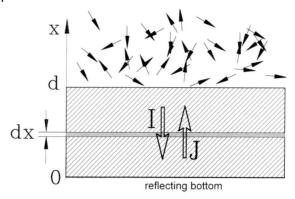

Figure 8.21 Derivation of the Kubelka–Munk equation. If a sample of thickness *d* and extended indefinitely horizontally is irradiated from the top with diffuse light of intensity *I*, a reverse flow *J* vertically to the top is measured (from Kortüm, modified). When the sample is not infinitely thick part of the incident light will be reflected from the bottom.

manganese carbonate or manganese oxide: $R_\infty(\lambda) - R_{Ref}(\lambda)$ {in its logarithmic form $\log [R_\infty(\lambda)/R_{Ref}(\lambda)]$}.

With these standards and in particular with the new materials used for Ulbricht's sphere we can obtain R_∞ values of better than 99 %. On the other hand, a perfectly black surface, which reflects as little light as possible, is much harder to manufacture (a fact which is a direct result of the Kubelka–Munk theory). We discussed this problem previously with respect to the correction of photoacoustic spectra by black reference samples covered with soot, the best known absorber (Section 7.4).

Assuming that the scattering coefficient *S* is independent of λ in the wavelength range measured (which usually holds true) in theory we are able to determine the concentration of a substance through reflection measurements by replacing "*K*" [Eq. (8.44)] by $2.3026 \times \varepsilon \times C$, where *C* is the molar concentration and ε the molar extinction coefficient [Eq. (5.5)]. For fine powdered substances with grain sizes close to 1 μm and a thickness of at least 1 mm Eq. (8.44) yields a fairly linear relationship. The correction and plots using the Kubelka–Munk-function are easily performed with on-line computers. For a non-reflecting background we have $R_g = 0$ and $R \equiv R_0$ and the theory provides for a sample thickness *d* the scattering coefficient *S*,

$$S = \frac{2.303}{d} \frac{R_\infty}{1 + R_\infty^2} \times \log \frac{R_\infty(1 - R_0 R_\infty)}{R_\infty - R_0} \tag{8.45}$$

and the absorption coefficient *K*:

$$K = \frac{2.303}{2d} \frac{1 - R_\infty}{1 + R_\infty} \times \log \frac{R_\infty(1 - R_0 R_\infty)}{R_\infty - R_0} \tag{8.46}$$

8.5.2
Some Practical Results

The absorption signal of an opaque and solid sample such as a young and very fine seed leaf is significantly magnified if it is covered by a strongly scattering (however not absorbing!) slurry of $CaCO_3$ (see Section 5.2.8) of approximately 1 mm thickness. Firstly, the normally collimated measuring light beam is thus diffused [cf. Eq. (8.42)], secondly, part of the reflected light is redirected again in the direction of the photodetector. Thus, signal magnification by a factor of three is easily obtained. An Ulbricht sphere is extremely useful for reflection measurements (Section 3.5.6), as it diffuses both the incoming and the reflected light almost perfectly.

Amplification of the optical signal in the absorption spectroscopy of highly dilute and liquid samples is very important is biochemistry. For example, the previously mentioned universal plant pigment *Phytochrome* is found only in extremely small, hormone-like concentrations. Normal absorption spectroscopy of solutions yields no signal at all. Concentration procedures are required. However, mixing the crudely extracted, without any further enrichment, sample with a slurry of (inert) $CaCO_3$, gives possible signal magnifications of between $\beta = 100$ and 150. For β the Kubelka–Munk theory yields:

$$\beta = 2 \left[\frac{-4R_\infty^3}{(1 - R_\infty^2)^2\ Sd} + \frac{1 + R_\infty^2}{1 - R_\infty^2} \right] \tag{8.47}$$

For most white and scattering media the so-called *scattering power* ($S \times d$) is large enough that the first term in Eq. (8.47) can be omitted, leaving only the second term, thus simplifying the equation. However, if the grain size of the scattering material becomes significantly smaller than the wavelength λ, then β will also be smaller because the scattering factor S decreases (see Figure 8.11).

Of course, when measuring a series of, for example 100, samples we have to maintain a constant ratio of the liquid–$CaCO_3$ mixture (typically 1 mL per 1 g), because we are interested in relative values (e. g., in column chromatography). The fact that β is fairly independent of the wavelength λ is very important, and it allows the measurement of undistorted reflection spectra of opaque and strongly scattering samples. Scattering is measured separately, e. g., based on white, non-fluorescent paper of similar *scattering power* ($S \times d$) as the reference.

As an example, Figure 8.22 compares the absorption spectrum (1) of a didymium filter (BG 36, Schott, Mainz, Germany, $A_{580} > 2$) using various plots with respect to the reflection values. Figure 8.22(5) shows the reflection spectrum of a white, non-fluorescing filter paper [log (R_{ref}), *baseline*] with collimated incident light. Figure 8.22(4) is the reflection spectrum of the previously used didymium filter on top of white paper [baseline (5)] ($R_g \approx 100\%$). Spectrum (3) shows the conventionally corrected and normalized reflection spectrum log ($R_{ref}/R_{didymium}$). Deviations from absorption spectrum (1) are fairly strong. However, if we plot the same spectrum according to the Kubelka–Munk function [Eq. (8.44)], the differences are much smaller [Figure 8.22(2)]. This is not a com-

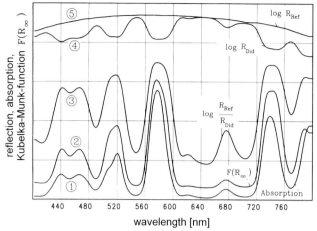

Figure 8.22 Comparison of the absorption spectrum (1) of a didymium filter with various logarithmic plots of the corresponding reflection values. Spectrum (5): reflection spectrum of a white, non-fluorescing filter paper (baseline) using collimated illumination. Spectrum (4): reflection spectrum of the didymium filter as used for spectrum (1), which had been put onto white paper ($R_g \approx 100\%$). Spectrum (3): reflection spectrum (4) corrected for the baseline (5) [log ($R_{ref}/R_{didymium}$)]. Spectrum (2): recalculated spectrum (3) according to the Kubelka–Munk function $F(R_\infty)$.

plete coincidence and is explained by the fact that the presumptions for application of the Kubelka–Munk theory are not perfectly met: this is mainly (1) a fine powdered sample, (2) diffuse irradiation and (3) independence of the scattering factor S from the wavelength λ.

Another example is shown in Figure 8.23, which compares the absorption spectrum of the green leaf of *Euphorbia pulcherima*, the corrected reflection spectrum and the reflection spectrum calculated on the basis of the Kubelka–Munk function. Again, obviously the Kubelka–Munk plot is preferred, reflecting the absorption spectrum much better than the normal reflection plot. Clearly, under specific circumstances reflection spectroscopy is capable of replacing absorption spectroscopy.

A further example is shown in Figure 8.24, where the corrected "absorption spectrum" of sporangiophores (SPPHs, carrying spores) of the fungus *Phycomyces b.* is plotted. This was obtained as a difference spectrum [2 × (2–1)] between the uncorrected reflection spectrum of sporangiophores (2) and a white reference material [ZnS, (1)]. The distance of the measuring head of the bifurcated light fiber from the sporangiophores was approximately 5 cm, allowing a non-invasive and contact-free measurement. Because of the filigree structure of the SPPHs normal absorption spectroscopy is not feasible.

A spectacular forgery was detected through using reflection spectroscopy. Up until the end of the 1980s the famous mosaic death mask from the Tikal tomb had been considered an exemplary artifact of Central American jade art, but

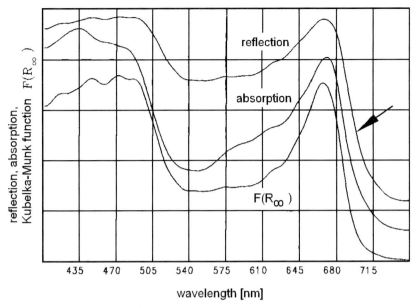

Figure 8.23 Comparison of the absorption spectrum of a green leaf of *Euphorbia pulcherima*, the corrected reflection spectrum and the reflection spectrum calculated on the basis of the Kubelka–Munk function.

then the mask material was recognized to be of inferior value; as only the earlaps are made from jade (Ward, 1987, measuring distance 30 cm).

Today remote reflection measurements are used to evaluate large woodlands either by plane or satellite. In particular, at the "red flank" (the Soret band between 690 and 730 nm) reflection spectra decline and healthy trees exhibit marked differences (see the arrow in Figure 8.23). This is used extensively in the cartography of forest damage.

Mimicry is a widely known phenomenon in biology. Imitating advantageous properties of a particular species allows other species to survive. Thus, for a long time why a particular bee (*Chelostoma fuliginosum*) preferred a certain blue bell-shaped flower (*Campanula persicifolia*), which offered a great deal of nectar could not be explained (Figure 8.25). However, the bee also visits a very different red orchid (*Cephalanthera rubra*), which in effect has no nectar. The solution to this inexplicable behavior came through reflection spectroscopy. The spectral properties of both flowers are practically identical in the UV-blue spectral region, which is the only range visible to insects. However, they are very different in the visible range where the human eye is most sensitive (Figure 8.26).

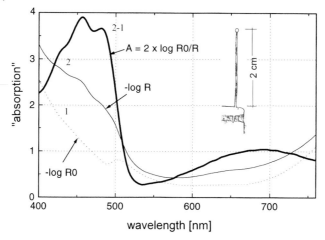

Figure 8.24 The first "absorption spectrum" of a very fine, filigree sporangiophore (SPPH) of the fungus *Phycomyces b.* as determined by simple and contact-free reflection spectroscopy *in situ*. The direct measurement of the absorption spectrum is impossible. About 1000 individual spectra of a "wood" of approximately 500 sporangiophores were measured within 20 s using a highly sensitive rapid scan spectrometer (RSS, Figure 5.51). This corresponds to a signal amplification of 5×10^5 compared with the conventional (which is not possible) measurement of a single sporangiophore (inset).

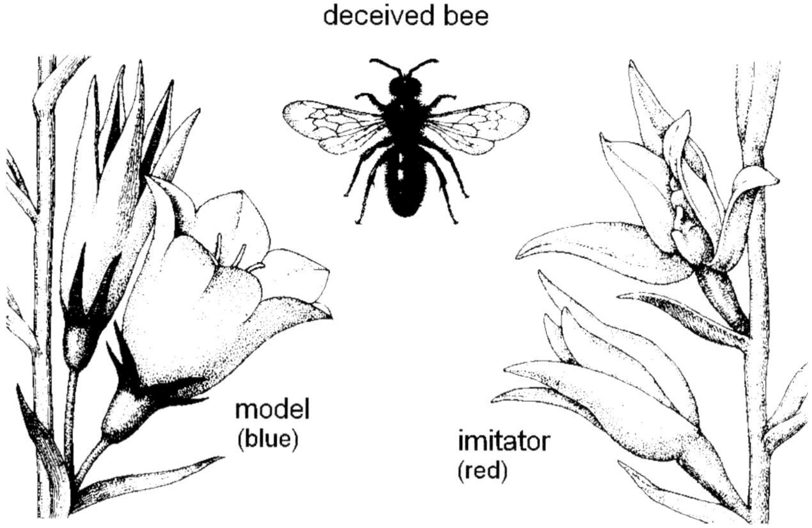

deceived bee

model (blue)

imitator (red)

Figure 8.25 A bee, *Chelostoma fuliginosum*, predominantly visits the blue bell-shaped flower *Campanula persicifolia*, which is rich in nectar and has a very different color compared with the red orchid *Cephalanthera rubra*, which has virtually no nectar. The bee and the two plants represent type of a triangular relationship (*Scientific American*) (from Barret, *Spektrum der Wissenschaft*, Nov. 1987, 102, modified).

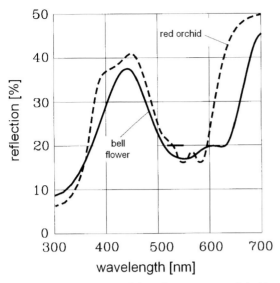

Figure 8.26 Comparison of the reflection spectra of the blue bell-shaped flower and the red orchid as shown in Figure 8.25 (from Barret, *Spektrum der Wissenschaft*, Nov. 1987, 102, modified).

8.6
Attenuated Total Reflection Spectroscopy (ATRS)

Although the principle of the total reflection of light that is observed on transition from an optically dense (higher refractive index *n*) to a less optically dense medium (lower refractive index *n*), including the theory of electromagnetism behind it, has been known since Newton's work, *internal reflection spectroscopy* was not developed not until 1947, by Goos and Hähnchen. They first recognized that at boundary layers the total reflection is always accompanied by a small displacement of the beam, which is about the same order as the wavelength [Figure 8.27(A)]. This is explained by Maxwell's theory and depends on a *damped surface wave* in the thinner medium between the entrance (K) and exit point (L): the light emanates slightly into the optically thinner medium – among other reasons as a function of the entrance angle – and thus transports energy across the phase border. Within the thinner medium a stationary field ***E*** is built up, the size of which declines exponentially with the distance from the interface:

$$\boldsymbol{E} = \boldsymbol{E} \times e^{\frac{-x}{d_p}} \tag{8.48}$$

The penetration depth d_p, is defined as the distance where ***E*** drops to the 1/e level, and is calculated by (Harrik, N. J., *Ann. N. Y. Acad. Sci.* 1963, *101*, 928):

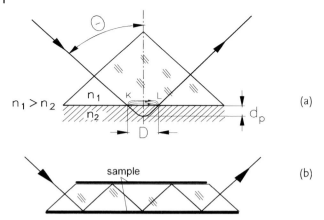

Figure 8.27 (a) The total reflection at boundary layers between media of different refractive indices is always accompanied by a small displacement of the beam in the order of a wavelength. This is explained by Maxwell's theory depending on a damped surface wave within the thinner medium between the entrance (K) and exit points (L): the light emanates slightly into the optically thinner medium where it is absorbed in the presence of a potential absorber (*attenuated total reflection*, ATR). (b) Using special prisms as shown, the absorption is significantly enhanced by multiple total reflection (up to 25 times).

$$d_p = \frac{\lambda_1}{2\pi[\sin^2 \alpha - (n_2/n_1)^2]^{1/2}}; \quad \alpha \geq \alpha_g \tag{8.49}$$

It is proportional to the wavelength λ_1 in the optically denser medium. By varying the entrance angle α it is possible – similar to modulation of the frequency f in photoacoustic spectroscopy (PAS) (Section 7.3.4) – to monitor depth profiles of absorption in thin layers, e. g., biological membranes.

If the optically thinner medium absorbs, the penetrating wave is attenuated (*attenuated total reflection spectroscopy*, ATRS). In the case of pure transmission and for a lower absorbance the following formula applies [$m_n < 0.1$; in dispersion theory m_n is the extinction module of the medium with a refractive index n and corresponds to the known product ac, Eq. (5.3)]:

$$\frac{I}{I_0} = e^{-m_n d} \cong 1 - m_n d_e \tag{8.50}$$

where d_e is defined as *effective thickness* of the layer. This is required in order to obtain the same extinction when measuring typical transmission spectra compared with total reflection measurements.

Similarly, for the reflectivity at small absorbencies we obtain

$$R \cong 1 - m_n d_e \tag{8.51}$$

With *N*-fold reflections [Figure 8.27(b)],

$$R^N \cong (1 - m_n d_e)^N \tag{8.52}$$

where d_e is a complex function of various parameters, however, for low concentrations it is independent of m_n.

The phenomenon of total reflection has found many different applications: precision measurements of angles, refractive indices, beam splitters, optical filters, light modulators, lasers, mirrors, measurements of reliefs and in absorption spectroscopy. This methods always becomes important if the more usual measurements are impossible or seriously restricted, e. g., because of extremely high absorbance of the sample within thin layers. The sample material is brought into contact with the optically denser but transparent material and the wavelength dependence of the total reflection starting within the denser medium is measured. The resulting spectrum is the result of interaction of the stationary field with the sample and resembles its transmission spectrum. However, theory shows that – in contrast to normal transmission spectroscopy – at certain angles of incidence the absorption of the sample dramatically increases. Therefore the *American Society for Testing and Materials* (ASTM) gave this technique the term "*attenuated total reflection*" (ATR) spectroscopy. It was originally suggested and developed independently by the Dutch Shell Laboratories (Fahrenfort) and the Philips Laboratories (Harrik).

Because of its fingerprint characteristics, today ATR is predominantly used in the infrared range, for example for determining fats. A series of different set-ups are on the market – depending on the envisaged application. Along with single reflection [Figure 8.27(a)], cuvettes (i. e., crystals) with multiple reflections (25 reflections and more) and the corresponding signal enhancement are available [Figure 8.27(b)]. While a traditional ATR cell requires a measuring volume of 1–2 mL, today diamond cells with sample volumes of only 1 µL are available. Diamond with a refractive index of $n = 2.4$ provides a four-fold higher reflectivity than glass ($n = 1.3$, Section 3.6.2). Depending on the application and required wavelength range (in µm) there are several crystal materials available, such as thallium bromoiodide (0.6 to 30 µm, soluble in bases, not in acids, higher refractive index than diamond), calcium fluoride (0.13–12 µm, hard but sensitive to thermal shock), sodium chloride (0.25–16 µm, soluble in water, low cost).

8.7
Selected Further Reading

Azzam, R. M.A., Bashara, N. M. *Ellipsometry and Polarized Light*, North Holland, New York, **1979**.

Bohren, C. F., Huffman, D. R. *Absorption and Scattering of Light by Small Particles*, John Wiley & Sons, New York, **1998**.

Campbell, I. D., Dwek, R. A. *Biological Spectroscopy*, The Benjamin Publishing Company, Menio Park, CA, London, Amsterdam, Sidney, **1984**.

Earnshaw, J. C., Steer, W. D., (eds.), *The Application of Laser Light Scattering to the Study of Biological Motion*, Plenum Press, New York, London, **1982**.

Frei, R. W., MacNeil, J. D. *Diffuse Reflectance Spectroscopy in Environmental Problem-Solving*, CRC Press, Boca Raton, FL, **1973**.

Harrik, N. J. *Internal Reflection Spectroscopy*, Interscience, New York, **1968**.

Hesse, M., Meier, H., Zeeh, B. *Spektroskopische Methoden in der organischen Chemie*, Thieme Verlag, Stuttgart, New York, **1984**.

Kortüm, G. *Reflexionsspektroskopie*, Springer, Berlin, **1969**.

Mirabella, F. M., ed. *Internal Reflection Spectroscopy*, Marcel Dekker Inc. NY., **1992**.

Newton, R. G. *Scattering Theory of Waves and Particles*, Springer, Berlin, **1982**.

Pecora, R., ed. *Dynamic Light Scattering: Applications of Photon Correlation Spectroscopy*, Plenum, New York, **1985**.

Schmidt, W., *Die Biophysik des Farbensehens*, 5th edn. Color, Regenbogenverlag, Klaus Stromer, Konstanz, **2000**.

Schrader, B., ed. *Handbook on Infrared and Raman Spectroscopy*, Wiley, New York, **1995**, p. 787.

Ward, F., Nat. Geo. Sept. 1987, 282–315.

Wendlandt, W. W. *Modern Aspects of Reflectance Spectroscopy*, Plenum Press, New York, London, **1968**.

Wiese, H. *Lichtstreuung und Teilchengrößenmessung (Light Scattering and Particle Size)*, *GIT Fachz. Lab*, **1992**, three subsequent articles, 4/92, 385–389; 7/92, 762–768; 10/92, 1029–1032.

9
Circular Dichroism and Optical Rotation

> Relata refero
> (Herodot)
>
> *I only report facts that have already been reported previously*

Circular dichroism (*CD*) and optical rotation (**o**ptical **r**otatory **d**ispersion, *ORD*) are two related optical phenomena that are closely correlated with the spatial structures of molecules. When discussing polarization fluorimetry in Section 6.4 we learnt about various analytical methods based on electronic excitation with polarized light: the depolarization of the emitted light was used to gain information about the motility of excited molecules, about intra- and intermolecular energy transfer and the electronic dipole moments of fluorophores. Advancing the application of plane polarized light (a two-dimensional property) by utilization of circularly or elliptically polarized light with different rotational directions of the electric and magnetic field vectors (different spatial properties), we are able to gain even more information about the spatial structures of molecules.

9.1
Polarized Light: a Generalization

With plane polarized light the **E**-vector vibrates in a plane that is perpendicular to the direction of propagation (usually drawn in the z-direction: Figures 2.1 and 8.1). Theoretically, plane polarized light is equivalent to the vector sum of two field components E_L and E_R, which rotate in opposite directions (Figure 9.1): we refer to elliptically or circularly polarized light. This finding can be verified readily by experiment. The term *interference* relates to the polarization properties of light with sufficient *coherence length* (Section 2.2): two plane polarized light beams with **E**-vectors that are perpendicular and possess the same amplitude and phase result in plane polarized light with the **E**-vector tilted by 45° to the vertical (Figure 9.2). The relative phase is shifted by π, and the resulting polarization plane is consequently tilted by −45°.

Optical Spectroscopy in Chemistry and Life Sciences. W. Schmidt
Copyright © 2005 WILEY-VCH Verlag GmbH & Co. KGaA, Weinheim
ISBN 3-527-29911-4

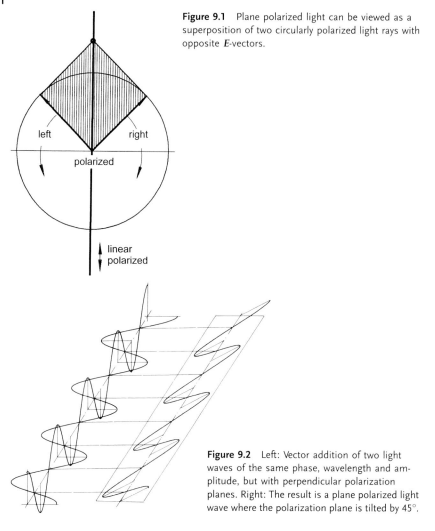

Figure 9.1 Plane polarized light can be viewed as a superposition of two circularly polarized light rays with opposite *E*-vectors.

Figure 9.2 Left: Vector addition of two light waves of the same phase, wavelength and amplitude, but with perpendicular polarization planes. Right: The result is a plane polarized light wave where the polarization plane is tilted by 45°.

If one of the two beams is shifted in phase by $\pi/2 = 90°$ with respect to the other, combination of the two will yield *circularly polarized* light. The rotational direction depends on which of the two light beams is delayed with respect to the other (Figure 9.3).

The spiral resulting from the linear propagation of the light wave is *not* viewed as being screwed into the space, however, it moves forward as a rigid body without rotation. A line where the spiral intersects with a spatially fixed surface that is perpendicular to the direction of propagation rotates circularly against the direction of polarization (with a rotating spiral the intersection would be a fixed line in space). According to convention a light wave is referred to as right polarized if an observer towards whom it is running sees it rotating to the right, i. e., clockwise, and vice versa.

Figure 9.3 Left: Vector addition of two light waves of the same wavelength and amplitude, but with perpendicular polarization planes. The interfering light waves are phase shifted by 90°. Right: The result is a circularly polarized light wave.

If the two waves, again have the same phase but differ in their amplitudes, they are recombined to a plane polarized wave where the **E**-vector is now tilted to a certain extent (Figure 9.4). A shift in the relative phase leads to a light wave with an elliptically circulating **E**-vector (Figure 9.5). *Elliptically polarized* light is defined by the long and short axis of the ellipse, its spatial tilt and by the direction of circulation of the **E**-vector (see Sections 9.3 and 9.7).

In practice, a well defined phase shift is realized through birefringent quartz crystals (Section 3.5.7). If a light beam impinges onto a plane parallel quartz plate, according to its polarization properties it will be split into an ordinary (o) and an extraordinary beam (e). Both beams experience different refractive indices within the crystal and thus propagate with different velocities $v = c/n$. Consequently, on exiting the crystal, which has a thickness d, there will be a phase shift $\Delta\Phi$ between both partial rays with

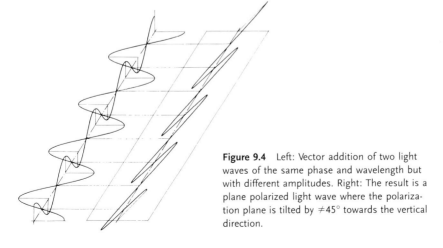

Figure 9.4 Left: Vector addition of two light waves of the same phase and wavelength but with different amplitudes. Right: The result is a plane polarized light wave where the polarization plane is tilted by $\neq 45°$ towards the vertical direction.

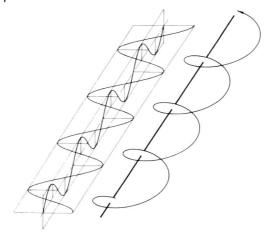

Figure 9.5 Left: Vector addition of two light waves of the same wavelength but with different amplitudes and where the polarization planes are perpendicular to each other. Right: The result is an elliptically polarized light wave.

$$\Delta\Phi = \Delta\Phi_e - \Delta\Phi_o = \frac{2\pi d}{\lambda}(n_e - n_o)$$

By changing the thickness of the crystal it is possible to generate any phase shift for each wavelength. Defined phase shifts of $\Delta\Phi = \pi/2$ (quarter wave plate) are technically interesting for the transformation of plane into circularly polarized light as are $\Delta\Phi = \pi$ (half-wave plate) for the simple rotation of the plane of plane polarized light. However, because they are extremely thin, in practice such wave plates of only 10 to 50 μm would be useless. Therefore, typical thicknesses of the wave plates are multiples of lambda, e. g., $10 \times \frac{1}{4}\,\lambda$ rather than $\frac{1}{4}\,\lambda$. Therefore they are known as *multiple order plates* (MO). However, as they are highly sensitive to changes in wavelength and temperature, so-called *zero order plates* (ZO) are utilized. These consist of two crossed MOs of, e. g., 10 λ and $10 \times \frac{1}{4}\,\lambda$ yielding a phase shift of $\pi/4$.

9.2
Optical Rotatory Dispersion (ORD)

We will now focus on the action of circularly polarized light on certain molecules. In Chapter 8 we studied scattering and diffraction as a consequence of various velocities in different media. Both phenomena depend directly on the polarizability α, i. e., the distortion of the electron cloud by the electric field vector of an external light field. Now, many organic molecules exhibit a significant "handedness", similar to right and left gloves: their structures can be transformed into each other only by a mirror (they are mirror image isomers of each other, i. e., enantiomers). The same holds true for circularly polarized light: a left-handed screw remains a left screw, even if we turn it upside down (see Figure 9.3).

It should not surprise us that enantiomeric molecules exhibit *optical activity*, i. e., their polarizability α depends on the direction of rotation of the incident circularly polarized light beam. The refractive indices of left or right polarized light are different (n_L and n_D, respectively; the subscripts are derived from the Latin and represent L for *laevus* = left and D as *dexter* = right). Van't Hoff (1852–1911) discovered that this effect is caused by the so-called *asymmetric carbon atoms*. Carbon can bind four different residues, as, for example, in lactic acid molecules (Figure 9.6). An incident, plane polarized light ray "sees" different refractive indices for its left (n_L) and right (n_D) circularly polarized components (see Figure 9.1), which therefore propagate with different velocities. After exiting a cuvette of length l the phases of the right and left circulating components will be shifted with the result that after the last vector addition the light is again plane polarized, however with a tilted plane of polarization (Figure 9.7). Even if only two residues at the carbon atom are identical the molecule loses optical activity and n_L and n_D are identical, as can easily be verified (Figure 9.6). Depending on whether the right or left circulating component is delayed, we observe a rotation to the left or the right side. Thus, we refer to left or right rotating lactic acid, so-called stereo or mirror image isomers.

Before the absolute configurations of optically active molecules were known a "relative configuration" was used. As a reference compound the right rotating glycerin aldehyde was chosen, and this was attributed the configuration "D" (as in "right"). Other optically active molecules were referred to this compound through methods of synthesis. Unfortunately, because of this the right rotating lactic acid was assigned as the L-form: L-(+)-lactic acid. A clockwise rotation of the polarization plane is defined by "+", a left rotation by "–", again relative to the direction of propagation of light.

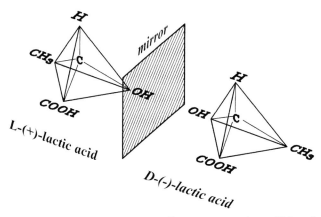

Figure 9.6 Spatial representation of lactic acid molecules with an asymmetric C atom causing *optical activity*. The mirror image represents the *stereoisomer*. The rotation direc- tions shown (right = "+" and left = "–", seen against the direction of propagation) are obtained by experiment. The terminology D and L refer to the *absolute* spatial configuration.

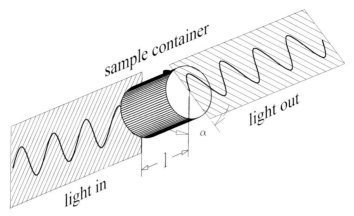

Figure 9.7 Rotation of the polarization plane by the angle α by an optically active molecule in an *ORD*-experiment. The sample has the thickness l (typically measured in decimeters).

The measured rotational angle α_{obs} is obtained by

$$\alpha_{obs} = [\alpha]_\lambda^T \times l \times c \tag{9.1}$$

where α is in degrees, the length l in decimeters (!) and c in g ml^{-1} (the unusual units are chosen for historical reasons). $[\alpha]_\lambda^T$ is the specific rotational angle and a constant for the measured compound, depending on temperature T and the wavelength λ. The molar rotation $[\Phi]_\lambda^T$ is defined by

$$[\Phi]_\lambda^T = \frac{[\alpha]_\lambda^T M_w}{100} = \frac{\alpha_{obs} \times M_w}{100lc} \tag{9.2}$$

Taking the effect of polarizability α of the solvent with refractive index n of the specific molecule into account, we have to correct $[\Phi]_\lambda^T$ by multiplication with $3/(n^2 + 2)$. For polymers in biochemistry it is not particularly meaningful to refer to the molar mass (molecular weight) of the whole molecule but rather to the molar mass of the polymerization unit, e.g., the medium mass of all amino acids of a protein (mean residue weight, *MRW*):

$$[m]_\lambda^T = [\alpha]_\lambda^T \frac{MRW}{100} \tag{9.3}$$

Table 9.1 shows some specific rotation angles of optically active components at 20 °C; the subscript D relates to the Na D-line at 589 nm (cf. Figure 2.13).

The rotation of the polarization plane is – in contrast to the CD signal, *vide infra* – fairly large and therefore detectable through a simple assay, particularly if we restrict the measurement to just one particular wavelength λ. Thus, the concentrations of aqueous solutions of cane sugar with a relatively large $[\alpha]^{20}{}_D = 66.4°$ are readily measured by the clockwise rotation of the plane polarized light with a simple, small polarization device, a so-called *saccharimeter*.

Table 9.1 Specific rotation angles of optically active components at 20 °C; the subscript D refers to the Na D-line at 589 nm (cf. Figure 2.13).

Molecule	Solvent	$[\alpha]^{20}_D$
D-Lactic acid	Water	−2.3
L-Lactic acid	Water	+ 2.3
Hexose ($C_7H_{14}O_4$)	Water	−11
D-Glucose ($C_6H_{13}NO_5$)	Water	+ 19
L-Mannose ($C_7H_{14}O_5$)	Water	+ 30
D-Ribose ($C_5H_{11}NO_4$)	Hydrochloride	−24.6
Hexose (amino sugar, $C_8H_{17}NO_3$)	Hydrochloride	+ 49.5
L-Alanine	Water	+ 2.7
Cane sugar	Water	+ 66.4
Cholesterol	Ether	−31.5
Vitamin D_2	Ethanol	+ 102.5

9.3
Circular Dichroism (CD)

Diffraction and scattering are the result of forced, light-induced dipole vibrations of the outer shell electrons of molecules. Forced oscillations depend characteristically on the exciting frequency and the resonance frequency of the induced oscillator. The frequency dependence of the amplitudes and phases can be visualized through a mechanical analogue, the so-called *Barton* pendulum system. This consists (see Figure 9.8) of a heavy energy-exciting pendulum and a series of coupled and comparable light and excitable pendulums of various lengths, which all show forced vibrations of the same frequency but different phase and amplitude. If the exciting pendulum is activated, the medium–light pendulum 3 of the same length is excited most by resonance – with a phase shift of $\pi/2$. Its amplitude, which in principle would increase infinitely compared with the exciting pendulum, is limited in under experimental conditions. Because of the ever present damping processes, the so-called resonance catastrophe does not occur. The pendulum absorbs energy. The shorter pendulums 1 and 2 vibrate with smaller amplitudes, in phase with the large pendulum, the longer pendulums 4 and 5 also show a smaller amplitude, however with opposite phase (phase shift of π).

Comment: For practical demonstration purposes the exciting pendulum should be at least 100 times heavier than the smaller pendulums, otherwise unwanted interferences obscure the expected result.

The following analytical approach adopted from theoretical, classical mechanics provides sufficient insight into the theory of the optical activity of chiral mole-

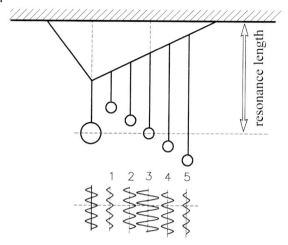

Figure 9.8 The *Barton* system of pendulums with an energy inducing pendulum (large sphere) and excited pendulums (small spheres). Phase and amplitude relationships are shown at the bottom (see text for details).

cules. A general differential equation for forced and damped vibrations of an oscillating mass m [Figure 9.9(a) and (b)] can be written as:

$$m \ddot{x} = - kx - f\dot{x} + F_0 \cos(\omega t) \tag{9.4}$$

where $\omega = 2\pi\nu =$ angular frequency, $-kx =$ restoring force, $-f\,dx/dt$ is the attenuating force and $F_0 \cos(\omega t)$ is the exciting force.

One or two dots on top of the extension parameter x define the first and second derivatives with respect to time, i. e., velocity and acceleration: dx/dt and d^2x/dt^2, respectively.

The general solution to this equation is:

$$x = \dot{x} \cos(\omega t) + \ddot{x} \sin(\omega t) \tag{9.5}$$

where

$$\dot{x} = F_0 \frac{m(\omega_0^2 - \omega^2)}{m^2(\omega_0^2 - \omega^2)^2 + f^2\omega^2} \tag{9.6}$$

and

$$\ddot{x} = F_0 \frac{f\omega}{m^2(\omega_0^2 - \omega^2)^2 + f^2\omega^2} \tag{9.7}$$

Plots of the functions in Eqs. (9.5) and (9.6) are shown in Figure 9.9(c) and (d). These are completely analogous to spectra from *optical rotatory dispersion* (ORD) and from the *circular dichroism* (CD) discussed here.

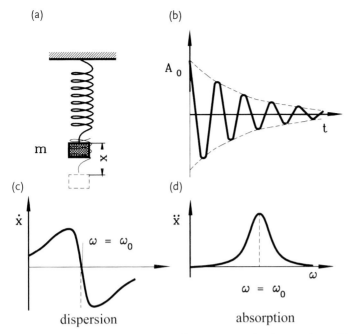

Figure 9.9 (a) A mass *m* is fixed to an elastic spring and is free to vibrate. (b) Diagram of damped vibration (kinetics). (c) Dispersion and (d) absorption properties of the system that are the direct result of the general solution to the explanatory differential equation.

Figure 9.10 shows the schematic dependency of the normal refractive index *n* for non-polarized light of a hypothetical molecule as a function of the frequency of light. We can recognize a general increase in the refractive index with increasing frequency, as already seen in Figure 3.34 (Section 3.9.1) when discussing prism dispersion. We term this *normal dispersion*.

If a measuring parameter such as the refractive index *n* changes with wavelength λ, we refer to this behavior as *dispersion*, which is defined by $dn/d\lambda$.

In the regions of the absorbances – of whichever type, rotatory, vibratory or electronic – according to Figure 9.10 we observe a "hump" in the otherwise smooth frequency dependence [cf. dispersion curve in Figure 9.9(c)]: the molecule shows resonance. If the refractive index *increases* with a *decreasing* frequency of the light we refer to this as *anomalous* dispersion. Anomalous dispersion is always observed near to absorption bands (e. g., with Fuchsin/Schiff's reagent between 460 and 560 nm), where the dispersion $dn/d\lambda$ becomes positive within small spectral ranges.

If an absorption band at frequency v_0 is approached from a longer wavelength, the incident, exciting light of frequency *v* and light velocity *c* is of lower frequency than the resonance frequency v_0 of the absorber. Thus,

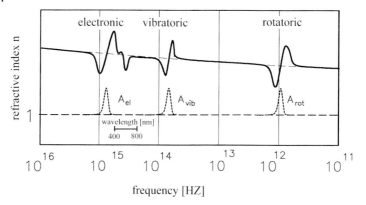

Figure 9.10 Decrease in the refractive index n of a hypothetical molecule with decreasing frequency of the light. *Anomalous dispersion* takes place in the vicinity of absorbances (electronic, vibratory or rotatory). The corresponding absorption bands are indicated by dashed lines.

the phase of the forced dipole oscillation is delayed compared with the exciting light [Figure 9.8, see also the discussion of Eq. (8.10) in Section 8.2.2]. As a result, the vector addition of the scattered and the incident light wave is again a sine oscillation, which, however, is somewhat phase-delayed compared with the incident light. This finally means a lower velocity of light for the resulting wave ($c_n = c/n$), equivalent to an increased refractive index. The contrast also holds true if we approach the absorption band from the shorter wavelength side.

According to the so-called *Drude equation*, at $\lambda = \lambda_0$ the previously mentioned resonance catastrophe would take place ($[\Phi]_\lambda^T \to \infty$):

$$[\Phi]_\lambda^T = \frac{96\pi N}{hc} \frac{R\lambda_0^2}{\lambda^2 - \lambda_0^2} \tag{9.8}$$

where h is Planck' constant, c the velocity of light, N the Loschmidt number and λ_0 the medium frequency of the band. R describes the so-called *rotational strength*, which is proportional to the integral of $\Delta\varepsilon$ across the individual band, which corresponds to the dipole or oscillatory strength f of an isotropic absorption band, which, in turn, is proportional to the integrated absorption:

$$f \propto \int_{band} \varepsilon(v)dv \tag{9.9}$$

As expected, enantiomeric molecules not only show different refractive indices, but also different absorption coefficients for left or right polarized light ($\Delta\varepsilon = \varepsilon_L - \varepsilon_D$): this is what is known as *circular dichroism* (CD) and the whole phenomenon of absorption and refraction of circularly polarized light by optically active molecules is the *Cotton effect*. As a function of wavelength we define the absorption difference as

$$\Delta A(\lambda) = \Delta\varepsilon(\lambda) \times c \times l = A_L(\lambda) - A_D(\lambda) \qquad (9.10)$$

where A_L and A_D are the absorptions of the left and right circularly polarized light, respectively.

Typically, $\Delta\varepsilon$ is only very small with values ranging between 10^{-3} and 10^{-6} M^{-1} cm^{-1}, while ε values can be found between 10^2 and 10^5 (see Table 5.1). For optically active molecules we define the *dissymmetric ratio* $g = \Delta\varepsilon/\varepsilon$ with values ranging from 10^{-3} down to less than 10^{-6}. Thus, experimentally we have to determine a very small difference between two large numbers; we will discuss the method in more detail in Section 9.5.

According to dispersion theory, ORD, CD and absorption spectra (A) are closely correlated, as shown for the hypothetical molecule in Figure 9.11. The molecule can have three absorption bands and two of them may be optically active and exhibit a positive or a negative circular dichroism, respectively. Naturally, the CD bands coincide with the common absorption bands (i.e., for non-polarized light). The following relationship holds true:

$$\varepsilon = \frac{1}{2}(\varepsilon_L + \varepsilon_D) \qquad (9.11)$$

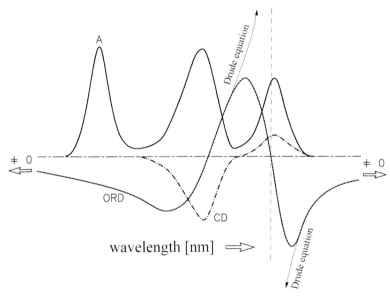

Figure 9.11 Comparison of ORD, CD and absorption spectra (A) of a hypothetical molecule. The molecule shows three absorption bands. Two of them are optically active and show positive and negative circular dichroism. The ORD bands extend far beyond the actual absorption range. According to the Drude equation the "resonance catastrophe" is indicated.

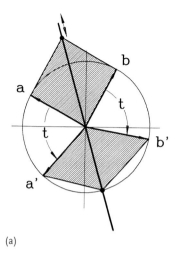

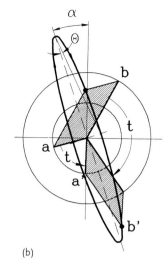

(a)   (b)

Figure 9.12 (a) For a pure *ORD* band left and right plane polarized light experiences different delays when passing through the sample, but on exiting they recombine to give plane polarized light but with the polarization plane rotated. (b) For a *CD* band one component is compared with the other stronger absorbing one, resulting in elliptically polarized light. The ratio of the short to the long axis of the ellipse is defined as a tangent of a certain angle θ, the so-called *ellipticity*.

The *ORD* bands are much less pronounced than the *CD* bands and show band boundaries that are "spread out", which is a measuring advantage (see below).

For a pure *ORD* band the left and right polarized components of plane polarized light only experience different delays during transmission through the sample, and after exiting they recombine to give plane polarized light, however, with a tilted polarization plane [Figure 9.12(a), cf. Figure 9.4]. For a *CD* band one component is absorbed more strongly than the other with the opposite polarization. Consequently, the exiting components combine to give elliptically polarized light [Figure 9.12(b), cf. Figure 9.5]. We can now define the ratio of the short to the long axis of the ellipse as the tangent of the angle θ, the so-called ellipticity.

For the purpose of demonstration the ellipticity in Figure 9.12(b) is greatly exaggerated. Owing to the typically small dissymmetry ratio of $<10^{-3}$ the true ellipse would show up only as a straight line – corresponding to the long axis and as indicated in Figure 9.12/a a vanishing short axis.

In a similar way to that shown in Eq. (9.1) for *ORD* we can now define the *specific ellipticity* by

$$[\Psi]_{\lambda}^{T} = \frac{\Psi_{obs}}{c \times l} \tag{9.12}$$

where Ψ_{obs} is the measured ellipticity, or the molar ellipticity,

$$[\Theta] = [\Psi]\frac{M_{\mathrm{w}}}{100} \qquad (9.13)$$

For biopolymers (e. g., proteins), again we refer to the average molar mass (molecular weight) of optically active residues (e. g., amino acid residues) (MRW = mean residue weight):

$$[\Theta]_{MRW} = [\Psi]\frac{MRW}{100} \qquad (9.14)$$

The signal strength of CD spectra is either presented as absorption difference $\Delta\varepsilon$ or as the ellipticity $[\theta]$. Molar ellipticity and circular dichroism ($\Delta\varepsilon = \varepsilon_{\mathrm{L}} - \varepsilon_{\mathrm{D}}$) are easily inter-converted using the following formula (given here without the derivation):

$$[\Theta] = 3300\ \Delta\varepsilon \qquad (9.15)$$

where $[\theta]$, again, is measured in (degrees cm^2 dmol^{-1}).

9.4
Theoretical Basis of the Cotton Effect

As already discussed in Chapter 6, electronic transitions result in charge displacement accompanied by an electrical transition dipole moment μ_e. These charge displacements are not necessarily strictly linear but often include – typically for $n \rightarrow \pi^*$ transitions – a circular component: upon excitation the electron cloud is also "twisted". The previously mentioned Maxwell equations of electrodynamics tell us that circular currents, in turn, generate a *magnetic* transition moment μ_m, perpendicular to the circular motion corresponding to μ_e. The collinear presence of both components μ_e and μ_m (or projections of them) is a necessary and minimum condition for optical activity. According to the theory, the "strength" of the transition corresponds to the imaginary part of the scalar product $[Im\ \mu_e \cdot \mu_m]$ (Figure 9.13). Even if the individual molecules are not highly ordered as shown in Figure 9.13 but move and rotate freely in space, e. g., in solution, this nevertheless does not change their "handedness" (parity), which can be reversed only by using a mirror (Figure 9.6: enantiomeric forms): the sample is optically active.

Optical rotational diffusion (*ORD*) and circular dichroism (*CD*) only appear to be different phenomena, however they are manifestations of one and the same physical principle of dispersion theory:
- Optical rotation (*ORD*) is the consequence of different refractive indices of right and left polarized light.
- Circular dichroism (*CD*) is the consequence of different absorbances for right and left polarized light.

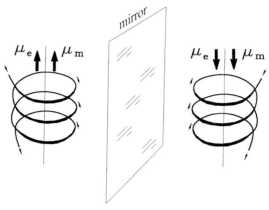

Figure 9.13 Optically active molecules posses some type of "handedness", which can be reversed only by a mirror. Electronic transitions of these molecules simultaneously exhibit translatory and rotatory charge displacements. In the final effect, the electric current rotates in a spiral. In addition to the *electric transition dipole moment* μ_e, according to the rules of electrodynamics a *magnetic transition dipole moment* μ_m is generated and has the same direction as the first one. This *colinearity* of a magnetic and electric dipole moment is the necessary minimum condition for optical activity.

Indeed, the so-called *Kronig–Kramer* relationship relates both phenomena to each other. It allows an *ORD* value at a certain wavelength λ to be deduced from the attributed *CD* spectrum:

$$[\Phi]_\lambda^T = 2.303 \frac{9000}{\pi^2} \int_0^\infty \Delta\varepsilon_{\lambda'} \frac{\lambda'}{\lambda^2 - (\lambda')^2} \, d\lambda' \tag{9.16}$$

The question arises as to why, in spite of this relationship, both measuring methods are utilized. The reason is: the mutual attribution of CD and ORD signals is not possible on a point by point basis. To obtain the optical rotation at a specific wavelength λ, the *complete* CD spectrum has to be taken into account. For practical work, however, it is often sufficient to include only the strongest CD bands in the vicinity of the corresponding wavelength of the ORD band in question. Similar to the relationship between the correlation and Fourier analysis (Section 5.4.4), today these calculations are performed quickly and reliably using an on-line PC. However, each of these measuring methods has their technical advantages and disadvantages, with the consequence that neither can be disregarded. (1) CD bands are attributed much more easily than ORD bands as they do not change sign and are much more sharply pronounced (Figure 9.10). On the other hand, point by point measurements of ORD values are performed with very simple and cheap devices (saccharimeters). (2) CD measurements are very precise, because they are measurements of alternating currents (ac), however, they are technically demanding, and the great advantages of lock-in detection can be utilized (cf. Section 5.5).

It is interesting to note that ORD bands attributed to sharper CD bands exhibit "long tails" that are spread out, which extend far beyond the range of the measurable absorption bands (Figure 9.11). This leads to the important consequence that even molecules that are far outside their technically more difficult to attain UV absorption bands can be investigated (see Table 9.1, where the NaD-line, which is located in the middle of the visible wavelength range, is utilized).

9.5
The CD Spectrometer

Modern *CD* spectrometers cover the whole wavelength range from the far *UV* (185 nm) up to the near-infrared (1000 nm). As mentioned earlier, the essential feature of this type of spectroscopic analysis is the measurement of very small differences between signals on top of large background signals. This requires an extremely low stray light level, relatively high measuring intensities and excellent efficiency. Thus, CD spectrometers are large, difficult to handle and expensive. In addition, because they are rarely fully utilized by the particular group working in this area, in contrast, for example, to absorption spectrometers or fluorimeters, a *CD* spectrometer is typically a joint investment for several groups. Figure 9.14 gives the basis of the measuring principle.

The light from a strong *UV–VIS* source is dispersed by a scanning monochromator, typically a double monochromator (see Section 5.2.5) and subsequently plane polarized by a polarizer (Nicol's prism), where the polarization plane is tilted by 45° with respect to the subsequent electrooptical modulator (Kerr or Pockel cell, cf. Figure 3.28). The modulator acts as phase shifter and thus generates left or right polarized light (see Figure 9.3), with a frequency in the kilohertz range, which finally hits the sample and is then detected as transmitted light by a photomultiplier. The subsequent *lock-in* amplifier, which obtains its reference signal from the driver unit of the modulator, finally generates the difference between the signals of the absorption of the left and right polarized light, i. e., the *CD* sig-

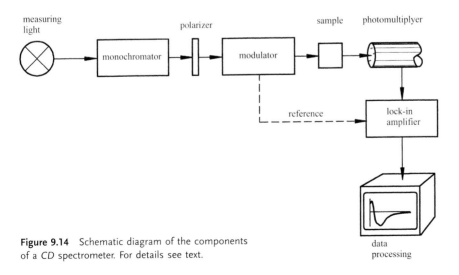

Figure 9.14 Schematic diagram of the components of a *CD* spectrometer. For details see text.

nal. By variation of the relative delay of the two partial rays by the modulator the *CD* difference signal is optimized (maximized). The final data processing, presentation and storage is performed by a normal *on-line* computer (*PC*).

We are dealing with a typical single beam reference procedure of the highest sensitivity – similar to dual wavelength spectroscopy as discussed in Section 5.5. Therefore the *baseline*, i. e., the dependence of efficiency of the spectrometer on the wavelength, has to be eliminated using the usual methods (see Section 6.3.2). Similar to normal absorption spectroscopy, the optimum signal to noise ratio (SNR) is obtained at an absorption of $A = 0.869$ (see Section 5.2.6). In *CD* spectroscopy special care has to be taken when choosing the solvent, because many solvents show strong absorption particularly in the interesting far-UV range. The sample thickness often has to be decreased to as little as $l = 0.1$ mm.

9.6
Applications

Chiroptical methods are used to study structures, particularly of macromolecules. In the early days of this type of analysis measurements were restricted to the determination of the helix content of polymers by the use of *ORD* spectroscopy (Linus Pauling, 1948), other conformations that also showed any optical activity were largely ignored. Except for special applications, today *CD* spectroscopy is the favored chiroptical method. Of course X-ray analysis is the most important method for investigating the structure of (bio-) polymers; nevertheless, chiroptical methods provide supporting, supplementary or information not achievable by any other route. They are relatively simple, fast, the temporal expense is small and they allow the direct observance of the *kinetics* of conformational changes, which is not feasible with X-ray structure analysis.

CD spectra of reference molecules with known conformations have been proven to be indispensable in *polarimetry* (the general terminology used for measurements with circularly polarized light). On the basis of known algorithms these allow the "mathematical structure analysis" of unknown molecules. Polypeptides consisting of identical units of amino acids exhibiting a well defined conformation are usually used as standards. Figure 9.15 shows examples of reference *CD* spectra of poly-L-lysine in various conformations: α-helix, β-sheet and coil. The differences are obvious. Figure 9.16 shows the dependence of the *CD* spectrum of insulin, a protein with many tyrosine residues, on temperature. With increasing temperature the molecule is converted increasingly from the α-helix form into the denatured, unordered coil form, possibly with some parts being of a β-structure. In spite of a great similarity of reference spectra with the spectra to be analyzed caution is required. Even very small helical fragments or β-sheets generate secondary effects that are difficult to analyze. Comparison with standard data from *CD* spectra of proteins, where the structure is has been determined by X-ray analysis, clearly increases the reliability of the interpretation.

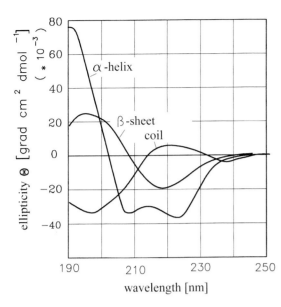

Figure 9.15 Typical reference CD spectra, *poly-L-lysine* in various basic conformations is shown here: α-helix, β-sheet and coil (from Ettinger, M. J., Timasheff, S. N. *Biochemistry,* 1971, *10*, 824–830).

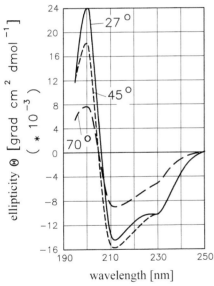

Figure 9.16 Dependence of the CD spectrum of insulin, a protein with numerous tyrosine residues, on temperature (°C) (from Ettinger, M. J., Timasheff, S. N. *Biochemistry,* 1971, *10*, 824–83).

9.7
Ellipsometry

Ellipsometry is a technique that is more than 100 years old (Drude, P., Über Oberflächenschichten, *Göttinger Nachrichten,* 14th July 1888). It is an optoanalytical procedure where elliptically or plane-polarized light waves are reflected by a substrate. Through this reflection the polarization state of the light wave is changed, a process which is highly sensitive to the material layered on the reflecting sub-

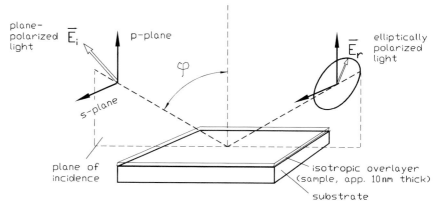

Figure 9.17 The so-called p–s coordinate system is used to describe an ellipsometric experiment. The s-plane is perpendicular to the direction of propagation and parallel to the sample surface. The p-plane is also perpendicular to the direction of propagation and contains the plane of incidence.

strate. In particular the ellipticity Θ and the orientation of the ellipsis α are influenced (cf. Figure 9.12) and can be measured with high precision and extremely good reproducibility ($<5 \times 10^{-4}$). However, two possible problems have to be taken into account: the cuvette windows themselves and the anisotropy of the sample. Ellipsometry has been used to study antigen–antibody reactions or lipid monolayers, and also marker free (!) deoxyribonucleic acid (DNA) (detection limit approximately 100 pg DNA), which has to be coupled to a substrate (Azzam, Bashara, 1979), in addition to the study of biological membranes. When connected to a laser as the light source the ellipsometer is predominantly used for kinetic investigations at a given wavelength.

The principle set-up including the terminology normally used is depicted in Figure 9.17. The polarization status of the incident light beam and the angle of incidence φ are variables. The measured values are expressed as γ_{obs} and Δ. These values are related to the ratio of the Fresnel reflection coefficients, R_p and R_s for p- and s-polarized light, respectively.

$$\tan (\Psi_{obs})\, e^{i\Delta} = \frac{R_p}{R_s} \tag{9.17}$$

Because ellipsometry measures the ratio of two values it can be very accurate and highly reproducible. From Eq. (9.17) the ratio is seen to be a complex number, thus it contains phase information in Δ. According to Fresnel's formulae (Section 3.5.2) and Figure 9.17, for any incident angle φ greater than 0° and smaller than 90°, p-polarized and s-polarized light will be reflected differently. The polarization state after reflection is determined by a rotating polarization filter. Concluding, the advantages of ellipsometry are the high accuracy and reproducibility, no reference sample is required and ellipsometry is not susceptible to scattering, lamp or purge fluctuations.

9.8
Closing Comments

Molecules are optically active if they show different interactions with left and right polarized light. These interactions are either detected as relative phase shifts between the left and right circularly polarized light (optical rotational dispersion, *ORD*), or as different absorbances of two oppositely polarized measuring lights (circular dichroism, *CD*). However, only a few chromophores that exhibit optical activity are actually optically active themselves. In the overwhelming number of cases the optical activity, i.e., the chirality, is generated by an asymmetric micro-environment, by asymmetrically positioned molecular groups. As would be expected, optical activity can also be induced by external magnetic fields (magnetic circular dichroism, *MCD*). The main applications of *CD* spectroscopy are the spectral and/or kinetic analysis of the secondary structures of proteins. The ellipsometry measures the change of ellipticity and spatial orientation of the polarization ellipse and is used particularly in the kinetic detection of even the smallest changes to substrate-bound biomolecules.

9.9
Selected Further Reading

Auzinsh, M., Ferber, R. *Optical Polarization of Molecules,* Cambridge Monographs on Atomic, Molecular, and Chemical Physics, No. 4, Cambridge University Press, Cambridge, **1995.**

Azzam, R. M.A., Bashara, N. M. *Ellipsometry and Polarized Light,* North Holland, Amsterdam, New York, **1979.**

Barron, L. D. *Molecular Light Scattering and Optical Activity,* Cambridge University Press, Cambridge, **2004.**

Baszkin, A., Norde, W., eds. *Physical Chemistry of Biological Interfaces,* Marcel Decker, New York, **2000.**

Berova, N., Nakanishi, K., Woody, R. W., eds. *Circular Dichroism: Principles and Applications* 2nd Edition, Verlag Chemie (VCH), Weinheim, **1994.**

Brittain, H. G. *Analytical Applications of Circular Dichroism,* Techniques and Instrumentation in Analytical Chemistry, Vol. 14, Elsevier, Amsterdam, **1993.**

Brosseau, C. *Fundamentals of Polarized Light: a Statistical Approach,* John Wiley, New York, **1998.**

Charney, E. *The Molecular Basis of Optical Activity: Optical Rotary Dispersion and Circular Dichroism,* John Wiley, New York, **1985.**

Fasman, G. D., ed. *Circular Dichroism and the Conformational Analysis of Biomolecules,* Sibirian School of Algebra and Logic, Kluwer Academic Publishers, **1996.**

Huard, S., Cacca, G. (translator) *Polarization of Light,* John Wiley, New York, **1997.**

Kliger, D. S., Lewis, J. W., Randall, D. A. *Polarized Light in Optics and Spectroscopy,* Academic Press, London, Sydney, Tokyo, **1990.**

Lightner, A. Gurst, J. E. *Organic Conformational Analysis and Stereochemistry from Circular Dichroism Spectroscopy,* Wiley-VCH, Weinheim, **2000.**

Nakanishi, K. et al. *Circular Dichroism,* Wiley-VCH, Weinheim, **2000.**

Rodger, A., Norden, B. *Circular Dichroism and Linear Dichroism,* Oxford Chemistry Masters, Oxford University Press, Oxford, **1997.**

Striebel, C., Brecht, A., Gauslitz, G. *Biosens. Bioelectron.* **1994,** 9(2), 139-46.

Sokolov, V. I. *Chirality and Optical Activity in Organometallic Compounds,* Gordon and Breach, New York, **1991.**

Thulstrup, E. W., Michl, J. *Elementary Polarisation Spectroscopy*, Verlag Chemie (VCH), Weinheim, **1997**.

Tompkins, H. G., McGahan, W. A. *Spectroscopic Ellipsometry and Reflectometry: a Users Guide*, John Wiley, New York, **1999**.

10
Near-infrared Spectroscopy

Legere enim et non intellegere neglegere est
(Cato)

Reading without understanding is idleness

10.1
Introduction

In 1800 the English–German physicist and astronomer Sir William Herschel discovered near-infrared radiation (NIR). Although today UV–VIS and classical infrared spectroscopy are routinely applied analytical methods, near-infrared spectroscopy (NIRS) is still only utilized for very specific applications. It is estimated that only around 1% of all current analyses are based on NIRS, but this does not reflect its analytical potential. Almost all organic substances and particularly water exhibit pronounced and specific absorptions in the spectral range between 1000 and 2500 nm (Figure 10.1). In addition, polymers of all types, such as polyethylene, polypropylene, polystyrene or polyamide, are readily discernible by NIRS.

The most important advantages of NIRS are:
- In contrast to the usual IR spectroscopy, a "normal" sample thickness (10 mm) and quartz cuvettes, as used in UV–VIS

Sir William Herschel (1738–1822)

Optical Spectroscopy in Chemistry and Life Sciences. W. Schmidt
Copyright © 2005 WILEY-VCH Verlag GmbH & Co. KGaA, Weinheim
ISBN 3-527-29911-4

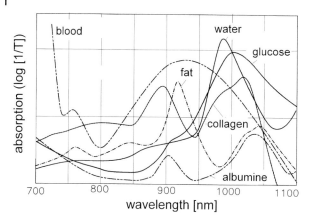

Figure 10.1 Nearly all substances show pronounced absorption in the near-infrared region, as shown here for water, blood and various blood components.

spectroscopy, can be utilized, and sample preparation is not required.

- Use of chemometric procedures, such as the so-called *principle component analysis* (PCA), as methods of multiple regression analysis allow the routine application of process control for a large number of samples.
- NIR light fibers are transparent up to 2500 nm (Figure 3.25) and allow measurements to be made at a central location for analyses of "inaccessible" samples at distances up to kilometers away, which is particularly useful for industrial applications.
- In contrast to other spectroscopic procedures, "non-spectroscopic" properties such as grain size, pH value, NaCl content, water of crystallization, the octane number for gasoline, the age of (perishable!) samples in blood-banks or just the temperature of biological tissues are also easily measured. In addition, NIRS has been successfully applied in snow and glacier analysis, in vegetational cartography and in research into forest decline using remote sensing from airplanes or satellites.

The range of applications is virtually unlimited. The first part of the present chapter considers the theoretical basis and methodology of near-infrared spectroscopy (NIRS). In the second part, several selected examples from various areas of science and technology exemplify the potential and advantages of NIRS. In 1987 the German Council for Near-infrared Spectroscopy was founded. The international *Council for Near-infrared Spectroscopy* (CNIRS) has been in existence since the early 1990s and in 1993 publication began of a specialist NIR journal, which gives information on the latest advances, in addition there are regular international conferences on NIRS. Currently NIRS is experiencing a boom with a world-wide growth of 15 % per year in instrument sales.

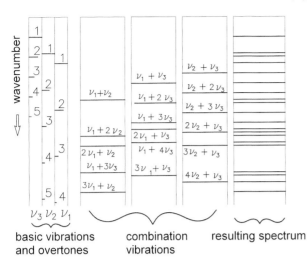

Figure 10.2 The basic vibrations ν_1, ν_2 and ν_3 of a three-atomic molecule are located in the mid-infrared region. The intensity of their overtones decreases with increasing order and increasing wavelength. Taking the selection rules into account, a multiplicity of combination bands is observed in the near-infrared spectral range, which results in the final spectrum (for simplicity only the wavelengths, not their intensities, are indicated).

basic vibrations and overtones   **combination vibrations**   **resulting spectrum**

10.2
Theory of Near-infrared Spectroscopy

While atoms exclusively show only electronic energy levels and thus sharp spectral lines, molecules exhibit additional vibrational and rotational energy levels. Consequently, their spectral properties are different from those for atomic line spectra: there are individual lines, bands and even systems of bands of great complexity. All are intimately connected with the so-called *valence electrons* that are responsible for the chemical properties of all molecules, which, in turn, also explains the general interest in NIRS. Infrared *emission* is practically insignificant, even if it could be measured: emissions from excited *vibrational* states can be observed only from very hot samples, and even then they are extremely weak.

Although electronic excitations are limited to the wavelength range between around 100 and 800 nm, the energy in the vibrational and rotational bands is also broadened (spread out) (Chapters 1 and 2). As discussed earlier, basic vibrations start at above 2.5 μm and extend up to 25 μm, while pure rotational bands with wavelengths between about 45 and 1000 μm are found in the far-infrared range. At first glance, the spectral range between 800 and 2500 nm is spectroscopically *no man's land*. Indeed, the first edition of the German standard textbook on infrared spectroscopy by Günzler and Böck, in 1975, contains a statement that NIR spectra are fairly useless.

While only a few molecules are colored and thus absorb in the visible wavelength range, in the NIR spectral range we are confronted with the problem that virtually all organic molecules, with their dominant CH-, NH- and OH-groups, are present spectrally. However, there are some exceptions, such as 1,2-dibromo-1,1,2,2-tetrachloroethane ($C_2Br_2Cl_4$), carbon disulfide (CS_2) or tribromo-boron (BBr_3), which do not show any NIR absorbance. Depending on the number of different vibrational modes and because of intermolecular interactions, we often find very sharp (CO_2), but sometimes also broad and faint absorption peaks (H_2O), although frequently there is a complex mixture of both. We are dealing with "molecular fingerprints" that are more pronounced than in the UV–VIS range, but less significant than in the IR region: the range is from simple molecules of organic acids, sugars or fats up to large molecules such as polymers, carbohydrates and proteins. Spectral libraries, particularly for the NIR range, allow fast and computer-assisted identification.

Where does this great spectral variation come from? Answer: it is caused by physical overtones, harmonics of the known basic infrared vibrations, and most importantly, all types of concomitant interfering *combination vibrations*. This can be explained by the three basic vibrations ν_1, ν_2 and ν_3 (two stretching and one bending vibration) of the three-atomic molecule CO_2, which are found in the mid-infrared region (Figure 10.2). The intensity of the overtones decreases with increasing order. However, liquids with sample thicknesses from 1 mm up to 1 cm are spectroscopically readily accessible, yielding strong signals. Taking the selection rules into consideration, we obtain a multitude of combination bands in the near-infrared range.

The number of combination vibrations increases rapidly with the number of atoms in a molecule. Even for simple molecules the total density of the vibrational bands is estimated to be in the order of about 10^3 per wavenumber! The individual vibrational modes within the molecule interfere with each other, however not in a stochastic manner. There are two types of infrared vibrations: on excitation, if the dipole moment is changed, the band is observed in the IR spectrum (*IR active*), however, if the polarizability α is changed, the band is *Raman* active. With respect to the basic vibrations, for combination vibrational modes, symmetry-related selection rules are valid. The number of combinations is greatly increased by a further quantum-mechanically based phenomenon, which determines the appearance of the NIR spectra: in cases of accidental degeneracy, i.e., when vibrations of different origin have the same energy, so-called Fermi coupling leads to energetic splitting of the former degenerate levels. Because of the high density of vibrational bands such degeneracies are more often the rule than the exception. Last but not least, there are frequently several or even many different molecules within a sample that lead to a spectral situation which appears to be analytically inaccessible – however it only *appears* to be inaccessible, as, in contrast to UV–VIS spectroscopy, a very different, i.e., a "statistical", approach yields excellent results.

10.3
The Infrared Spectrometer

While optical components such as mirrors, lenses, gratings or incandescent lamps and even light fibers are utilized in the NIR range, the classical and oldest detector in NIR is a lead sulfide resistor (PbS, 1000–4000 nm). Because of its benefits, in recent years a gallium indium arsenic photodiode (InGaAs, a semi-conductor) has become the detector of choice. Cooling is not necessarily required, but extends the accessible spectral range and improves the dynamic properties of all NIR detectors. However, cooling is indispensable with the new NIR photomultipliers (Section 3.7.1). All classical and also novel types of monochromators can be used:

- Photometer with a filter wheel (3–6 filters): the simplest low cost system.
- Classical grating monochromator with an NIR grating and corresponding order filters.
- Diode array spectrometer, which is suitable for the NIR range: InGaAs arrays are still fairly expensive (a factor of five more expensive compared with diode arrays in the UV–VIS).
- Fourier spectrometer: the multiplex advantage, throughput advantage, Connes advantage (i.e., laser calibration). Measurement in the seconds time range and high precision, however, expensive.
- Rapid scan spectrometer (RSS) with vibrational or rotational mirrors.
- Opto-acoustically adjustable gratings (TeO_2 crystal, which is sintered onto piezo-electric plates).

With respect to the UV–VIS range, there are several calibration lamps (*penlights*) available with sharp atomic emission lines for argon, xenon or krypton. For immediate calibration of absorption/reflection spectra in the NIR range, oxides of rare earths such as didymium and erbium or silicone oil and polystyrene are used, even if their spectral resolution is much lower than that of the emission lines of interest. The company Brucker (Germany) has adopted a special trick for calibrating their Fourier–NIR spectrometers: water vapor, always present as a "contamination", exhibits extremely sharp and energetically well defined rotational lines in addition to the high order vibrations observed in the NIR. These provide an excellent and automatic wavenumber/wavelength calibration (Figure 10.5).

Of course, even in the visible and lower NIR range between 600 and 1000 nm, there are even higher order vibrations (overtones) of various molecules that can be utilized (such as by Futurex, USA, for measuring blood sugar). The advantages of the classical (more sensitive and highly dynamic) UV–VIS photomultiplier and also of the low-cost and sensitive silicon detectors can be utilized, but this is at the expense of the disadvantages of lower absorbance and even higher complexity of spectra, and interferences with electronic spectra in the visible wavelength range.

10.4
Presentation of Near-infrared Spectra

There are various methods of presenting NIR spectra: firstly, plotting the ("real") absorption as deduced from transmission of light through liquid and gaseous samples; secondly, the (linear) non-corrected transmission presentation (T). NIR absorption spectra are usually presented as "A" in the reflection mode: $A = \log (R_0/R) = \varepsilon \times x \times c$, where, according to convention, the concentration is given in mol L^{-1} and the dimensions of ε in (mol^{-1} cm^2). The reason for this is the low dynamic range of NIR detectors (1–2 orders), while photomultipliers offer 6–7 orders (decades). However, there are many advantages to using reflection spectroscopy. It is non-invasive, requires non-transparent samples and allows spectral measurements of all samples without exception – even of completely opaque samples, and thus offers further possibilities.

While traditionally in "normal" IR spectroscopy the transmission is plotted versus wavenumber k (cm^{-1}) (i. e., energy $E = khc$), in NIR spectra, as in UV–VIS-spectra, the absorbance is often plotted as a function of wavelength (1000–2500 nm corresponds to 10000–4000 wavenumbers). The reason is the use of largely identical techniques in NIR and UV–VIS spectroscopy, in contrast to IR spectroscopy.

10.5
Algorithms for the Analysis of Near-infrared Spectra

In addition to the great complexity of NIR spectra even for small molecules, real samples often consist of complex mixtures where, for example, the water, fat, starch, sugar and protein content have to be measured at the same time. Thus, it is virtually impossible to break down such a complex NIR spectrum, similar to an NMR (nuclear magnetic resonance) or an IR spectrum, into its individual components. The information regarding structure and composition of NIR spectra is completely hidden, as if it were encoded, and a classical interpretation of the spectra is not feasible. A meaningful quantitative and qualitative analysis is only possible on the basis of a statistically based method, i. e., "chemometrics".

Procedures for identifying the features of large data sets are well documented and have been available in the statistical literature for a long time. The general principle of the various and often mathematically complex methods (*multiple linear regression, partial least squares, P-matrices procedures*) can be explained briefly by an example based on the statistical technique of *principle component analysis* (PCA), often adopted in NIR analysis today. To help in the understanding of the method, we will first focus on only two wavelengths (Figure 10.3). However, PCA is a so-called full-spectra method, where the final whole spectrum, which includes, for example, about 500 measuring points, is utilized. The first step is the measurement of so-called *training samples*: NIR reflection spectra of several identical samples, e. g., $m = 50$, are measured. Subsequently the measured values E (only E_1, E_2 in this instance) of all individual training samples (here 50) are

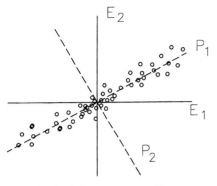

$$P_1 = C_{11} E_1 + C_{12} E_2$$
$$P_2 = C_{21} E_1 + C_{22} E_2$$

$$C_{11}^2 + C_{21}^2 = 1$$
"weights"

each sample corresponds to one individual point in n-dimensional space (e.g. n = 500)

P = "principle component"

E = measured values

Figure 10.3 Explanation of chemometrics, a widely used statistical method in NIR spectroscopy, using *principle component analysis* (PCA) as an example. This example is simplified greatly by using only two (rather than the 500 used in practice) wavelengths, λ_1 and λ_2, with the measured values E_1 and E_2.

plotted in a 500-dimensional (only 2-dimensional in this instance) coordinate system. As principle components are concerned only with the variation in the data, the reflectance energies E are shown only as *deviations* from their mean values. A rigid rotation of the original axes E_1 and E_2 then takes place, which maximizes the variation along one new axis. This axis, P_1, is the "first principle component". The second new axis, P_2, which by definition lies at right angles to the first, becomes the second principle component. The new axes can be expressed as linear combinations of the original axes:

$$P_1 = C_{11} E_1 + C_{12} E_2 \tag{10.1}$$

$$P_2 = C_{21} E_1 + C_{22} E_2 \tag{10.2}$$

The coefficients $C_{11}, \dots C_{nn}$ are referred to as "weights" and are always normalized so that the sum of the squared weights for any single component is unity:

$$C_{11}^2 + C_{12}^2 = 1 \tag{10.3}$$

$$C_{21}^2 + C_{22}^2 = 1 \tag{10.4}$$

The weights of any principle component are derived with reference to a rotation, which defines a new axis. In the example given, each sample is represented as a single point in two dimensions. Measuring a full spectrum, e.g., for 500 points, the diagram in Figure 10.3 becomes 500-dimensional, within which each sample shows up as a single point. Each PC is therefore defined in terms of the original reflectance values by

$$P_n = C_{21} E_1 + C_{22} E_2 + \ldots + C_{500} E_{500} \tag{10.5}$$

where E_i denotes the reflected energy at wavelength i and C_i denotes the coefficient or weight for the energy values at wavelength i.

$$C_1^2 + C_2^2 + \ldots + C_{500}^2 = 1 \tag{10.6}$$

Mathematically, principle components are the eigenvectors of the covariance matrix S of the reflected energies, and the variations along each of the component axes are the eigenvalues of S (for more details see Selected Further Reading). Of course, the largest variations in the spectrum will occur at wavelengths where the individual components are at their maxima. Each PC shows a smaller variation than its predecessor. Practical work shows that in most cases it is just the first six PCs (up to 99.8%) that describe the variation of the spectra almost completely.

PCs are only "constructed" spectra without any immediate value, however, due to their mathematical origin, they do represent the main individual components of a sample fairly well. For each chemical component the optimum measuring wavelength and corresponding coefficient for calculating the concentration can be calculated on the basis of these PCs. Finally, a calibration diagram is obtained by determining "once and for all" the concentration of l components, on the basis of m calibration samples, using methods other than NIRS, preferably wet-chemical procedures.

Routine measurements, particularly in the chemical and pharmaceutical industry, require fast spectral monitoring. The central feature of the quantitative chemometric procedures is the possibility of calibrations for analytically interesting components without any prior knowledge of the sample and independent of the remaining composition. The total composition of a sample is known only in rare cases, however, in chemometrics individual components are *uncoupled*. The most appropriate wavelengths are selected by the mathematical algorithm. Another significant aspect of chemometrics is the reliable evaluation of the standard deviations involved, as these are provided automatically by the algorithm, i. e., the computer program. Therefore, routine NIR analyses can be readily performed, particularly by non-specialist personnel.

10.6
Applications

The scope for NIR applications is virtually unlimited and is only confined by the imagination of the user. To date NIR spectroscopy has been used for a range of very different measuring tasks in industry and research, and in the future it will surely replace the more expensive and time consuming classical analytical methods, as exemplified by following examples:
- Separation and control of plastic waste, e. g., in the waste and recycling industry (polyethylene, polypropylene, polyamide, etc.).

- Iodine value (a measure of the number of double bonds in unsaturated fatty acids), OH-number.
- Aromatic content.
- High-pressure liquid chromatography (HPLC), thin-layer chromatography.
- Environmental analysis, vegetation, limnology, geology, soil analysis.
- Water and waters (lakes, rivers, etc.) analysis, snow analysis, salt content of oceans, pH values, water of crystallization content.
- Food analysis (meat, cakes and breads, juices, fruits, flour, coffee, nicotine content, cocoa, milk products, etc.).
- Petrochemistry (gasoline components, octane value).
- Biochemistry (proteins, carbohydrates, fats, water).
- Control of various raw materials, pharmaceuticals such as caffeine, theophylline, salicylic acid, paracetamol, novonal.
- Textile industry.

10.6.1
Medicine and Pharmacy

Figure 10.1 shows the NIR spectra of some blood components (glucose, albumin, fat, water). Non-invasive measurements of cholesterol, albumin, bilirubin, glucose and others, are of practical importance, in a similar manner to the non-invasive

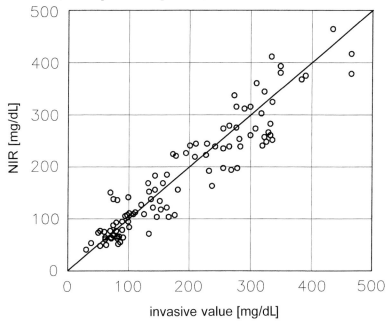

Figure 10.4 Comparison of the transcutaneous NIR measurement (of the left index finger of the patient) with the usual invasive measurement of vein blood taken immediately after. NIR full spectra were analyzed statistically (from Futurex Inc., USA).

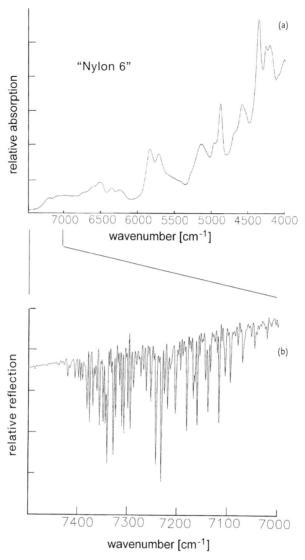

Figure 10.5 (a) Corrected NIR absorption spectrum (measured as a reflection spectrum) of "Nylon 6". (b) In the range between 7000 and 7400 cm^{-1} an overtone of a vibration band from the contaminating water vapor is located. This overtone, in turn, exhibits a large number of very narrow rotational lines, which are utilized for precise wavelength calibration (Brucker, Germany).

and reliable determination of blood oxygen using a pulse oximeter with a finger probe. Near-infrared spectroscopy has already found many applications in the pharmaceutical industry, and to a certain extent in medicine. Its principle advantage is the fast, contact-free and precise analysis of liquids as well as solids such as tissues, tablets or powders by untrained personnel. Figure 10.4 demonstrates the promising comparison of transcutaneous NIR measurement of glucose (using the left index finger of the patient) with the usual, invasive measurement of vein blood (with the same finger, modified from Futurex Inc., USA).

10.6.2
Plastics

Figure 10.5(a) shows the corrected NIR absorption spectrum (measured as a reflection spectrum) of "Nylon 6", a common plastic (modified from Bruker Optics GmbH, Ettlingen, Germany). The smooth shape of the baseline throughout the whole NIR range is, however, in the range from 7000 to 7400 cm^{-1} somewhat affected by an overtone of a vibration band due to the presence of contaminating water vapor. The overtones of water, in turn, show a large number of very sharp, well defined rotational bands [Figure10.5(b)], which are utilized for wavelength calibration.

Various plastics such as the polymers polyamide (PA), polypropylene (PP) or polyethylene (PE) show specific and pronounced absorption spectra, which serve to distinguish between them distinctly. The high pressure polymerization of ethylene was discovered in 1933 and is, next to Haber's ammonia synthesis, the most important high pressure reaction performed on a large scale. In Europe alone a yearly volume of approximately three million tons of polyethylene is produced at pressures of 3 kbar and temperatures of between 150 and 300 °C. Process analysis using NIRS depends on the characteristic absorption spectra of ethylene and polyethylene. Figure 10.6 (modified from Cook and Jones, *FT-NIR Atlas*, Verlag Chemie, VCH, Weinheim) shows second-order vibrational spectra for ethylene and polyethylene. A similar picture is repeated, and is even more pro-

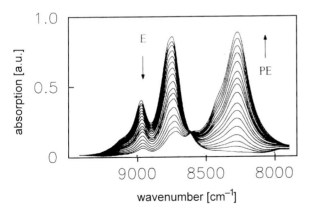

Figure 10.6 NIR absorption spectra as scanned during the polymerization process of free radicals of ethylene at 190 °C and 2630 bar in the region of the second overtones in the near-infrared (from Buback and Vögele, *FT-NIR Atlas*, Verlag Chemie (VCH), Weinheim, 1993). A clear-cut isosbestic point at a wavenumber of 8600 demonstrates the straightforward polymerization reaction.

nounced, in the third overtone between wavenumbers 10 000 and 12 000. The general rule holds true that spectra become more and more pronounced with an increasing number of overtones – and are increasingly separated; this is explained by the increasing inharmonicity with higher vibrational excitations (cf. Figure 2.23).

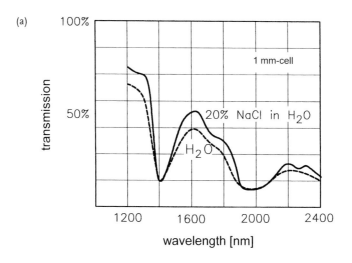

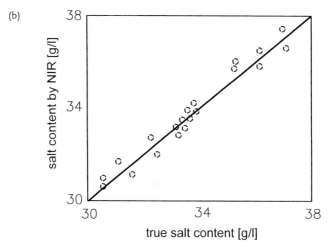

Figure 10.7 Determination of "non-spectroscopic" properties with NIR spectroscopy using as an example the salt content of sea water: (a) The NIR transmission spectrum of pure water and of water with a 20% salt content. (b) On the basis of chemometrics, as described in the text, a good correlation and proportionality between the real (as weighed) and NIR spectroscopy measured salt content is obtained.

10.6.3
Salt Content of Sea Water

At first glance the determination of the salt content of sea water by optical spectroscopy does not appear to be feasible, because sodium chloride is almost the archetypal infrared transparent material: hence this is a typical example of the spectral measurement of "non-spectroscopic properties" by NIRS. In this context the concept of an *infrared shift agent* becomes meaningful. Sodium chloride changes the structure of water by modification of hydrogen bridges and consequently the NIR spectrum [Figure 10.7(a)]. On the basis of one of the chemometric calculations currently being used, we obtain the calibration curve in Figure 10.7(b) (modified from Hirschfeld, Lawrence Livermore Laboratory, Livermore, CA, USA). This allows a very fast, reliable and low cost determination of the salt content of sea water, e. g., on-line on a research vessel. Similar to the salt content, other entities such as proton concentrations, pH values or even the just temperature can be measured easily, contact-free and precisely.

10.7
Summary

Even though NIR spectroscopy is considered by "pure spectroscopists" to be no more than a pseudo-science, due to the inherent complexity and problems concerned with the immediate interpretation of NIR spectra, for the practical spectroscopist it is only the results that count. What is the easiest, fastest, cheapest and possibly non-invasive way to obtain valuable analytical information, if possible *in vivo* and without a laborious wet-chemical preparation? It is precisely for these reasons that NIR spectroscopy provides unprecedented advantages and possibilities. Thus, the future of NIR spectroscopy is in very practical routine applications rather than in theoretical research.

10.8
Selected Further Reading

Batten, G. D., Flinn, P. C., Welsh, L. A., Blakeney, A. B. *Leaping Ahead with Near-infrared Spectroscopy*, NIR Publications, Chichester, **1995**.

Buback, M, Vögele, H. P. *FT-NIR Atlas*, Wiley-VCH, Weinheim, **1993**.

Burns, D. A., Ciurczak, E. W., eds. *Handbook of Near-Infrared Analysis, Practical Spectroscopy*, 2nd edn., Marcel Decker, New York, **2001**.

Cho, R. K., Davies, A. M. C., eds. *Near-infrared Spectroscopy*, Proceedings of the 10th International Conference, Chichester, UK, ISBN 0952 8666 33, NIR-Publications, **2002**.

Ciurczak, E. W., Drennen, J. K. *Pharmaceutical and Medical Applications of Near-infrared Spectroscopy (Practical Spectroscopy)*, Marcel Decker, New York, **2002**.

Conzen, J.-P., Schmidt, A. *Non-destructive Quality Control of Pharmaceutical Tablets by Near-infrared Reflectance Spectroscopy. Near-infrared Spectroscopy: The Future Waves*, Proceedings of the 7th International Conference on Near-infrared Spectroscopy, Montreal, Canada, **1995**.

Daehne, S., Resch-Genger, U., Wolfbeis, O. S. *Near-infrared Dyes for High Technology Applications*, Kluver Academic Publishers, **1998**.

Davies, A. M.C., Williams, P. C., eds. *Near-infrared Spectroscopy: The Future Waves*, NIR Publications, Charlton, Chichester, **1996**.

Fearn, T. *Chemometrics for Near-infrared Spectroscopy: Past, Present and Future*, Spectrosc. Europe **2001**, *13/2*, 10–14.

Gans, P. *Data Fitting in Chemical Sciences by the Method of Least Squares*, John Wiley & Sons, New York, **1992**.

Mardia, K. V., Kent, J. T., Bibby, J. M. *Multivariate Analysis*, Academic Press, New York, London, **1979**.

Mark, H. *Principles and Practice of Spectroscopic Calibration*, John Wiley & Sons, New York, **1991**.

Mark, H., Workman, J. *Statistics in Spectroscopy*, Academic Press, New York, London, **1991**.

Martens, H., Naes, T. *Multivariate Calibration*, John Wiley & Sons, New York, London, **1992**.

Massart, D. L., Vandeginste, B. G.M., Buydens, L. M.C., De Jong, S., Lewi, P. J., Smeyers-Verbeke, J. *Handbook of Chemometrics and Qualimetrics*, Data Handling in Science and Technology, Vol. 20, Elsevier, Amsterdam, **1998**.

McClure, W. F. *Near-infrared Spectrometry: Learning the Fundamentals*, book & disk edn., John Wiley & Sons, New York, **2006**.

Meier, P. C., Zund, R. E. *Statistical Methods in Analytical Chemistry*, John Wiley & Sons, New York, **1993**.

Morgan, E. *Chemometrics: Experimental Design: Analytical Chemistry by Open Learning*, John Wiley & Sons, New York, **1991**.

Morrison, D. F. *Multivariate Statistical Methods*, McGraw-Hill, New York, **1967**.

Murray, I., Cowe, I. A. *Making Light Work: Advances in Near-infrared Spectroscopy*, Verlag Chemie (VCH), Weinheim, **1992**.

Osborn, B. G., Fearn, T., Hindle, P. H. *Practical NIR Spectroscopy in Food and Beverage Analysis*, 2nd. edn., Longman, London, **1993**.

Otto, M. *Chemometrics*, Wiley-VCH, Weinheim, **1999**.

Raghavachari, R. *Near-infrared Applications in Biotechnology (Practical Spectroscopy)*, Marcel Decker, New York, **2000**.

Zupan, J., Gasteiger, J. *Chemometrics: Experimental Design*, Verlag Chemie (VCH), Weinheim, **1991**.

General: http://www.nirpublications.com (books, periodicals, web, CD-ROM, training).

Appendix

A1
Conversion Factors for Units of Energy

	J	erg	kWh	cal	eV	cm^{-1}	Hz
1 J	1	10^7	$2.777\,778 \cdot 10^{-7}$	$2.388\,459 \cdot 10^{-1}$	$(6.241\,8 \pm 3) \cdot 10^{18}$	$(5.034\,5 \pm 3) \cdot 10^{22}$	$(1.509\,3 \pm 1) \cdot 10^{33}$
1 erg	10^{-7}	1	$2.777\,778 \cdot 10^{-14}$	$2.388\,459 \cdot 10^{-8}$	$(6.241\,8 \pm 3) \cdot 10^{11}$	$(5.034\,5 \pm 3) \cdot 10^{15}$	$(1.509\,3 \pm 1) \cdot 10^{26}$
1 kWh	$3.600\,000 \cdot 10^6$	$3.600\,000 \cdot 10^{13}$	1	$8.598\,452 \cdot 10^5$	$(2.247\,1) \pm 10^5$	$(1.812\,4 \pm 1) \cdot 10^{29}$	$(5.433\,5 \pm 5) \cdot 10^{39}$
1 cal	$4.186\,8$	$4.186\,8 \cdot 10^7$	$1.163\,000 \cdot 10^{-6}$	1	$(2.613\,3 \pm 1) \cdot 10^{19}$	$(2.107\,8 \pm 1) \cdot 10^3$	$(6.319\,1 \pm 3) \cdot 10^{14}$
1 eV	$(1.602\,10 \pm 7) \cdot 10^{-19}$	$(1.602\,10 \pm 7) \cdot 10^{-12}$	$(4.450\,28 \pm 19) \cdot 10^{-26}$	$(3.826\,55 \pm 16) \cdot 10^{-20}$	1	$(8.065\,8 \pm 6) \cdot 10^3$	$(2.418\,0 \pm 1) \cdot 10^{14}$
1 cm^{-1}	$(1.986\,3 \pm 2) \cdot 10^{-23}$	$(1.986\,3 \pm 2) \cdot 10^{-16}$	$(1.5.517\,5 \pm 5) \cdot 10^{-30}$	$(4.744\,2 \pm 5) \cdot 10^{-24}$	$(1.239 \pm 1) \cdot 10^{-4}$	1	$(2.997\,925 \pm 3) \cdot 10^{10}$
1 Hz	$(6.625\,6 \pm 5) \cdot 10^{-34}$	$(6.625\,6 \pm 5) \cdot 10^{-27}$	$(1.840\,4 \pm 2) \cdot 10^{-40}$	$(1.582\,5 \pm 1) \cdot 10^{-34}$	$(4.135\,6 \pm 4) \cdot 10^{-15}$	$(3.335\,640 \pm 3) \cdot 10^{-11}$	1

A2
Important Natural Constants

speed of light in a vacuum	c	2.9979245	10^8 m s^{-1}
Planck constant	h	6.626176	10^{-34} J s
	$\hbar = h/2\pi$	1.0545887	10^{-34} J s
Stefan–Boltzmann constant	$\sigma = (\pi^2/60)k^4/\hbar^3\,c^2$	5.67032	10^{-23} J K^{-4}
Wien's displacement law constant	$\lambda_{max}\,T$	2.8978	10^{-3} m · K
Rydberg constant	$R_y = 2\pi^2 me^4/h^2$	1.0973731	10^7 m^{-1}
Bohr radius	a_0	5.29167	10^{-11} m
electron rest mass	m_e	9.1091	10^{-31} kg
neutron rest mass	m_n	1.67482	10^{-27} kg
proton rest mass	m_p	1.67252	10^{-27} kg
	(m_p/m_e)	1836.151	
elementary charge	e	1.602189	10^{-19} C
electron radius	r_e	2.81777	10^{-15} m
Bohr magneton	$\mu_B = e\hbar/2m_e$	9.2407	10^{-24} J T^{-1}
nuclear magneton	$\mu_N = e\hbar/2m_p$	5.0508	10^{-24} J T^{-1}
Avogadro constant	N_A	6.0225	10^{23} mol^{-1}
gas constant	R	8.31441	J mol^{-1} K^{-1}
Boltzmann constant	$(\varkappa =)k = R/N_A$	1.3806	10^{-23} J K^{-1}
molar volume of an ideal gas	$V_m = RT_0/p_0$	2.2413	10^{-2} m^3 mol^{-1}
	$(T_0 = 273.15\text{K},\; p_0 = 101325\text{ Pa})$		
Faraday constant	$F = N_A\,e$	9.6484	10^4 C mol^{-1}
gravitational constant	G	6.6720	10^{-11} Nm^2 kg^{-2}

A3
Manufacturers and Distributors (representative selection)

Listed are some representative (mainly German) contact addresses, manufacturers and suppliers of various spectrometers and components used in "Optical Spectroscopy". The suppliers given are also worldwide, and their details can be found by search engines such as GOOGLE or ALTAVISTA, where they give information on their individual programs. Additional sources are listed in various free laboratory journals (see Laboratory Journals below).

Optical spectrophotometers
(UV-VIS-NIR)

Bruker Optik GmbH
Rudolf-Plank-Str. 27
76275 Ettlingen, Germany
phone + 49 7243 504-600
fax + 49 7243 504-698
optik@bruker.de
info@brukeroptics.de
www.bruker.de
www.bruker.com (FT-IR/FT-NIR/FT-
Raman-spectrometer)

BÜCHI Labortechnik AG
Postfach
9230 Flawil, Switzerland
phone + 41 71 394 6363
fax + 41 71 394 6565
buchi@buchi.com
www.buchi.com
(only NIR-spectrometers)

getSpec.com, c/o Sentronic GmbH
Emil-Figge-Str. 76–80
44227 Dortmund, Germany
phone + 49 231 9742285
fax + 49 231 9742286
s.thomassen@getSpec.de
www.getSpec.de

Dr. Gröbel, UV-Elektronik GmbH
Goethestr. 17
76275 Ettlingen, Germany
phone + 49 7243 71839-0

fax: + 49 7243 71839-300
info@uv-groebel.de

Hamamatsu Photonics
Deutschland GmbH
Arzbergerstr. 10
82211 Herrsching, Germany
phone + 49 8152 3750
fax + 49 8152 2658
info@hamamatsu
www.hamamatsu.com

HORIBA Jobin Yvon GmbH
Chiemgaustr. 148
81549 München, Germany
phone + 49 89 462317-0
fax + 49 89 462317-99
info@jobinyvon.de
www.jobinyvon.de (UV-VIS)

HORIBA Jobin Yvon GmbH
Neuhofstrasse 9
64625 Bensheim, Germany
phone + 49 6251 8475-0
fax + 49 6251 8475-20
raman@jobinyvon.de
www.jobinyvon.de (Raman, Laser)

L. O.T.-Oriel GmbH & Co. KG
Im Tiefen See 58
64293 Darmstadt, Germany
phone + 49 6151 8806-0
fax + 49 6151 896667
info@LOT-Oriel.de
www.oriel.com

MIKROPACK GMBH
Maybachstraße 11
73760 Ostfildern, Germany
phone + 49 711 341696-0
fax + 49 711 341696-85
spectroscopy@mikropack.de
www.mikropack.de

m · u · t GmbH
Am Marienhof 2
22880 Wedel, Germany
phone + 49 4103 9308-435
fax + 49 4103 9308-7435
sjohannsen@mut-gmbh.de
www.mut-gmbh.de

Ocean Optics
Nieuwgraaf 108 G
6921 RK Duiven, The Netherlands
phone + 31 26 31905-00
fax + 31 26 31905-05
info@OceanOpticsBV.com
www.OceanOpticsBV.com
(diode arrays)

PerkinElmer LAS (Germany) GmbH
Ferdinand Porsche Ring 17
63110 Rodgau-Jügesheim, Germany
phone + 49 800 18100-32
fax + 49 800 18100-31
cc.germany@perkinelmer.com
www.instruments.perkinelmer.com
(AAS, ICP)

Polytec GmbH
Polytec-Platz 1–7
76337 Waldbronn, Germany
phone + 49 7243 604-0
fax + 49 7243 69944
info@polytec.com
www.polytec.com

Polytec GmbH
Vertriebs- und Beratungsbüro Berlin
Schwarzschildstraße 1
12489 Berlin, Germany
phone + 49 30 6392-5140

fax + 49 30 6392-5141
info@polytec.de

Shimadzu Deutschland GmbH
Albert-Hahn-Str. 6–10
47269 Duisburg, Germany
phone + 49 203 7687-0
fax + 49 203 766625
info@shimadzu.de
www.shimadzu.de
www.shimadzu.com

SI Scientific Instruments GmbH
Römerstr. 67
82205 Gilching, Germany
phone + 49 8105 77940
fax + 49 8105 5577
info@SI-GmbH.de
www.si-gmbh.de

SPECTRO Analytical Instruments
GmbH & Co. KG
Boschstraße 10
47533 Kleve, Germany
phone + 49 2821 892-2109
fax+ 49 2821 892-2210
info@spectro.com
www.spectro.com

tec5 Aktiengesellschaft
In der Au 25
61440 Oberursel, Germany
phone + 49 6171 9758-0
fax + 49 6171 9758-50
sales@tec5.com
www.tec5.com

Light sources (lamps, lasers, LEDs)

Hamamatsu Photonics
Deutschland GmbH
Arzbergerstr. 10
82211 Herrsching, Germany
phone + 49 8152 3750
fax + 49 8152 2658

info@hamamatsu
www.hamamatsu.com

Heraeus Noblelight GmbH
Heraeusstraße 12–14
63450 Hanau, Germany
phone + 49 6181 35-8492
fax + 49 6181 35-16 8410
hng-marketing@heraeus.com
www.heraeus-noblelight.com

L.O.T.-Oriel GmbH
Im Tiefen See 58
64293 Darmstadt, Germany
phone + 49 6151 8806-0
fax + 49 6151 896667
info@LOT-Oriel.de
www.oriel.com

MUT, Meßgeräte für
Medizin und Umwelttechnik
Hafenstraße 32–32a
22880 Wedel, Germany
phone + 49 4103 9308-15
fax + 49 4103 9308-99
uwachholz@mut-gmbh.de
www.mut-gmbh.de

Ocean Optics
Nieuwgraaf 108 G
6921 RK Duiven, The Netherlands
phone + 31 26 31905-00
fax + 31 26 31905-05
info@OceanOpticsBV.com
www.OceanOpticsBV.com

Polytec GmbH
Polytec-Platz 1–7
76337 Waldbronn, Germany
phone + 49 07243 604-0
fax + 49 07243 69944
info@polytec.com
www.polytec.com

Polytec GmbH
Vertriebs- und Beratungsbüro Berlin
Schwarzschildstraße 1
12489 Berlin, Germany

phone + 49 30 6392-5140
fax + 49 30 6392-5141
info@polytec.de

Roithner Lasertechnik
Schoenbrunenrstr. 7
1040 Vienna, Austria
phone + 43 1 586 5243-0
fax + 43 1 586 5243-44
office@roithner-laser.com
www.roithner-laser.com
(LEDs from 350 to 1650 nm)

tec5 Aktiengesellschaft
In der Au 25
61440 Oberursel, Germany
phone + 49 6171 9758-0
fax + 49 6171 9758-50
sales@tec5.com
www.tec5.com

Light detectors

L.O.T.-Oriel GmbH
Im Tiefen See 58
64293 Darmstadt, Germany
phone + 49 6151 8806-0
fax + 49 6151 896667
info@LOT-Oriel.de
www.oriel.com

MIKROPACK GMBH
Maybachstraße 11
73760 Ostfildern, Germany
phone + 49 711 341696-0
fax + 49 711 341696-85
spectroscopy@mikropack.de
www.mikropack.de

Ocean Optics
Nieuwgraaf 108 G
6921 RK Duiven, The Netherlands
phone + 31 26 31905-00
fax + 31 26 31905-05
info@OceanOpticsBV.com
www.OceanOpticsBV.com

Polytec GmbH
Polytec-Platz 5–7
76333 Waldbronn, Germany
phone + 49 7243 604-0
fax + 49 7243 69944
h.holbach@polytec.de

tec5 Aktiengesellschaft
In der Au 25
61440 Oberursel, Germany
phone + 49 6171 9758-0
fax + 49 6171 9758-50
sales@tec5.com
www.tec5.com

Optical/electronic/chemical components

BFI Optilas
Lilienthalstr. 14
85375 Neufahrn, Germany
phone + 49 8165-6173-0
fax + 49 8165-6916-67
RF.DE@bfioptilas.com
bfi.avnet.com/Germany/index.html
Laser, Optik, Sensor, Imaging, Test &
Measurement, Magnetic, Fibre Optic

Edmund Industrie Optik GmbH
Schoenfeldstr. 8
76131 Karlsruhe, Germany
infogmbh@edmundoptics.de
www.edmundoptics.de
(supplier of all optical components)

FEMTO Messtechnik GmbH
Paul-Lincke-Ufer 34
10999 Berlin, Germany
phone + 49 30 44693-86
fax + 49 30 44693-88
info@femto.de
www.femto.de
(light detectors, electronics)

Hamamatsu Photonics
Deutschland GmbH

Arzbergerstr. 10
82211 Herrsching, Germany
phone + 49 8152 3750
fax + 49 8152 2658
info@hamamatsu
www.Hamamatsu.com

Hellma Optik
Postfach 100403
07704 Jena, Germany
phone + 49 3641 609 814
fax + 49 3641 609 815
sales@hellmaoptik.com
www.hellmaoptik.com

Hellma Spectrooptics
Postfach 1163
79371 Müllheim/Baden, Germany
phone + 49 7631 182-0
fax + 49 7631 13546 (cuvettes)

HORIBA Jobin Yvon GmbH
Chiemgaustr. 148
81549 München, Germany
phone + 49 89 462317-0
fax + 49 89 462317-99
info@jobinyvon.de
www.jobinyvon.de (UV-VIS)

L.O.T.-Oriel GmbH
Im Tiefen See 58
64293 Darmstadt, Germany
phone + 49 6151 8806-0
fax + 49 6151 896667
info@LOT-Oriel.de
www.oriel.com

Melles Griot GmbH
photonic components
Lilienstraße 30–32
64625 Bensheim, Germany
phone + 49 6251 8406-0
fax + 49 6251 8406-24
USA: 16542 Millikan Avenue
Irvine, CA 92606, USA
phone + 1 800 835-2626
fax + 1 949 261-7790

Molecular Probes Inc.
29851 Willow Creek Road
Eugene, OR 97402, USA
phone + 1 541 465 8300
fax + 1 541 335 0504
order@probes.com
www.probes.com
(all types of fluorescence probes)

National Instruments
Corporate Headquarters
11500 N Mopac Expwy
Austin, TX 78795, USA
phone + 1 512 794 0100
fax + 1 512 683 9300
info@ni.com

Poerschke GmbH
PB 1226
64734 Höchst, Germany
phone + 49 6163 912130
fax + 49 6163 912132
poerschke@t-online.de
www.poerschke.de

SCHÖLLY FIBEROPTIC GMBH
Robert-Bosch-Str. 1–3
79211 Denzlingen, Germany
phone + 49 7666 908-0
fax + 49 7666 908-380
info@schoelly.de
www.schoelly.de

SCHOTT AG
Hattenbergstr. 10
55122 Mainz, Germany
phone + 49 6131 66-0
fax: + 49 6131 66-2000
info@schott.com
www.schott.com
(optical filters of all types)

Spectra-Physics
Guerickeweg 7
64291 Darmstadt, Germany
phone + 49 6151 708-0
fax + 49 6151 708-217
(lasers-photonics)

Sphere Optics Hoffman GmbH
88690 Uhldingen, Germany
phone + 49 7556 9299666
fax + 49 7556 50108
www.sphereoptics.com

MIKROPACK GMBH
Maybachstraße 11
73760 Ostfildern, Germany
phone + 49 711 341696-0
fax + 49 711 341696-85
spectroscopy@mikropack.de
www.mikropack.de

Ocean Optics
Nieuwgraaf 108 G
6921 RK Duiven, The Netherlands
phone + 31 26 31905-00
fax + 31 26 31905-05
info@OceanOpticsBV.com
www.OceanOpticsBV.com

Polytec GmbH
Polytec-Platz 5–7
76333 Waldbronn, Germany
phone + 49 7243 604-0
fax + 49 7243 69944
h.holbach@polytec.de

tec5 Aktiengesellschaft
In der Au 25
61440 Oberursel, Germany
phone + 49 6171 9758-0
fax + 49 6171 9758-50
sales@tec5.com
www.tec5.com

Laboratory Journals

Laboratory journals are an important source of information on the latest developments, trends and prices. In addition to advertisements of diverse opto-analytical devices they often offer interesting articles on topics relevant to the field, written by specialists. Typically they appear monthly and are usually supplied free of charge. Some laboratory journals that are important for the Optical Spectroscopist are listed below:

Electro Optics
Freepost
5 Tranquil Passage
Blackheath
London SE3 OBR, UK

GIT VERLAG GMBH
Rösslerstr. 90
64293 Darmstadt, Germany
phone + 49 6151 809-00
fax + 49 6151 809-46

LABO, Verlag Hoppenstedt GmbH
Havelstraße 9
Postfach 100139
64201 Darmstadt, Germany
phone + 49 6151 380-0
fax + 49 6151 380-360
info@hopp.de
www.hoppenstedt.com

Opto & Laser Europe (OLE)
WDIS Publishing House
Victoria Road
Ruislip, Middlesex
HA4 OSX, UK

Photonik, AT-Fachverlag GmbH
Saarlandstraße 28
70734 Fellbach, Germany
phone + 49 711 952951-0
fax + 49 711 952951-99

Spectroscopy Europe, joint publication of VCH Verlag GmbH
Postfach 101161
69451 Weinheim
Germany and IM Publications
6 Charlton Mill
Charlton, Chichester
West Sussex PO180HY, UK

In addition, various companies such as Hamamatsu, LOT, Molecular Probes, Shimadzu and Varian publish catalogues, CDs or company produced journals offering valuable information about the latest developments in Optical Spectroscopy.

A4
Periodic Table of the Elements

The * sign denotes a radioactive element. The number above the element symbol indicates the unified atomic mann in u, where 1 u is 1/12 of the mass of the carbon isotope ^{12}C. For naturally occuring elements this is the atomic mass of the normal isotopic mixture. For elements that can only be generated artivicially and radioactive elements that do not occur in nature, the mass number of the most stable isotope in the decay series is given. The number to the lower left of the element symbol is the atomic number. The small boxes below the element symbols show the detailed distribution of electrons within the various shells (K, L, M, N, O, P, Q) according to the Figure 2.9.

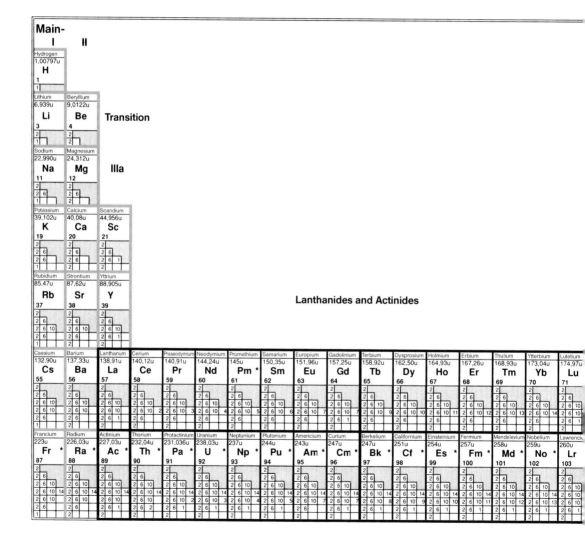

					-groups				
				III	IV	V	VI	VII	VIII/0

	III	IV	V	VI	VII	VIII/0	Period
						Helium 4,0026u **He** 2 2	**1.** Periode K(n=1)
Elements	Boron 10,811u **B** 5 2 2 1	Carbon 12,011u **C** 6 2 2 2	Nitrogen 14,007u **N** 7 2 2 3	Oxygen 15,999u **O** 8 2 2 4	Fluorine 18,998u **F** 9 2 2 5	Neon 20,183u **Ne** 10 2 2 6	**2.** Periode K(n=1) L(n=2)
Va Va VIa VIIa VIIIa Ia IIa	Aluminium 26,982u **Al** 13 2 2 6 2 1	Silicon 28,086u **Si** 14 2 2 6 2 2	Phosphorus 30,974u **P** 15 2 2 6 2 3	Sulfur 32,064u **S** 16 2 2 6 2 4	Chlorine 35,453u **Cl** 17 2 2 6 2 5	Argon 39,948u **Ar** 18 2 2 6 2 6	**3.** Periode K(n=1) L(n=2) M(n=3)

| nium
90u
Ti
23
2
2 6
2 6 2
2 | Vanadium
50,942u
V
23
2
2 6
2 6 3
2 | Chromium
51,996u
Cr
24
2
2 6
2 6 5
1 | Manganese
54,938u
Mn
25
2
2 6
2 6 5
2 | Iron
55,847u
Fe
26
2
2 6
2 6 6
2 | Cobalt
58,933u
Co
27
2
2 6
2 6 7
2 | Nickel
58,71u
Ni
28
2
2 6
2 6 8
2 | Copper
63,54u
Cu
29
2
2 6
2 6 10
1 | Zinc
65,37u
Zn
30
2
2 6
2 6 10
2 | Gallium
39,102u
Ga
31
2
2 6
2 6 10
2 1 | Germanium
72,59u
Ge
32
2
2 6
2 6 10
2 2 | Arsenic
74,922u
As
33
2
2 6
2 6 10
2 3 | Selenium
78,96u
Se
34
2
2 6
2 6 10
2 4 | Bromine
79,909u
Br
35
2
2 6
2 6 10
2 5 | Krypton
83,80u
Kr
36
2
2 6
2 6 10
2 6 | **4.**
Periode
K(n=1)
L(n=2)
M(n=3)
N(n=4) |

| onium
22u
Zr
41 | Niobium
92,906u
Nb
41 | Molybdenum
95,94u
Mo
42 | Technetium
98,906u
Tc *
43 | Ruthenium
101,07u
Ru
44 | Rhodium
102,90u
Rh
45 | Palladium
106,4u
Pd
46 | Silver
107,87u
Ag
47 | Cadmium
112,41u
Cd
48 | Indium
114,82u
In
49 | Tin
118,69u
Sn
50 | Antimony
121,75u
Sb
51 | Tellurium
127,60u
Te
52 | Iodine
126,90u
I
53 | Xenon
131,30u
Xe
54 | **5.**
Periode
K(n=1)
L(n=2)
M(n=3)
N(n=4)
O(n=5) |

| nium
3,49u
Hf | Tantalum
180,95u
Ta
73 | Tungsten
183,85u
W
74 | Rhenium
186,2u
Re
75 | Osmium
190,2u
Os
76 | Iridium
192,2u
Ir
77 | Platinum
195,09u
Pt
78 | Gold
196,97u
Au
79 | Mercury
200,59u
Hg
80 | Thallium
204,37u
Tl
81 | Lead
207,19u
Pb
82 | Bismuth
208,98u
Bi
83 | Polonium
209u
Po *
84 | Astatine
210u
At *
85 | Radon
222u
Rn *
86 | **6.**
Periode
K(n=1)
L(n=2)
M(n=3)
N(n=4)
O(n=5)
P(n=6) |

| erfordium
fu
Rf
4 | Dubnium
262u
Db *
105 | Seaborgium
Sg *
106 | Bohrium
Bh *
107 | Hassium
Hs *
108 | Meitnerium
Mt *
109 | *
110 | 111 | 112 | 113 | 114 | 115 | 116 | 117 | 118 | **7.**
Periode
K(n=1)
L(n=2)
M(n=3)
N(n=4)
O(n=5)
P(n=6)
Q(n=7) |

A5
Index

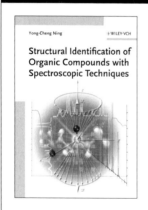

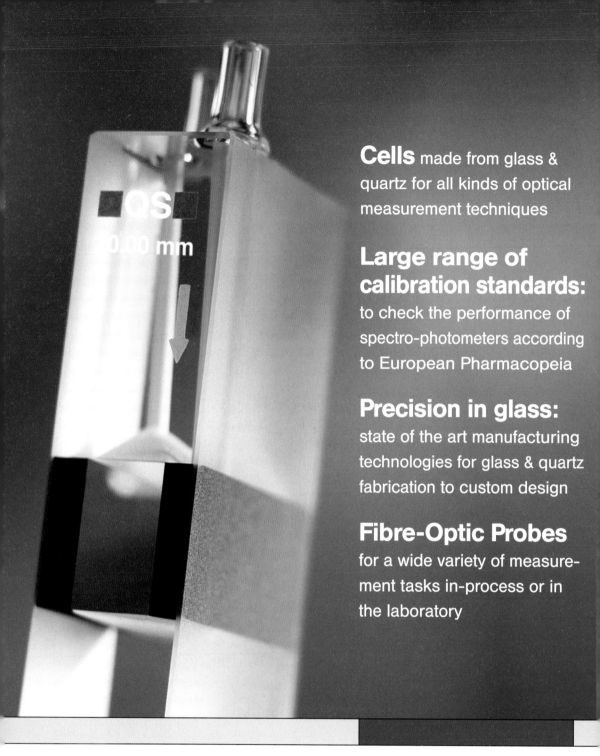